THE EARTH'S TECTONOSPHERE:

ITS PAST DEVELOPMENT AND PRESENT BEHAVIOR

THE EARTH'S TECTONOSPHERE:

ITS PAST DEVELOPMENT AND PRESENT BEHAVIOR

An analysis of the deep-seated global forces that have been operating within closely defined constraints over long periods of time to produce the Earth's surficial features.

J. H. TATSCH

TATSCH ASSOCIATES
Sudbury, Massachusetts 01776
1972

Printed in the United States of America. Copies are available from Tatsch Associates, Sudbury, Massachusetts, 01776, U.S.A.

Library of Congress Catalog Card Number: 76-190460.

International Standard Book Number: 0-912890-00-2.

TO HELEN

This book analyzes the evolution of the upper 1000 km of
the Earth. The basic premise of this analysis is that deep-
seated, global forces operating within closely-defined constraints
over long periods of time have caused the development of the
Earth's surficial features. A global analysis of these surficial
features is used to define the nature and probable origin of
the hypothesized deep-seated features.

Embodied in this analysis is a three-fold central concept:
(1) the tectonosphere of the Earth is a unified dynamic entity
that has been evolving to its present state and behavior since
the Earth began; (2) the evolution of the tectonosphere was
motivated through a single global mechanism operating continually
during the past 4.6 b. y.; and (3) evidence of this mechanism
exists today as surficial manifestations of the Earth's internal
behavior and structure. An important element of this concept
involves a tectonospheric Earth model derived from the dual
primeval planet hypothesis.

This hypothesis, based on fundamentals of solar-system
genesis and mechanics, establishes the tectonospheric Earth
model as a genetic solar-system entity. The driving mechanism
of this model uses energy from two sources that have existed
since primordial times: (1) the geogenetically disequilibrated
shape of the Earth; and (2) the flow of heat (and possibly of
volatiles) from the inner to the outer parts of the Earth. This
driving mechanism has produced, according to this concept, the
present tectonospheric structure as well as a panorama of tectonospheric behavior during the past 4.6 b. y. Consequently,
the Earth's surficial features are manifestations of the tectonospheric driving mechanism.

Using a multiple-hypothesis approach, the book provides a
judicious review of current hypotheses about tectonospheric
behavior, structure, and evolution, introducing the tectonospheric Earth model as an alternate working hypothesis in areas
where current hypotheses are weak, non-commital, or non-existent.
Being unified and global in both geometrical and mechanical
aspects and spanning a temporal frame of 4.6 b. y., this hypothesis provides answers within much of the spatio-temporal domain

applicable to the evolution of the Earth's tectonosphere during the past 4.6 b. y.

The book discusses correlations between observation and predictions of this hypothesis in areas related to tectonospheric evolution: continental and oceanic structure, seismicity, heat flow, magmatic activity, mountain building and related geosynclinal development, gravity anomalies, geomagnetism and polarity reversals, sea-floor spreading, continental drift, plate tectonics, and other surficial manifestations of the Earth's internal behavior.

TABLE OF CONTENTS

THE EARTH'S TECTONOSPHERE:

ITS PAST DEVELOPMENT AND PRESENT BEHAVIOR

Chapter 1
THE ORIGIN OF THE SOLAR SYSTEM

Introduction.

The Earth's tectonosphere comprises the crust and upper
mantle to a depth of about 1000 km. This book analyzes the
evolution of the Earth's tectonosphere by studying the surficial
manifestations of the Earth's internal behavior on a global
scale. Forces operating within closely-defined constraints
over long periods of time appear to have caused these manifestations.

A mountain, for example, suggests long-lived radially directed
forces. Sea-floor spreading and continental drift imply tangential

forces operating on a global scale. Earthquakes involve the accumulation of stresses from long-lived forces having both radial and tangential components. These and other observations imply that the internal forces producing the Earth's surficial manifestations are deep-seated, long-lived, and global.

Most hypotheses about the evolution of the Earth's tectonosphere embody the global and deep-seated nature of these internal forces but leave unanswered questions regarding how long these forces have been operating.

If the forces generating the Earth's internal behavior have existed since the Earth began, solar system evolutionary hypotheses should define the nature of these forces. If these causal forces originated after the Earth began, solar-system evolutionary hypotheses should indicate when they originated and how their deep-seated and global characteristics have been maintained.

To find answers to these and related questions, this chapter examines hypotheses regarding the origin and evolution of the solar system.

A Brief Description of the Present Solar System.

The Earth and other planets (except perhaps Pluto) have so many features and characteristics in common that they appear to have a common origin. The salient common characteristics, which must be taken into consideration in any discussion of the origin of the solar system, may be summarized as follows:

a. All the planets rotate about the Sun and describe almost circular orbits, in contast to most other celestial bodies which most usually describe ellipses of high eccentricity. The planetary orbits within the solar system are practically in the same plane, which is inclined about 6° to the equatorial plane of the Sun. All planets orbit the Sun in a similar direction, rotating about their respective axes in this same direction, except for Uranus whose equator is inclined at about 98° and possibly Venus which appears to rotate retrogradely.

b. The distances of the planets from the Sun form a series in which the separation between the planets increases with their distance from the Sun in a very nearly geometric progression. This relationship was discovered by Bode in 1772; it predicted the presence of another planet in the space between Mars and Jupiter. Thirty years later, the region was found to contain not a planet but a swarm of fragments, called asteroids, which may be the remnants of one or more former planets.

c. Bode's separation, between Mars and Jupiter, divides the planets into two distinct groups. The terrestrial planets (Mercury, Venus, Earth, and Mars) are all small, fairly dense, and close to the Sun. They rotate relatively slowly on their axes and possess only a few satellites. The Jovian planets (Jupiter, Saturn, Uranus, and Neptune) are all large and relatively distant from the Sun; they rotate rapidly on their axes and have many satellites. Pluto, the most distant, seems not to fit the general scheme and may perhaps be an erstwhil satellite of one of the other planets. Like the Earth, the other terrestrial planets are probably poor in gaseous elements, whereas the Jovian planets are relatively rich in gases, particularly methane, ammonia, and possibly hydrogen.

d. Although the Sun contains 99% of the total mass of the solar system, it contributes only 2% of the angular momentum, the other 98% being contributed by the planets, with Jupiter furnishing almost all of it.

Today, the solar system consists of the Sun, nine planets, many with extensive satellite systems, numerous smaller objects, such as asteroids and meteorites, a large family of comets, and a mixture of dust and gas in the space between the major members of the solar system. There is evidence to indicate that our solar system is not unique but that many other stars have planetary systems similar to that of the solar system. The birth of a star has, therefore, been of interest to astronomers for many years, since the origin of the solar system probably followed the same general steps as those which other stars appear to

4

follow in stellar evolution.

Briefly stated, most astronomers now believe that the evolution of a star may be described as follows: A mass of gas and dust in interstellar space is in motion, and the various forces acting on the gas and dust are in stable equilibrium. The primary forces postulated for such systems are (1) those due to the gravitational attraction between the particles and (2) the centrifugal forces due to rotation. At some stage, since matter in space is heterogeneously distributed, the balanced system would tend to become unbalanced according to the principal of universal heterogeneity. This, in turn, would cause the condensation to proceed, with the dust and gas aggregating toward the center of mass of the system. Such, briefly stated, is the consensus of how a star is born.

Although there is general agreement among astronomers about this as being the preferred general hypothesis for stellar formation, there is no general agreement regarding the details for such things as how a star acquires a planetary system. Consequently, a myriad of hypotheses has been proposed by astronomers, physicists, and mathematicians to provide an explanation for the evolution of the quasi-orderly planetary system surrounding our Sun.

Particularly during the past few hundred years, many noteworthy scientists have turned their attention to the problem of the origin of the solar system. As a result, more than thirty major hypotheses have been proposed, since the middle of the 18th century, to account for the origin of the solar system. All of them are open to criticism and none is completely convincing, but several of them appear to include some measure of truth and permanence.

It is not necessary to review all of the more than 30 hypotheses. However, representative ones will be summarized as a means of emphasizing the complexity of solar system evolution and the difficulty of fitting all observational evidence into a single, consistent, all-inclusive hypothesis.

All of the hypotheses have a common starting point for the substance of the planets as small particles. These particles must have been either solid or gaseous, or a combination of both. It is not likely, however, that they could have been liquid since they would soon have evaporated in the near vacuum of space. The first stage in all hypotheses is the agglomeration of the particles, and all agree that heat was involved in the process. Thus, a new planet would tend to warm up as long as the heat was not dissipated too quickly by radiation from its outer surface.

From the above, it is possible to see the basis for conflicting hypotheses regarding the origin of the solar system, since questions such as the following remained undetermined. Were the original particles gaseous, solid, or a combination of both? Were they hot or cold? Did they come from the Sun,

6

from an interloper, or from other cosmic matter which is known
to be scattered throughout all of space?

The ambiguities are unavoidable, because all cosmic matter
is of about the same basic composition regardless of where found.
Many moot questions must obviously remain, therefore. However,
it is possible to arrange the possible answers to questions
such as the above into two broad categories: (1) those definable
as uniformitarian, which postulate the formation of the planets
by a continuous, slow process; and (2) those definable as cata-
strophist, which postulate that the planets formed as the result
of some unusual and violent event. Some hypotheses of each of
these two major types are included in the following summaries.

The Nebular Hypothesis.

The first attempt to formulate a scientific hypothesis for the evolution of the solar system dates back to 1796, when the distinguished French astronomer, Simon Laplace, formulated the "nebular hypothesis" (Laplace, 1796). Actually, as is the case in almost all hypotheses, Laplace's idea was not entirely original but had been suggested, a few years earlier, by the famous German philosopher, Immanuel Kant. However, credit for the nebular hypothesis is usually given to Laplace since it was he, rather than Kant, who actually developed the idea into a form simple and lucid enough to permit other scientists to understand, test, and thereby be willing to accept it as a working hypothesis.

Laplace assumed that in the beginning of the solar system, all matter existed in the form of a hot, slowly-rotating, gaseous nebula, much larger than the present extent of the most remote planet. He then introduced simple physical principles to show how the primeval nebula evolved into the present solar system.

According to Laplace's concept, as the nebula lost heat to outer space it cooled and shrank in size. This, in turn, caused acceleration in the rotational velocity and gradually increased the centrifugal force in the equatorial plane. As a consequence, the nebula slowly evolved into a thin, lenticular shape. Eventually, the rotational velocity was such that the centrifugal force at the periphery equalled the force of gravity.

8

When this occurred, a peripheral belt of the lenticular nebula was thrown off as a gaseous annulus, while the rest of the nebula continued to shrink centrally away from it. The annulus continued to rotate about the primary with the angular velocity it had at the time of separation. Then, with continued shrinkage of the central nebula, a second ring was shed in a similar manner, followed by sufficient rings to allow one for each planet. Laplace then conceived that mutual attraction among the particles within each annulus eventually caused it to break and condense into a sphere that continued to orbit along the path previously occupied by the annulus.

Thus, in the nebular hypothesis, Laplace's simple scheme seemed to account logically for the gross features of the solar system as they were observed in the nineteenth century. In particular, this simple scheme could account for the spacing of the planets, their nearly circular and concentric orbits, their rotational motion in a more-or-less common plane, the gross organization of the satellite systems, and the residual heat of the Sun. However, even Laplace recognized certain shortcomings in his scheme, and he relegated his hypothesis to the position of a note in a larger work and presented it as having been arrived at "as a result of neither observation nor calculation". Nevertheless, the simplicity of the scheme gave it a certain attractiveness. In fact, it had such a wide appeal that it was not until over 100 years later that the more serious shortcomings of the nebular hypothesis were recogniz

One of the most obvious failings of the nebular hypothesis was that, if the present Sun were the residue of the original nebula and had shrunk to its present size since the formation of Mercury, the innermost planet, then the present Sun should be a thin, lenticular body spinning rapidly and on the verge of shedding another annulus. However, it is well known, of course, that in actuality the Sun is essentially spherical in shape and rotates slowly.

Perhaps even more critical is the problem of the angular momentum of the solar system. Since the angular momentum of a planet is, for all practical purposes, the force propelling it along its orbit, it is quantitatively approximated by the product of its mass and the area swept over, in unit time, by an imaginary line connecting it to the center of attraction about which it revolves. In this connection, Russell (1935) showed that the entire solar system does not possess enough angular momentum to have caused an annulus to have been thrown off at any stage of solar system evolution.

A third shortcoming of the nebular hypothesis involves the direction of rotation of the planets. Simply stated, if the planets had been formed according to Laplace's hypothesis, the direction of planetary rotation should be retrograde according to Kepler's laws, rather than direct as they are.

These three, plus other shortcomings, forced the abandonment of Laplace's nebular hypothesis at about the turn of the present century. However, as will be seen in a later section, certain features of this hypothesis were introduced into present hypotheses when other hypotheses developed even more critical shortcomings.

The Tidal Disruption Hypothesis.

Two of the scientists responsible for analyzing the short-comings of the nebular hypothesis, Chamberlin and Moulton, were responsible also for supplanting it by a more satisfactory hypothesis: the tidal disruption hypothesis (Chamberlin and Salisbury, 1928). They assumed that the Sun had traveled alone in space until it almost collided with another star, perhaps five or six billion years ago. At that time, as the two stars passed within a few million miles of each other, each was partly disrupted by tidal forces of mutual attraction. After the interloping star receded rapidly into space, trailing its fragmentary retinue, the pieces remaining near the Sun fell back into elliptical orbits about the Sun and condensed into the planets. Since the postulated fragments resembled tiny meteorites they were termed "planetesimals", and an alternate name for the tidal disruption hypothesis was the planetesimal hypothesis.

The planetary satellites, under this hypothesis, were assumed to be lesser condensations that were too far from the primeval planets to be intercepted but near enough to be captured.

One of the more serious shortcomings of this hypothesis involved the almost circular orbits of the planets. To rectify this, Chamberlin and Moulton postulated that a large part of the matter ejected from the Sun dissipated into a vast cloud which revolved about the Sun in a spherical configuration.

Consequently, the planetesimal condensations also traveled in circular orbits.

However, aside from the question of the adequacy of the above ad hoc explanation, the tidal disruption hypothesis has other shortcomings. For example, an analysis by Spitzer (1939) indicated that the entire disrupted matter should have dissolved into a vast cloud of dust and gas and that it would never have condensed into planetesimals. Also, of course, the Jovian planets are largely gaseous and thereby difficult to derive from planetesimals except under special assumptions.

In view of the above and other considerations, Jeffreys (1918) and Jeans (1919) modified the basic tidal disruption hypothesis by placing greater emphasis on the Sun's enormous gravity field. However, a more serious problem was faced by the basic tidal disruption hypothesis of Chamberlin and Moulton and its several modifications, including the Jeans–Jeffreys gaseous–tidal concept. Briefly stated, this problem arises from the fact that the angular momemtum of Jupiter exceeds that of all the rest of the solar system; that is, it would have required more momentum to launch Jupiter into its present orbit than to place all the other present planets into theirs and to give the Sun its present spin.

For these and other reasons, the basic idea of all tidal disruption hypotheses is now considered inadequate to account for the present organization and behavior of the solar system.

13

As a result, most newer hypotheses for the origin of the solar
system have reverted to a modification of concepts similar to
those embodied in the original nebular hypothesis of Laplace.
One of these, the dust cloud hypothesis, is described in the
following section.

The Dust Cloud Hypothesis.

Astronomers have known for years that much of outer space is filled with dust clouds. It was only natural, therefore, that two renowned astronomers should combine their talents into a "dust cloud" hypothesis for the origin of the solar system.

Accordingly, in 1944, C. F. von Weizsäcker (1944) began the development of a hypothesis based on the premise that the solar system has evolved from a dust cloud. Basically, von Weizsäcker postulated that, in the remote past, the Sun was at the center of such a dust cloud which was rotating slowly and extended out to beyond the present orbit of Pluto. Under the von Weizsäcker concept, individual particles in the dust cloud revolved about the Sun in "free" or Keplerian orbits. That is, their angular velocity differed progressively with distance from the Sun. Von Weizsäcker postulated that this created turbulent eddies arranged in concentric zones about the Sun.

On the basis of the above, von Weizsäcker proceeded to develop protoplanetary bodies from the concentric, turbulent eddies (See Fig. 1-5-1). Briefly stated, this meant that the condensations began in a postulated roller-bearing type structure and gradually grew until the structure developed gravitative fields capable of capturing all the material within the counter-rotating eddies. Mutual attraction among the eddies within

15

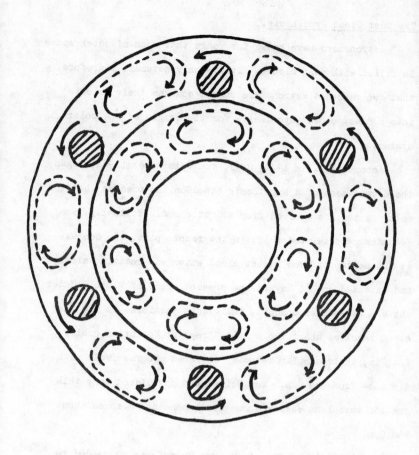

Fig. 1-5-1. Schematic diagram of von Weizsäcker's hypothesis, showing concentric rings of clockwise eddies (dashed) and counterclockwise "roller-bearing" eddies in a solar nebula.

16

a single zone of the configuration was unopposed and gradually
caused them to coalesce into a protoplanet much larger than
the present planets and much more diffuse. Under this concept,
satellites would be formed when the larger protoplanets developed
zones of turbulent eddies.

Von Weizsäcker's concept was elaborated by G. P. Kuiper
(1951), who added theoretical and mathematical aspects. Among
other things, Kuiper postulated that the solar system evolved
from a gaseous nebula of cosmic composition, i. e., with a predominance of hydrogen and helium but with much lesser amounts
of the heavier elements and of atomic particles of many kinds.

Under the Kuiper concept, the initial nebula was cold;
but, as it shrank from mutual attraction of the individual
particles, it became more dense at its center and progressively
more rarefied at its periphery. Then it began to rotate slowly.
As it shrank further, its angular momentum increased with the
consequence that centrifugal force in the equatorial plane
gradually transformed the dust cloud to a lenticular shape.
However, unlike Laplace, Kuiper postulated that centrifugal
force and internal gravity came into balance at all depths
within the nebula, so that, by the time it had shrunk to the
diameter of Pluto, it became stable and rotated as a unit.

Then, in accordance with Kepler's third law, the period
of revolution varied with distance from the center, such that
the internal parts rotated much faster than the outer. These

17

angular velocity differences created vast turbulent eddies. But, in contrast to von Weizsäcker, Kuiper postulated that turbulent eddies were not necessarily generated in concentric zones but could occur anywhere within the cloud. Then, both large and small eddies formed and reformed until a few finally achieved a stable configuration which in turn grew at the expense of the smaller, less-stable eddies. As a consequence, the nebula finally assumed a configuration composed basically of several large, rotating gaseous masses, one for each planet.

However, there are still unresolved problems. One of these concerns direction of rotation. If the young protoplanet was traveling in a roughly, circular orbit and sweeping up planetesimals also moving in circular orbits, the resulting motion should have been retrograde. But not all observers agree with such an analysis; Whipple (1964), for example, says that, if the young protoplanet was spiraling either in or out through a field of solid particles moving in circular orbits, it should have developed forward rotation. The answer probably lies somewhere within these two schools of thought.

To account for the present satellite system, the Kuiper-von Weizsäcker hypothesis assumes that each protoplanet was originally much larger than its present counterpart, both because it was much less dense and also because it still contained many gaseous elements which have since been lost. Then, as the primordial protoplanet contracted, it developed turbulent

18

eddies which condensed into the present satellite systems, through a process similar to that which formed the protoplanets but on a smaller scale.

As the central part of the nebula continued to shrink, it generated heat until it reached a thermonuclear temperature. This, in turn, caused radiant energy to pour outward in sufficient quantity to dispel into space most of the gases originally settling around the protoplanets. Since this effect would be largest nearest the center, the nearer planets (Mercury, Venus, Earth, and Mars) lost most of their gas constituents and condensed into solid bodies. At the same time, the remote planets (Jupiter, Saturn, Uranus, and Neptune) were little affected. In fact, it is entirely possible that their gaseous envelopes may easily have gained in size through capture of some of the gases blown outward from the inner planets.

From the above, it may be seen that the dust cloud hypothesis explains many of the observed features and phenomena of the solar system. Among these are the spacing of the planets, their almost circular orbits, their coplanar rotation, the general organization of the larger satellite systems, and the high temperature of the Sun. Also, it offers an explanation for the differences between large and small planets. It explains Jupiter's large angular momentum as being derived from the protosun, i. e., the original rotating dust cloud.

Like all hypotheses, however, the Kuiper-von Weizsäcker

hypothesis has certain shortcomings. For example, the asteroids are too small to represent condensations from even the smallest of the hypothesized eddy systems. In fact, one of the nearest and largest asteroids, Eros, is not even roughly spherical (as would be expected of an eddy-system condensate) but it is a tumbling fragment of rock of very rough and unequal dimensions. These two reasons are among those which indicate that the asteroids are fragments of a planet (or planets) that originally formed between Mars and Jupiter but later disrupted.

Similar arguments apply also to meteorites, which do not fit easily into the basic eddy-system evolutionary scheme of the Kuiper-von Weizsäcker hypothesis. The subject of meteorites is discussed in greater detail in a later chapter of this book. Suffice it here to say that their mineralogy indicates that meteorites crystallized under great pressure and relatively high temperature, i. e., they appear to be fragments of a celestial body of planetary size.

Another problem faced by the Kuiper-von Weizsäcker hypothesis concerns the retrograde motion of 6 of the 32 known planetary satellites of the solar system. Under the hypothesis, they were originally all formed into direct orbits, but assumed retrograde orbits after subsequently leaving their erstwhile parents and then returning either to the planets from whence they came (but into retrograde orbits) or else straying to some other planet (where they assumed retrograde orbits).

20

This is discussed in greater detail in a following section, entitled The Dual Primeval Planet Hypothesis.

The Moon is unique in that it is relatively large compared with its primary, the Earth. For this, and other reasons, it poses a special problem for the Kuiper-von Weizsäcker hypothesis, as well as for all other hypotheses. Because of the importance of this problem, the details are discussed in a separate chapter, entitled The Earth-Moon System.

At this point, it might be well to quote a well-worded summary written by Sir Harold Spencer Jones (1958):

"The problem of formulating a satisfactory theory of the solar system cannot be regarded as yet solved, though important advances have been made in recent years. The theories that involved the close approach of another star to the Sun, or an actual collision, have been abandoned, and it seems reasonably certain that an explanation has to be sought in the evolution of an extended system containing gas and solid particles. - - - The theory developed by Kuiper, though not entirely free from objection, appears at present (1958) to provide the most satisfactory explanation of the main features of the solar system and may, with some modification, prove capable of leading to a theory that can be generally accepted."

The Accretionary Hypothesis.

Urey (1959) has proposed that the planets were built up by the accretion of small particles or meteorites. The principal attraction of this hypothesis is that it avoids the problems arising from the differences in chemical composition between the Sun and the planets. On the other hand, it fails to provide an explanation for the origin of the small, meteorite type bodies. At the present time, the Earth is bombarded by about 500 kg of meteorites per day; at this rate, the mass of the Earth would increase by only one ten thousand millionth, or about one millimeter of depth, in a period of 3 b. y. Obviously, then, under Urey's concept, there would have to be an extremely large source of small bodies during the early stages of the formation of the Earth. If these could be provided, then Urey's concept would have certain very attractive features, particularly when the combined Earth-Moon origin is considered as a single system. Urey's hypothesis will, therefore, be reconsidered briefly in the next chapter when the Earth-Moon system is discussed.

Unanswered Questions Regarding the Origin of the Solar System.

A review of the above representative hypotheses reveals that there is no completely satisfactory hypothesis for the origin and evolution of the solar system. Therefore, in analyzing the evolution of the Earth's tectonosphere, certain assumptions must, of necessity, be made regarding the "missing parts" in schemes for the evolution of the solar system. Preliminary to making such assumptions, however, it is well to stop and review briefly some of the unanswered major questions regarding the origin of the solar system.

Three such unanswered major questions come to mind almost immediately: (1) the observed retrograde motion of 6 of the 32 natural satellites within the solar system; (2) observational evidence supporting the existence of a complete spectrum of iron-to-stony meteorites; and (3) observational evidence supporting certain unexplained differences between the composition, density, and other characteristics of the Earth and Moon. Although it is true, of course, that not all of these unanswered major questions have a _direct_ bearing on the evolution of the Earth's tectonosphere, it is important to realize, nevertheless, that any solar system evolutionary scheme that is to be used as the basis for the evolution of the primordial Earth should at least attempt to answer these three major questions, since the tectonosphere appears to have been evolving to its present state since almost the birth of the primordial Earth.

The author invites the reader to consider possible answers
to the above three major unanswered questions. Specifically,
an attempt will be made to answer them on the basis of the
"dual primeval planet hypothesis", which is discussed in the
next section.

The Dual Primeval Planet Hypothesis.

Having reviewed representative classical hypotheses regarding solar system origin and having noted certain major questions left unanswered by them, it is well now to consider what logical assumptions should be made in those areas where present hypotheses are inadequate, either from the standpoint of logic or of the details necessary for even a preliminary analysis of the evolution of the Earth's tectonosphere during the past approximately 5 b. y. Therefore, the reader is invited to consider certain features of the dual primeval planet hypothesis which the author has devised as a "working hypothesis" for supplying answers where none exist in classical schemes for the origin of the solar system.

From the preceding review of representative hypotheses regarding the origin of the solar system, it may be seen that, generally speaking, most solar system evolutionary models are based on a simple monistic configuration, such as may be represented schematically as shown in Fig. 1-8-1. Regardless of the ultimate source of the material from which the planets are assumed to be condensed, the monistic concept provides that condensation shall be basically into one planet or protoplanet in each circumsolar orbit. (See, for example, Kuiper, 1951, 1956; Urey, 1951, 1952, 1956, 1959, 1960; Alfven, 1954; Levin, 1958; Schmidt, 1958; Lyttleton, 1961).

Although basic concepts other than the monistic are feasible

25

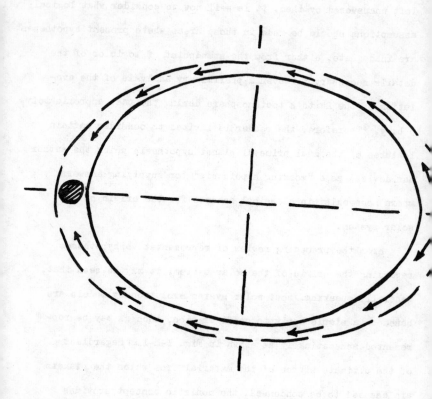

Figure 1-8-1. Formation of the planets from matter surrounding the primordial Sun: basic monistic concept. (Schematic; not to scale).

from elementary considerations, the monistic is normally used, partly perhaps because of its simplicity and partly because all present known planetary configurations are of course monistic. However, in a basic analysis of the evolution of the solar system, it might be well to consider briefly also certain more general "polyistic" concepts. Under such concepts, the initial planetary or protoplanetary condensation, in any particular circumsolar orbit, might not necessarily be into single bodies, as shown in Fig. 1-8-1, but might be rather into a multiplicity of planetary or protoplanetary condensations in any given circumsolar orbit, as shown schematically in Fig. 1-8-2.

Obviously, the exact number of such planetary or protoplanetary condensations assumed to have existed in any given primordial circumsolar orbit is more or less arbitrary and, in fact, inconsequential with the final results (so long as it is greater than one) since it can be shown that any polyistic planetary configuration tends to degenerate, with time, into a dualistic configuration and, eventually, into a monistic one. Fig. 1-8-3 shows schematically one way in which a multiplicity of planetary condensations in a given orbit might degenerate into a dualistic configuration.

Certain features of the actual process of the further degeneration of the resulting dualistic planetary configuration into a monistic one are of interest since they have a bearing on the evolution of the Earth's tectonosphere when the dual pri-

27

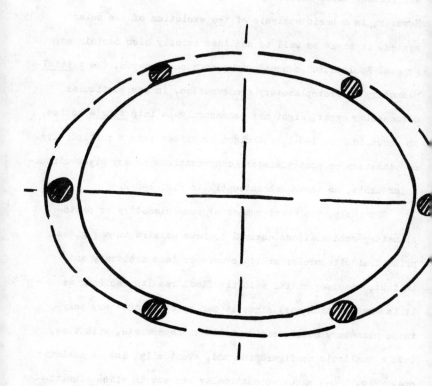

Figure 1-8-2. Hypothetical condensations within one planetary orbit from matter surrounding the primordial Sun: basic polyistic concept (Schematic; not to scale).

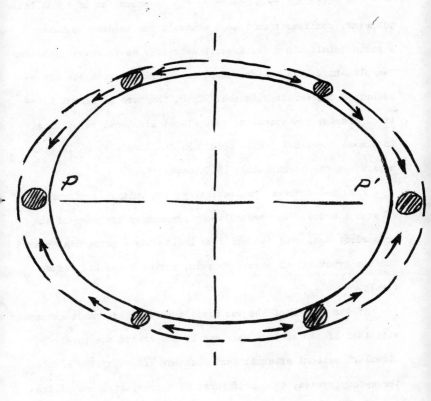

Figure 1-8-3. Formation of the planets from matter surrounding the primordial Sun: basic dualistic concept. (Schematic; not to scale).

meval planet hypothesis is applied to the primordial Earth.

The basic analysis regarding the degeneration of a dualistic planetary configuration into a monistic one is done by using a basic dualistic model described briefly as follows: Consider two planetary-size bodies formed in accordance with the above-described degeneration scheme. Thus, they are of almost identical masses, are traveling with almost identical velocities in almost identical orbits about the Sun, and with an initial separation of approximately 180° between them.

Fig. 1-8-4 shows, schematically, the initial configuration of such a model. The central body represents the mass of the primordial Sun. The two orbiting bodies are a primordial planet and its hypothetical prime, or twin, marked P and P', respectively.

If the masses of the two orbiting bodies are small compared with that of the central mass and if the orbits are purely circular, without external perturbations affecting the simple three-body system, this configuration approximates one of the "stable" Lagrangian configurations. (See, for example, Lagrange, 1772; Moulton, 1914; Whittaker, 1937; Finlay-Freundlich, 1958; Kurth, 1959; Brouwer and Clemence, 1961; Blanco and McCuskey, 1961; Danby, 1962).

Assuming that the two bodies, P and P', are of planetary size and, as indicated above, that their respective masses, orbits, and velocities are not necessarily exactly identical,

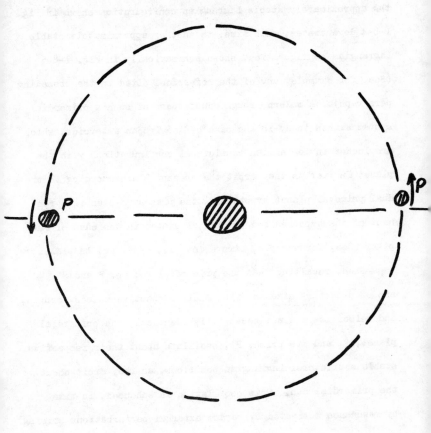

Figure 1-8-4. Initial dualistic planetary configuration in the dual primeval planet hypothesis. (Schematic; not to scale).

31

the approximately stable Lagrangian configuration shown in Fig. 1-8-4 degenerates, with time, to another approximately stable Lagrangian configuration, shown schematically in Fig. 1-8-5 (See, for example, any of the references cited in the preceding paragraph). A modern, rough counterpart of such a primordial system exists today in the case of the Trojan asteroids, which are locked in Lagrangian equilateral configurations with the planet Jupiter as they orbit the present Sun; according to the dual primeval planet hypothesis, the present system is a remnant of the hypothesized primordial system in the case of Jupiter (See, for example, Struve, et al., 1959, pp. 143-144).

Then, recalling that the primordial bodies, P and P', are of planetary size and that their respective masses, orbits, and velocities are not necessarily identical, the primordial planet, P, and its prime, P', oscillate about their respective stable equilateral Lagrangian positions, as they orbit about the primordial Sun. This oscillation is enhanced, in time, by resonance motivated by random external perturbations arising from the extreme heterogeneity and turbulence existing within the solar system during the early, formative stages of planetary evolution (See, for example, Kuiper, 1951). The resulting resonance causes the primordial planet, P, and its prime, P', eventually to fracture (but, of course, without necessarily causing either of them to separate into the fracture parts, since gravitational and other forces tend to hold the parts

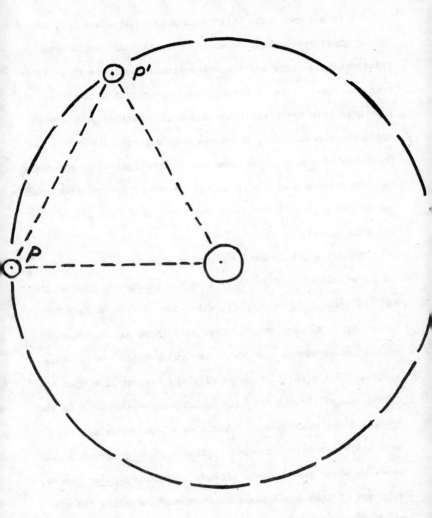

Figure 1-8-5. Temporarily stable equilateral Lagrangian configuration in the dual primeval planet hypothesis. (Schematic; not to scale).

33

together).

It is assumed that, other things being equal, a body fracturing under resonance does so harmonically and, therefore, fractures into parts whose configurations duplicate that of the original body. Thus, a tetrahedron, under such conditions, fractures into tetrahedra; and so forth. Similarly, a sphere, under these conditions, fractures into octants, since octantal fragments of a sphere are more nearly spherical than any other regular fragments of a sphere, the center of gravity of a spherical octant being only about 13% removed from that of a sphere of the same volume.

Since the configuration shown in Fig. 1-8-5 is not perfectly stable, the distance between the primordial planet, P, and its prime, P', eventually, with time, decreases, and the bodies approach each other. When this happens, the "weaker" of the two separates along the previously described fracture planes. (Since one of them is slightly less massive then the other, one of them is slightly weaker than the other. In the special case, wherein such difference might be negligible, both bodies would, of course, separate, more or less simultaneously, along the fracture planes. But in the general case, only one of them separates into its fracture parts, the other remaining intact but fractured).

Before considering the general case, it might be of interest that, according to the dual primeval planet hypothesis,

34

the asteroidal belt between Mars and Jupiter constitutes what remains of the hypothetical planet, Aster, and its prime, Aster Prime. This is discussed in more detail in Chapter 19 (Asteroids, Meteorites, and Tektites).

Also, before considering the details of the general case, wherein only _one_ of the primordial bodies, P and P', separates into its fracture parts while the other remains intact but fractured, it should be pointed out that certain simplifications will be made in connection with the following analyses. For example, the initial analyses need consider only the case in which separation is into **perfect** octants of a **perfect** sphere. Although the approximations resulting from such simplifications are adequate for this analysis, the reader should realize, of course, that in more advanced analyses, such simplifying approximations are not used, since the fracture products of a non-spherical, non-homogeneous body, such as the primordial Earth, could not, of course, be perfect octants of a perfect sphere. For the purpose of this analysis, however, only the case of separation into perfect octants need be considered. To help visualize the geometry and mechanics involved in the general case, the accompanying sequence of two-dimensional schematic diagrams has been prepared to show the initial path of each of the spherical octants of the hypothetically separating body, i. e., the primordial body, P'. This is further simplified by successively considering each of several possible modes of separation.

Thus, Fig. 1-8-6 shows the paths of the spherical octants of a non-rotating, fractured sphere which is assumed to be separating in the absence of any significant field. In this simple case, the spherical octants move radially outward relative to their center of mass (which, of course, continues to move along its approximately circular orbit about the primordial Sun).

Fig. 1-8-7 shows the paths of the spherical octants for the same situation as that of Fig. 1-8-6, except that the separating body is assumed to be rotating at the time of separation into the octantal fragments. In this case, the eight spherical octants move outward along spiral paths from the moving point which defines their center of mass (which, as before, continues to move along its approximately circular orbit about the primordial Sun, of course).

Fig. 1-8-8 shows the paths of the spherical octants for the same situation as that of Fig. 1-8-7, except that the results include the effects from the significant fields of the Sun, of the unseparated body, and of the spherical octants of the separating body. As indicated in the two-dimensional diagram, when this ten-body problem is solved by the methods of celestial mechanics, the four spherical octants, P_1' and P_2', of the separating body, P', assume direct orbits about the unseparated body, P. (The actual analysis, facilitated by high-

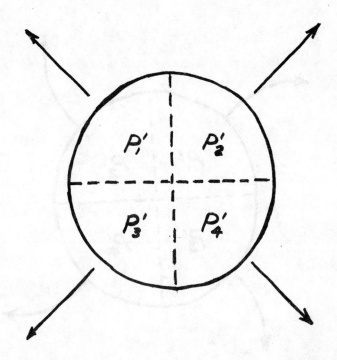

Figure 1-8-6. Paths of the initial fragments of a non-rotating, fragmentizing body in the sbsence of a significant external field, with initial fragments assumed to be into octants. (Schematic; not to scale).

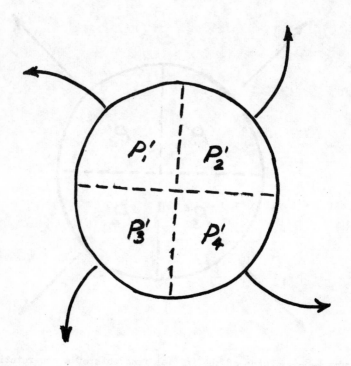

Figure 1-8-7. Paths of the initial fragments of a rotating, fragmentizing body in the absence of a significant external field, with the initial fragmentation assumed to be into octants. (Schematic; not to scale).

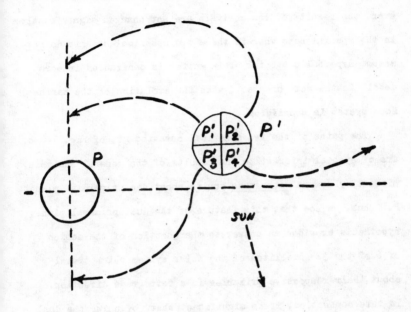

Figure 1-8-8. Paths of the initial fragments of a rotating, fragmentizing body, P', in the presence of a significant external field due to the Sun, to the body, P; and to the fragments of the fragmentizing body, P'. (Schematic; not to scale).

39

speed computers, is normally done in three dimensions, of course; but, in this two-dimensional schematic diagram, half of the spherical octants are hidden behind those in view).

In the general case, the two spherical octants, marked P_4', assume hyperbolic orbits and are normally lost to P. However, the results of the analysis are not changed significantly in the special case wherein these two spherical octants do <u>not</u> assume hyperbolic orbits. This matter is considered in more detail in the next chapter , when the evolution of the Earth-Moon system is considered.

The paths of the two spherical octants, P_3', of Fig. 1-8-8, are of special interest to the results of the general analysis, since they assume <u>retrograde</u> orbits about the primordial planet, P. Thus, we see that this feature of the dual primeval planet hypothesis provides an answer to the question of the motion of 6 of the 32 satellites of the solar system which travel about their respective primaries in a retrograde direction. In this connection, it is significant that, by using the dual primeval planet hypothesis, the motion of the retrograde satellites does not depend upon any special reversing mechanism but is simply a consequence of the <u>same</u> mechanism which explains the direct satellites.

By making appropriate adjustments in the initial values of the variable parameters of the model, each of the several satellite systems of the present solar system is duplicated.

The asteroids evolve, of course, in the special case mentioned earlier, wherein both the primordial planet and its prime separate, more or less simultaneously, into their respective fracture parts. The evolution of the meteorites and comets, similarly, is a consequence of the basic dual primeval planet hypothesis. Also, the various orbital and axial obliquities of the planets and their satellites are accounted for. The hypothesis also provides a mechanical framework for studying other aspects of the solar system, such as, for example, the possibility of the existence of intra-Mercurial planets and extra-Plutonian planets, as well as for investigating certain aspects of possible planetary systems for stars other than the Sun.

However, for the purpose of this book, which is to analyze the evolution of the Earth's tectonosphere, we are concerned with only three aspects, or predictions, of the dual primeval planet hypothesis. These are (1) the evolution of the Earth-Moon system; (2) the evolution of the primordial Earth; and (3) the evolution of a complete spectrum of iron-to-stony meteorites. Each of these aspects of the dual primeval planet hypothesis is considered, together with appropriate other hypotheses, in the following chapters, beginning with an analysis of the evolution of the Earth-Moon system in Chapter 2.

REFERENCES

Alfven, H., 1954. _On the Origin of the Solar System_.
Clarendon, Oxford, 191 pp.

Blanco, V. M., and McCuskey, S. W., 1961. _Basic Physics of the Solar System_. Addison–Wesley, Reading, pp. 179 et seq.

Brouwer, D., and Clemence, G. M., 1961. _Methods of Celestial Mechanics_. Academic, New York, pp. 260 et seq.

Chamberlin, T. C., and Salisbury, R. D., 1928. _The Two Solar Families_. Univ. of Chi. Press, Chicago, 156 pp.

Danby, J. M. A., 1962. _Fundamentals of Celestial Mechanics_. Macmillan, New York, pp. 198 et seq.

Finlay-Freundlich, E., 1958. _Celestial Mechanics_. Pergamon, New York, pp. 33 et seq.

Jeans, J. H., 1919. _Problems of Cosmogony and Stellar Dynamics_. Cambridge Univ. Press, London, 121 pp.

Jeffreys, H., 1918. On the early history of the solar

system. _Mon_. _Not_., _Roy_. _Astron_. _Soc_., 78: 424–441.

 Jones, H. S., 1958. The origin of the solar system.

Endeavour, 171 (67): 140–144.

 Kuiper, G. P., 1951. On the origin of the solar system.

In: J. A. Hynek (Editor), _Astrophysics_. McGraw-Hill, New

York, pp. 357–427.

 Kuiper, G. P., 1956. The formation of the planets, part

3. _J_. _Roy_. _Astron_. _Soc_. _Can_., 50: 158–173.

 Kurth, R., 1959. _An Introduction to the Mechanics of_

the Solar System. Pergamon, New York, pp. 136 et seq.

 Lagrange, J. L., 1772. _Essai sur le Probleme du Trois_

Corps. Paris Academy (Collected Works, vol. 6), pp. 229 et seq.

 Laplace, S. P., 1796. _Exposition du Systeme du Monde_.

Paris.

 Levin, B., 1958. _The Origin of the Earth and Planets_.

Foreign Languages Publishing House, Moscow, 88 pp.

Lyttleton, R. A., 1961. An accretion hypothesis for the origin of the solar system. _Mon. Not. Roy. Astron. Soc._, 122: 399–405.

Moulton, F. R., 1914. _An Introduction to Celestial Mechanics_ Macmillan, New York, pp. 277 et seq.

Russell, H. N., 1935. _The Solar System and Its Origin._ Macmillan, New York, 235 pp.

Schmidt, O. J., 1958. _Theory of the Earth's Origin._ Foreign Languages Publishing House, Moscow, 139 pp.

Struve, O., Lynds, B., and Pillans, H., 1959. _Elementary Astronomy._ Oxford, New York, 396 pp.

Spitzer, L., Jr., 1939. Dissipation of planetary filaments. _Astrophys. J._, 90: 675–688.

Urey, H. C., 1951. The origin and development of the Earth and other terrestrial planets. _Geochim. Cosmochim. Acta_, 1: 207.

Urey, H. C., 1952. *The Planets, Their Origin and Development*. Yale, New Haven, 178 pp.

Urey, H. C., 1956. Diamonds, meteorites, and the origin of the solar system. *Astrophys. J.*, 124: 625.

Urey, H. C., 1959. Primary and secondary objects. *J. Geophys. Res.*, 64: 1721-1737.

Urey, H. C., 1960. Criticism of the melted moon theory. *J. Geophys. Res.*, 65: 358-359.

Von Weizsäcker, C. F., 1944. Evolution of the solar system from a dust cloud. *Z. Astrophys.*, 22: 319.

Whittaker, E. T., 1937. *Analytical Dynamics*. Cambridge, London, pp. 406 et seq.

Chapter 2

THE EARTH-MOON SYSTEM

Complexity of the Earth-Moon System.

The Earth-Moon system may be defined as consisting basically
of the Earth, the Moon, and such meteorites and "interlopers"
as may have come under the influence of the Earth's gravitational
field during the past 4.6 b. y. This chapter will, therefore,
discuss those general features, phenomena, and interrelationships
of the Earth, the Moon, and the meteorites as might be necessary
for an understanding of the evolution of the Earth's tectonosphere.
Because the Earth and the meteorites are discussed in other
chapters, this chapter will be devoted primarily to the Moon
and its possible genetic relationships with the evolution of
the Earth and the meteorites.

Broadly speaking, the Earth's tectonosphere appears to be
composed of meterial very much like that of the Moon. In other
respects, however, there are gross differences between the Earth
and Moon. The complexity of some of these differences was first
suspected, in 1610, when Galileo pointed his telescope at the
Moon for the first time. Before long, scientists were asking
questions about the apparent similarities and differences in
the Earth and Moon. In a never-ending attempt to answer basic
questions about such apparent similarities and differences,

46

more and more observations and analyses were made about the Moon.

Unfortunately, however, many of the observations and analyses tended to complicate the problems, rather than to simplify them. For example, in 1959, the initial photographs of the farside of the Moon suggested that it has a basically different topography than does the nearside. Ten years later, the results from the Apollo flights further complicated the problem.

Gross Characteristics of the Moon.

On the basis of presently available information, the gross characteristics of the Moon may be summarized as follows:

a. The Moon is a very old body, probably as old as the Earth, i.e., about 4.6 b. y.

b. Like the Earth, the Moon is severely depleted in alkalis.

c. The mineralogy of the lunar maria indicates high-temperature crystallization under anhydrous conditions, and all present evidence is contrary to any significant role for water in either lunar minerals or topography.

d. The composition of the lunar surface minerals clearly suggests magmatic differentiation as the basic lunar genetic process, but the ages of the lunar rocks indicate that such differentiation occurred at a very early time in the history of the Earth-Moon system.

e. Cosmic-ray exposure studies of lunar crystalline minerals suggest either (1) a constant eroding of the lunar surface by a process similar to sandblasting, or (2) a sudden ejection of these rocks from, perhaps, impact craters.

f. There are basic differences in lunar and terrestrial interior structure (See, e. g., Schreiber and Anderson, 1970).

g. There are many lunar features which appear to have been produced by quasi-endogenous tectonism rather than

48

by impact (See, e. g., Ronca, 1965, 1966).

h. The lunar interior presents a somewhat paradoxical aspect: in certain respects, the lunar structure appears strong; in other respects, it appears fractured (See, e. g., Vrebalovich and Jaffe, 1967).

i. Certain linear features of the Moon appear to be genetically endogenous (Ronca, 1965, p. 293), caused either (1) by a reactivation of the deep-seated lunar grid system, or (2) by the escape of the volatiles from the subsurface along fractures related thereto.

j. The Moon has a prominent pear-shape (See, e. g., Michael, 1966), suggesting a very unusual internal structure, with the surface rising 2.2 km to form an equatorial bulge.

k. The basic seismic behavior of the Moon is suggestive of the response expected from a "sphere with a mushy core", or from a "small planet with a double crust", with the top of the inner crust at a depth of several hundred km (Schreiber and Anderson, 1970).

l. The Moon does not appear to be layered like the Earth but rather to be deeply fractured and heterogeneous.

m. The lunar magnetic field is only about 30 gammas, or more than three orders of magnitude less than the Earth's field of 35,000 gammas (See, e. g., Helsley, 1970a, 1970b).

n. The remanent magnetism of some lunar breccias

49

includes <u>two</u> components: (1) a large <u>viscous</u> component with a time constant of several hours; and (2) a high-coercivity remanence, possibly acquired by <u>impact</u> processes on the lunar surface (Doell et al., 1970).

o. The bulk of the mechanical, seismic, gravimetric, and magnetic evidence for the Moon indicates that the lunar interior is quite unlike that of the Earth.

p. Lunar mascons seem to be quite different from similar features on the Earth, in that the lunar mascons appear correlatable with topographic <u>deficiencies</u> rather than with <u>excesses</u> and, presumably, with <u>ancient</u> geology (i. e., the ringed maria), rather than with <u>recent</u> geology as is the case on the Earth (Kaula, 1969).

q. It is difficult to determine exactly <u>what</u> material constitutes the lunar <u>interior</u>. However, several analyses seem to suggest that the volcanic rocks on the lunar surface may have cooled, for some days, at temperatures above the solidus, <u>under vacuum conditions</u>, with consequent loss of the alkalis and other volatile oxides (See, e. g., O'Hara et al., 1970). Thus, according to that concept, the material collected by the Apollo flights would represent the residue of <u>fractional</u> <u>sublimation</u>, superimposed upon the effects of fractional crystallization, thereby suggesting that the lunar interior is composed of clino-pyroxenite or amphibole peridotite or similar material.

Lunar Craters.

In its usual definition, a crater is a circular, polygonal, or elongate topographic depression, generally with steep inner slopes. Lunar craters range in size from the very small to more than 100 km in diameter.

Almost continual bombardment of the Moon, particularly during the Moon's early history, has served to give the lunar surface a "well-cratered look", even to the unaided eye. Such bombardment usually arises from one or more of the following causes: (1) ions from cosmic rays, solar wind, or solar storms; (2) gases; (3) meteoroids; (4) the nuclei of comets; and (5) asteroids (Whipple, 1967).

Some observers interpret disconnected crater "chains" to be evidence of lunar volcanism along linear traces of subsurface fractures (See, e. g., Kosofsky and El-Baz, 1970, p. 100).

Some irregular depressions in crater floors may indicate subsidence of a solid crust into a weak substratum or possibly a fractured substructure (Ibid., p. 101).

Lunar craters appear to fall into two distinct categories: (1) circular and (2) relatively non-circular (Ronca and Salisbury, 1966; Ronca, 1967). The relatively non-circular category is composed entirely of craters older than the maria. These un-usual, non-circular craters may be explained in several ways: (1) that they were formed by a different process than were the ordinary, circular craters, or (2) that they have the same origin as the circular ones but were deformed during a period of compressive stress in the lunar crust (Ronca, 1967).

The thickness of rubble (1 to 10 m) on some mare surfaces implies that most major craters were formed early in lunar history, followed by lava extrusion (Short, 1970).

Analyses of the lunar craters and other effects of the bombardment of the lunar surface have served not only to explain what gave the Moon its cratered appearance but also have helped to explain the nature of the lunar rocks (Fredriksson et al., 1970).

Faults, Rilles, and Domes.

Faults, on the Moon as on the Earth, are planes along which rock masses have shifted horizontally, vertically, or diagonally, apparently in response to subsurface readjustments or equilibrations. At places, the lunar surface appears to have faults with escarpments as high as a quarter of a km (See, e. g., Kosofsky and El-Baz, 1970, p. 103).

Rilles, unique to the lunar surface, bear some resemblance to terrestrial valleys. They may be rectilinear, sinuous, or composite. In most cases, the sinuosities vary only slightly from a straight line. Many are suggestive of grabens and may be 10 km wide and half a km deep. They may extend for hundreds of km across the lunar surface with complete disregard for other surficial features, thereby suggesting a genetic relationship with subsurface fractures. Particularly noteworthy in this respect is a sinuous rille, 150 km long and 1 km wide, whose average direction is perfectly rectilinear for its entire length.

Domes are raised, circular structures up to 20 km across and several hundred km high, suggestive of terrestrial mesetas. They abound on the lunar surface, particularly in the mare regions. Since some steep-sided domes are similar to terrestrial volcanic domes, some scientists associate lunar domes with volcanic activity (Kosofsky and El-Baz, 1970, p. 103).

Maria and Terrae.

Since the 17th century, when they were actually considered
to be oceans, the dark areas of the Moon have been called maria,
from the Latin word for sea. One of the most unusual observations
regarding the maria is that they are far more numerous and deeper
on the nearside than on the farside of the Moon. Generally speaki
they constitute the dark, relatively flat and smooth areas of the
lunar surface. Many of the maria are circular. The sizes of
maria range up to 1300 km (Mare Imbrium). Some are surrounded
by arcuate cliffs as high as 1.3 km. Many maria contain long,
linear wrinkle ridges extending diagonally across them, suggestive
of a subsurface readjustment along a fracture plane. Some maria
have long (200 km), narrow (8 to 10 km) ridges, as well as elongat
rectilinear features which appear to be fractures (Kosofsky
and El-Baz, 1970, p. 41).

Lower parts of maria may be composed of eclogite (Ringwood
and Essene, 1970). There is an indication that the low viscosity
of the lunar parent magmas (possibly an order of magnitude less
than that of terrestrial olivine basalts) is related to low
concentrations of Si and Al (Weill et al., 1970). As a consequence
of this, lunar lavas, in spite of lower gravity, could have been
capable of faster and more extensive flow than their terrestrial
counterparts, say the primordial flood basalts. Thus, before
solidifying, the primordial lunar lavas could easily have flowed

sufficient distances to fill even the largest lunar basins to produce the maria during primordial times.

In contrast to the maria, the relatively <u>bright</u>, <u>rough</u>, and <u>elevated</u> lunar surfaces are referred to as terrae, or highlands, to distinguish them from the relatively darker, smoother, and lower areas of the lunar surface (maria). Much of the highlands area, though rough, is actually a highly-modified plain. Morphological features of the highlands, although originally sharp, have become subdued with time. In part, this subdual may be the result of the movement of loosely consolidated material from higher positions to lower levels on the surface, possibly under the influence of lunar gravity, aided by moonquakes (Kosofsky and El-Baz, 1970, p. 52). Others could have resulted from more general causes, such as gravitational sliding associated with isostatic adjustment of a fractured, heterogeneous understructure. The resulting texture, in either case, would tend to make the lunar topography more nearly coincide with the selenoid (the gravitational shape of the Moon).

Mountain ranges, perhaps more genetically related to the basins than to the highlands, form arcuate patterns around most major circular basins. Some of these mountains lie within the highlands but they appear to be concentric with nearby basins. Many mountains of lesser magnitude appear in the highlands as isolated blocks or peaks.

Most Apollo results seem to indicate that the highlands
are _older_ than the maria. At many places, the interface between
maria and terrae is quite unique, with a convex profile of certain
terrae appearing to bulge out over the adjacent maria (See, e. g.,
Kosofsky and El-Baz, 1970, p. 60). In some such cases, a depressi
or trench at the foot of the slope suggests that the contact
at this point is _controlled_ _by_ _faulting_. Compositionally, it
appears that the maria contain more iron (and related chemical
elements) than do the highlands.

The Lunar Farside and Nearside Features.

The lunar farside is markedly different from the nearside in that it has fewer very large basins and virtually no maria comparable in either extent or volume with those of the nearside. Mare Orientale, the largest of the circular features on the farside, is only partially filled with mare material. Mare Moscoviense, also on the farside, is comparable with Mare Nectaris, one of the smaller nearside circular maria. Although the _floors_ of a _few_ individual craters, most notably Tsiolkovsky, are substantially covered with _smooth_, dark mare material, the predominant impression is of the scarcity of such material on the farside. In this respect, _much_ of the farside could be compared, in gross morphology, with the southern highlands of the nearside (Kosofsky and El-Baz, 1970, p. 124). The latter region's large basins with light-toned floors, exemplified by the craters Baily and Clavius, have about a score of farside counterparts, most of which are larger.

The pattern which radiates from, or is concentric with, Mare Orientale dominates the eastern part of the lunar farside: ridges and furrows extend radially over 1000 km from this multiringed circular basin, while secondary craters, believed to have originated from Orientale ejecta, overlie older features on much of the eastern farside.

As is the case in the southern highlands on the Moon's

nearside, the structural and textural characteristics of the farside are complex. This is largely due to the existence of countless craters of various ages and sizes, each modified by younger ones.

Three concentric circular scarps ring the inner basin of Mare Orientale (21°S, 85°W). The outermost and best developed, the Cordillera Mountains scarp, is almost 900 km in diameter. The Cordillera and Rook Mountains, which form the middle ring, are among the highest mountain chains on the Moon. Each rises more than 3 km above the adjacent terrain.

The inner basin is nearly, but not completely, filled with dark material such as is characteristic of mare areas. The unfilled portions, which retain the features of the original basin floor, provide important new data on the likely early configuration on the other older large basins. Surrounding the _basin_ _center_ to distances up to 1000 km is a _coarsely_ _braided_ blanket of material that clearly covers an older, cratered terra surface. The unusually well preserved textures on this surface, and the sharpness of the mountain rings indicate that Mare Orientale may be among the youngest of the Moon's large, circular basins (21°S, 85°W).

The dark material covering the inner Orientale basin (on the farside of the Moon) may be quite shallow, since there are numerous "islands" of material that project through it. The circular, rectilinear, and irregular depressions are probably "collapse" features, formed under control of subsurface structures which also controlled the complex pattern of rilles and wrinkle ridges. All these structures appear to have formed _after_ the flooding of the central depression (24°S, 96°W).

The 200-km crater Tsiolkovsky (20°S, 129°E) is unique in that its floor is _partially_ filled with smooth, _very_ dark material, through which protrudes a prominent central peak. The diagonal banding, which gives the rim crest a "swept" appearance, is the surficial manifestation of a mass of material that may have been formed by an enormous downslope movement comparable in magnitude with a terrestrial avalanche.

Three other observations serve to emphasize the contrast between the lunar farside and nearside: (1) some evidence seems to indicate that the nearside is _flatter_ than the farside, i. e., the nearside is more planar, whereas the farside appears to be more rounded; (2) some evidence indicates that the crust of the nearside is _thinner_ than that of the farside; and (3) almost all of the many basins on the nearside appear to be more-or-less filled whereas those few which are on the farside appear to be "dry" or empty.

Kosofsky and El-Baz (1970) have reported a number of probable igneous intrusions within the lunar highlands on both the nearside and the farside. Here again, the contrast between nearside and

farside is evident: there are fewer such probable intrusions
on the farside than on the nearside.

Other interesting contrasts between the lunar farside and
nearside are discussed in Markov (1962), Barabashov et al. (1961),
and Kosofsky and El-Baz (1970).

Composition of Lunar Rocks.

To appreciate the complexities of the evolution of the Earth-Moon system, it is necessary to have an understanding of the lunar rocks as they are now, plus some indication of how and to what extent present lunar rocks are related to primordial lunar rocks, as well as how and to what extent they may be related to terrestrial and meteoritic rocks, both present and primordial.

Since an understanding of present rocks is pivotal to an understanding of how rocks evolved within the Earth-Moon system, a few of the characteristics of present lunar rocks which may reveal some of the characteristics of primordial lunar rocks are listed:

a. Moon rocks, even when compared with their closest terrestrial analogues, show significant peculiarities, indicating a decidedly different history of geochemical evolution for the Earth and for the Moon.

b. Variations in ratios such as Al/Ca, Si/Mg, Na/K, and in the amounts of Ti and P emphasize the diversity of the igneous processes that produced the Earth rocks and the Moon rocks in their present forms and compositions.

c. Evidence indicates that the major process in operation on the Moon was the crystallization of silicate melts in a highly reducing (oxygen depleted) environment.

61

d.   An iron analogue of pyroxmangite, $FeSiO_2$, is found on the Moon but does _not_ occur naturally on the Earth. The same is true of even more complex lunar-unique minerals such as Ti-Fe-Zr silicates with small concentrations of Ca and Y and lesser amounts of 8 other elements including Al and Na. It is significant that all such lunar-unique rocks contain _iron_.

e.   Some lunar samples contain pure iron, whereas in terrestrial rocks, iron _never_ occurs in the pure form but always as an oxide.

f.   The concentration of Ti is not only greater in Moon rocks than in Earth rocks but also it varies significantly from mare to mare.

g.   The relative abundances of O^{16} and O^{18} isotopes in lunar rocks appear to match those of terrestrial rocks, thereby suggesting that the Earth and Moon probably formed in the _same part_ of the solar system. Although this may appear to be somewhat at variance with other evidence, indicating that the evolutionary processes for the Earth and Moon differed significantly, it is not necessarily contradictory to it.

h.   A small, lemon-size rock from Oceanus Procellarum resembles an ordinary piece of terrestrial granite, but it is _compositionally_ and age-wise quite different. It contains 20 times as much radioactive U, Th, and K, and is a billion years older than the oldest of known terrestrial granites.

i. Most lunar rocks contain many of the elements with high boiling points (e. g., Ti, Zr, Cr), but very few of the elements with low boiling points (e. g., Pb, K, Na).

j. Both lunar and terrestrial basalts are basically the same, but there are significant differences in the proportions of individual minerals, i. e., _more_ pyroxenes and ilmenite (rich in Ti and Fe) but _less_ plagioclase and ferric irons in the lunar samples.

k. Compared with lunar analogues, the basalts from the terrestrial oceanic crust, ridges, and rises contain approximately the _same_ concentrations of Mg, Ca, K, P, U, Th, Y, and Yb as do the lunar samples but _far_ less Fe, Ti, Ba, and Zr, and _higher_ concnetrations of Si, Al, and Na.

l. Since the Moon's primordial radionuclide composition (U, Th, and K) is significantly different from that of either the Earth or of meteorites, it appears that the Moon has gone through a substantially different geochemical history than either the Earth or the meteorites.

m. No more than about 3% of the present lunar soil is composed of chondritic meteorite contamination.

n. Except for their high ilmenite content, the lunar rocks resemble the Ca-rich achondritic meteorites (eucrites and howardites) in both composition and structure. In this connection, Mason et al. (1970) concluded that the principal lunar rock types can be broadly grouped into ilmenite basalts and breccias.

o. Ulbrich (1970) concluded that lunar and meteoritic rocks may resemble terrestrial rocks texturally and quasi-mineralogically, but there is no chemical similarity in many of them.

p. A pseudo-similarity between the Moon and Earth may appear to lie in the fact that each is deficient in siderophiles, heavy metals that presumably sink to the core of a planet during its molten phase, leaving rock slag making up the crust and mantle. This seems to be what happened when the Earth formed but why should the Moon, which appears to have little or no metal core now, show this same marked shortage of siderophiles? Could this possibly mean that the Moon, as we now know it, once had a core? If so, what happened to it?

q. Many lunar samples appear to be the result of limited partial fusion of material similar to the brecciated eucritic meteorites (See, e. g., Philpotts and Schnetzler, 1970).

r. Certain lunar rocks are depleted one to two orders of magnitude in Ag, Au, Zn, Cd, In, Tl, and Bi, suggesting loss by high-temperature volatilization _before_ or _after_ accretion of the Moon (See, e. g., Keays et al., 1970).

s. Some lunar igneous rocks appear to be the product of _extensive_ igneous fractionation (See, e. g., Gast and Hubbard, 1970).

t. Hurley and Pinson (1970) concluded that Rb depletion in lunar igneous rocks occurred during or shortly after accretion of the terrestrial planets.

u. Douglas et al. (1970) concluded that the overall homogeneity, igneous textures, and absence of xenoliths in the lunar crystalline rocks indicate that they were derived from a common Ti-rich magma by internal, anorogenic volcanism rather than by impact. In their model, crystallization conditions allowed strong compositional variations in pyroxenes, olivine, and plagioclase, as well as the growth of certain lunar-unique minerals.

v. Lunar glasses occur primarily (1) as spheres that are associated with lunar basalts, or (2) as coating glasses that compositionally approximate basalts and microbreccias (Argell et al., 1970).

In addition to the above listed fundamental differences and similarities between lunar, terrestrial, and meteoritic minerals, there are also individual differences within each of these. For example, there is a complete spectrum of meteorites, ranging from pure stone, through stony-iron, to pure iron; these are discussed in Chapter 19 (Asteroids, Meteorites, and Tektites). Similarly, there is a broad spectrum of terrestrial minerals; these are discussed in Chapter 10 (Intrusive and Extrusive Activity).

Also, as would be expected, the rocks in one area of the Earth are somewhat different from those of other areas. The same is true for different parts of the Moon. The most pronounced differences in rocks on the Moon are found between the lunar highlands and the maria. However, there are gradations as one goes from one highland to another, and from one mare to another. As one example, salient differences between the minerals at Mare Tranquillitatis (Apollo 11) and those at Oceanus Procellarum (Apollo 12) may be summarized as follows (See, e. g., LSPET, 1970):

a. Both are of about the same age, with the OP (Oceanu Procellarum) rocks averaging about 5% to 10% younger, perhaps.

b. The OP material contains about an order of magnitude less microbreccia than does the MT (Mare Tranquillitatis) material.

c. The regolith (an incoherent covering above the underlying rocks) at OP is only about half as thick as that at MT, with a possibility that this item is related to the previousl indicated differences in degree of brecciation at the two sites.

d. The amount of solar wind material in the OP fines is considerably less than that in the MT fines.

e. The crystalline rocks in the OP collection display a wide range in both modal mineralogy and primary texture, in contrast to the apparent uniformity of the MT rocks.

f. The "non-earthly" chemical character of OP samples (high in refractory and low in volatile concentrations) is shared by the MT samples in an even more pronounced manner.

g. The chemical composition of fine material at the OP site is the same as that of the breccias, but it is different from that of the crystalline rocks; this relationship is also true for the MT site but to a lesser degree.

h. In the OP breccias (which are similar to those from MT), the average mineral composition is _less_ olivine-rich than that of the majority of the rocks collected.

i. Impact metamorphism similar to that found in MT samples occurs also in OP samples.

Age of the Moon.

In analyzing the evolution of the Earth-Moon system it is well to compare the age of the Moon with that of other parts of the system. For example, if all parts of the Earth-Moon system are of about the same age, a different type of evolutionary hypothesis might be applicable than if some parts of the Earth-Moon system are not of the same age.

Here it is understood that the "age" intended is that which dates back to the time when the rocks being studied were last crystallized. On this basis, there are some lunar, terrestrial, and meteoritic rocks which have about the same age, i. e., they were crystallized from a molten state at about the same time in the history of the Earth-Moon system. This does not mean that such simultaneous crystallization necessarily occurred at the same place within the Earth-Moon system. Nor is this necessarily applicable to more than a relatively few of the lunar, terrestrial, and meteoritic materials. Nor would it necessarily by fair to say that all, or even most, lunar, terrestrial, and meteoritic materials were once of the same "age". However, it may be said that the "younger" ones are those which crystallized most recently. Obviously, then, a rock which has just recently crystallized is the "youngest" of all, regardless of its previous history. From this it follows that some rocks have had several "ages", the number being the same at the number of times the rock has been crystallized. Thus, if a rock was crystallized (or recrystallized) 4.6, 3.6, 2.7, and 1.8 b. y. ago, it is now 1.8 b. y. old, but had previous "ages" of 2.7, 3.6, and 4.6 b. y.

The above indicates that, particularly when analyzing the evolution of a large system such as the Earth-Moon system during the past 4.6 b. y., it is essential that we consider the different mechanisms which might have changed the "age" of a rock (i. e., crystallized it from a melt or recrystallized it from an already crystalline state). Perhaps even more important is the question of what was the causal mechanism or device which operated to fix the age (or perhaps to change the "age") of a given rock, or of a given group or suite of rocks. Why, for example, are the "oldest" terrestrial rocks only about 3.6 b. y. old? What was the causal mechanism or device which operated to fix (or to change the age) of most of the Earth rocks sometime during the past 3.6 b. y.? Why did not a similar causal mechanism operate to change the age of most of the lunar rocks sometime during the past 3.6 b. y.?

These and other questions must be answered by any successful hypothesis for the evolution of the Earth-Moon system. It will have to show why certain lunar soil, breccia, and rocks appear to have remained unmolten during the past 4.6 b. y. It will have to show also why the oldest meteorites landing on the Earth are about this same age but that no such meteorite has ever been found with a chemical composition exactly like that of lunar rocks of similar age. Yet, although they differ in age, some of the oldest lunar rocks are similar in appearance to rough, marbelized granite found near meteorite craters on the Earth. What exactly were the spatial and temporal characteristic of the geometry and mechanics which permitted such well-constrain behavior during the evolution of the Earth-Moon during the past 4.6 b. y.?

Reynolds et al. (1970) analyzed a Mare Tranquillitatis rock 4.1 b.y. old which reached the lunar surface 35 to 65 m. y. ago and lay amid soil whose particles have typically been within a meter of the surface for a b. y. or more. In this connection, although Mare Tranquillitatis may have formed as recently as 3.6 b.y. ago, the soil and breccia there are about 4.6 b. y. old, indicating that the time of initial differentiation of the lunar crust was about 4.6 b. y. ago (Albee et al., 1970). Again, we may ask what were the temporal and spatial mechanics and geometry of the Earth-Moon system at that time which permitted it to perform so precisely and within such well-defined constraints?

From the above it appears that, unlike that of the Earth, most of the Moon's activity took place during the first part of its history. Any successful hypothesis for the evolution of the Earth-Moon system would have to explain this, of course.

Needless to say, these and other observations serve to place very strict constraints on the mechanics and geometry of lunar evolution. For example, unless evidence regarding the age of the lunar material is proved erroneous (and it is not likely that it will be), then it appears that one can presume either (1) that the Moon was accreted from _already_ differentiated material, or (2) that, _early_ in the history of the Earth-Moon system, some device such as a large number of impacts played a major role in creating superheated melts over considerable portions of the lunar surface (See, e. g., Fredriksson et al., 1970). This process seems to have been somewhat selective, because it appears that lunar activity might have occurred more profusely on the _nearside_ than on the farside, for some reason of other. Even when both of the processes suggested by Fredriksson et al. (1970) are assumed to have operated in conjunction with each other, a hypothesis with nicely constrained mechanics and time constants is almost essential if one wishes to explain the observations regarding the early evolution of the Earth-Moon system.

Evidence for Hot Lunar Evolution.

One of the major unanswered questions about the Moon conce
its origin. In an attempt to answer this basic question, astro
and physicists have tried to determine, on the basis of both
direct and indirect evidence, to what extent vulcanism and related magmatic processes have occurred on the Moon. If this
could be determined, it would be easier to determine how the
Moon originated and evolved to its present state and behavior.

If a body as large as the Moon compositionally resembled
either the Earth or chonodritic meteorites, then it follows
that it must have undergone substantial heating at some time
in its history. This, of course, leads to questions as to how
it was heated, when and for how long the heating lasted, and to
what extent heat still exists within the Moon.

The Moon rocks analyzed from the Apollo flights proved
to have surprisingly low ratios of K/U compared with meteorites
and terrestrial rocks of similar bulk composition. Yet, the
lunar rocks very definitely contain enough K, U, and Th to prod
a thermally active Moon if it is assumed that the lunar rocks
examined are representative of the composition of the entire
Moon. However, the Apollo results indicate that the lunar surficial rocks have not been heated above 250° C during the past
3.6 b. y. This means, simply stated, that if the igneous rocks
from Mare Tranquillitatis and other maria are volcanic eruptives
from a hot interior, then this would require that such occurred

72

as a very early thermal phase followed by at least 3.6 b. y. of relatively cool conditions on the Moon.

The surprisingly old age found for the lunar maria challenges many carefully conceived hypotheses for the origin of the Moon as well as for the surficial topography. Most observers now feel that the lunar maria are younger than the highlands (See, e. g., Kosofsky and El-Baz, 1970). On the assumption that the maria are indeed 3.6 b. y. old, then the highlands must include regions that are even older, perhaps as old as the solar system itself. In comparison, most terrestrial meteorites appear to have crystallized about 4.6 b. y. ago, and the oldest known terrestrial rocks appear to have crystallized about 3.6 b. y. ago.

Arguing for a "hot" Moon, many scientists have, for years, insisted that thermal activity has been an important element in lunar history and that lunar topography is governed mainly by _internal_ forces. In their view, the lunar surface consists primarily of volcanic cones and craters, ash flows or fissure eruptions, fault scarps, and grabens. Another group of "hot" Moon proponents postulates that the relatively large lunar relief of almost 10 km is mostly isostatically adjusted at depth; i. e., high elevations have deep roots. According to such a concept, the lunar crust, particularly in the highlands, consists of a relatively low-density siliceous material overlying a denser more-mafic substratum.

73

Although there is no visible evidence of large-scale vulcanism on the surface of the Moon, Kozyrev reported in 1958 that a gaseous hydrocarbon was observed in the crater Alphonsus. Since then, similar activity has been reported in Aristarchus and other craters. Also, infrared maps of the Moon contain "hot" areas, suggesting thermal anomalies within the Moon.

Most present observers, however, prefer to believe that the Moon shows more evidence of having been cold during most of its existence. But, although there are less "hot" Moon proponents today than there once were, there are still many who claim that much evidence points to the necessity for a lunar evolution involving a "hot" Moon. As will be seen from the next section, there is even more evidence indicating that the lunar evolution might have involved both a hot and a cold Moon.

Evidence for Cold Lunar Evolution.

The previous section has summarized evidence supporting a hot Moon. However, there is much evidence that the Moon has been cold for at least the last 3.6 b. y. For example, although the previous section considered the possibility that the lunar craters are volcanic in origin, a more plausible explanation may be that they resulted from meteoritic impact (See, e. g., Baldwin, 1963).

Lunar crater diameters range from millimeters to hundreds of kilometers, in sharp contrast to terrestrial craters which range from fractions of km to only about 30 km, making the smallest lunar craters smaller that the smallest terrestrial craters and the largest lunar craters larger than the largest terrestrial craters. Even more surprising, almost all the volcanic activity on the Earth is limited to well-defined, linear belts, whereas the lunar craters appear to be more-or-less randomly distributed.

The total number of lunar craters is almost unbelievable, there being over 300,000 with diameters larger than 1 km on the nearside of the Moon. Proponents of the impact hypothesis consider the lunar maria as being flows of molten material resulting from local, temporary heating triggered by high-energy impacts. Under such a concept, the reported gas discharges from Alphonsus and Aristarchus, as well as the infrared hot spots, are interpreted as being "residual" effects from fairly recent meteoritic impacts.

Aside from the craters and related topographic aspects of
the lunar surface, there are other observations about the Moon
which support a "cold" Moon evolution. For example, the Moon
has no appreciable magnetic field, suggesting that it does not
now have a metallic core and that the Moon might possibly always
have been in a cold, unmolten state. Such is not necessarily
the only interpretation, of course, since there is other evidence
indicating that the Moon might at one time have had a core.

Another physical characteristic of the Moon suggests a
"cold" Moon evolution: the Moon's orbital motions show that
it has a disequilibrated shape, suggesting that the interior
is strong enough to have resisted long-term stress and must,
therefore, have been cold since it was formed, 4.6 b. y. ago,
or at least during the last 3.6 b. y. Specifically, the Moon's
shape is not that of a smooth spheroid of revolution but rather
a triaxial ellipsoid with an apparent bulge of several km toward
the Earth. Although this bulge is interpreted from the Moon's
gravity field, it has been widely assumed to represent an <u>actual</u>
topographic bulge on the lunar <u>surface</u>. Some observers feel
that it was formed and "frozen" in at some remote time when the
Moon was closer to the Earth.

Alternatively, of course, the gravity "bulge" may represent
major inhomogeneities within the <u>interior</u> of the Moon, rather
than an <u>external</u> topographic bulge. In either case, of course,
the preservation of either a fossil tidal bulge of an asymmetric

distribution of lunar mass would be difficult to explain on the basis of a "hot" Moon history, since the magnitude of the bulge is such that a near-zero viscosity is indicated for the Moon for at least the past 3.6 b. y.

Recent data show that instead of a single mass inhomogeneity on the nearside of the Moon, there may be several mass concentrations (also called "mascons"), suggested by individual positive gravity anomalies as large as 230 mgal, localized roughly over certain ringed maria (See, e. g., Muller and Sjogren, 1968). The relative size and importance of the mascons may be appreciated when one recalls that the gravity anomaly of the Mare Imbrium mascon is equivalent to that of an iron ball about 100 km in diameter lying at a depth of about 50 km; that of the Mare Serenitatis mascon is equivalent to an iron meteorite 37 km in diameter buried at a depth of 270 to 400 km, or to a lenticular-shaped stony meteorite 600 km in diameter and 8 km thick.

Since the lunar maria cover about 35% of the nearside of the Moon but only about 2% of the farside, the question arises, of course, as to why. If the maria and the associated mascons are due to deep-seated magmatic activity, why are they so asymmetrically limited to only one side of the Moon? Even if it is assumed that the maria resulted from meteoritic impacts, their asymmetric distribution on the far and near sides of the Moon is just as puzzling. It is generally agreed by astronomers and physicists that all meteorites striking the presently airless

Moon tend to blast craters and largely destroy themselves in the process. Has such always been the case?

In one hypothesis, large meteorites or asteroids up to 100 km in diameter are postulated to have come in at relatively low velocities and were, therefore, not destroyed on impact but became the mascons. Since only a satellite in _Earth_ orbit could attain so low a velocity, it is postulated that the impacting bodies were orbiting the Earth in primordial times and that the Earth-Moon separation was much less, but it was increasi The impacting bodies, therefore, collided with the Moon's primord _farside_ which, having been made more massive thereby, shifted to the _nearside_ as a result of the Earth's gravitational pull. It is to be recalled here that low-velocity impacts would not have produced excessive melting and, therefore, the maria could _not_ have been created by such low-velocity impacts.

Whatever the origin of the lunar maria and the associated mascons, they present a strong argument for a strong _outer_ shell of the Moon for at least the past 3.6 b. y. They need not necessarily tell us anything about _inner_ parts of the Moon, nor about _any_ part of the Moon _prior_ to 3.6 b. y. ago.

Depending upon what assumptions one wishes to make, the absence of an atmosphere on the Moon may be interpreted as meaning that there has been no continuous outgassing such as occurs from the warm interior of the Earth, thereby possibly indicating that the Moon's interior is not warm or at least that its internal

78

structure is different from the Earth's or that of any other body with an atmosphere that must be continuously replaced from internal sources. With an escape velocity of only 2.38 km/sec, the lunar gravity field is weak, but it is adequate to hold an atmosphere for hundreds of millions of years. Yet at present the Moon has neither atmosphere nor ionosphere. In fact, the lunar surface vacuum is more nearly perfect that any reproducible in a laboratory. Neither a cold interior nor the Moon's low gravity can completely account for such an extremely low surface pressure; however, the absence of a lunar atmosphere favors a "cold" Moon rather than a "hot" one. But when all the evidence is considered, it appears that the lunar evolution might have involved both hot and cold processes.

<u>Evidence</u> <u>for</u> <u>Catastrophic</u> <u>Events</u> <u>in</u> <u>the</u> <u>History</u> <u>of</u> <u>the</u> Earth-Moon System.

In analyzing the evolution of any large system, it is helpful to determine at the outset whether <u>any</u> of the activity contributing to the evolution of the system was catastrophic. That is, was <u>all</u> of it such that it may be characterized as having been purely uniformitarian? If any of it appears to have been catastrophic, then the evolutionary hypothesis for the system should, of course, answer questions regarding the <u>cause</u> and nature of the suspected catastrophe or catastrophes. Also, the evolutionary hypothesis should predict or determine <u>when</u> the catastrophe occurred and <u>where</u>, including the role played by the then extant environment in which the catastrophe, or catastrophes, must have occurred.

In the case of the Earth-Moon system, there are certain indications that at least one, and possibly two, such catastrophes have occurred during the past 4.6 b. y. For example, many meteorites appear to have been involved in at least one catastrophe, some in several. Can the same be said of the rest of the Earth-Moon system? If so, then what evidence do we find that such might have occurred on either the Earth or the Moon, or both, during the past 4.6 b. y.? If the <u>entire</u> Earth-Moon system was <u>not</u> involved in the catastrophes that affected the meteorites, <u>how</u> did the Earth and Moon escape their effects?

There are some indications that the abundance of U, Th, and

K in lunar rocks can best be explained by an "event" that occurred about 4.6 b. y. ago, probably crystallization and melting caused by tremendous heat. Obviously not part of uniformitarian evolution, such an "event" falls within the definition of a catastrophe. But what caused it?

Another perplexing problem is that the lunar maria contain material spatially separated by only a few centimeters and yet temporally by a b. y. Most of the maria rocks are about 3.6 b. y. old, whereas the dust particles (and some small rock fragments) are about 4.6 b. y. old or a b. y. older than the maria rocks.

Thus, preliminary examination of the evidence would seem to indicate that the Moon might have undergone at least two catastrophes, one at its "creation" 4.6 b. y. ago and another about a b. y. later. As for the nature of the "catastrophes", it is still difficult to determine whether all lunar magmatism came from a hot lunar interior or whether all of it was created by heat from the impact of large meteorites.

If any of the lunar magmas came from the lunar interior, did they rise up slowly as they do on the Earth or did they come out violently or catastrophically? If the lunar magmas rose up slowly, as they still do on the Earth, why do they no longer do so on the Moon? Even more importantly, why does it appear that the lunar magmas suddenly ceased flowing about 3.6 b. y. ago? Why did not a similar thing happen on the Earth 3.6 b. y.

ago? Of course, to suddenly _stop_ such a flow would require energy of catastrophic proportions and such would also have to be explained. If any of the lunar magmas came out of the lunar interior by some catastrophic means, what was its cause?

Faced with these and similar questions, some observers prefer to postulate that the lunar melting was caused by a deluge of large meteorites which rained down on the lunar surface until about 3.6 b. y. ago. An extension of such a postulate explains the lack of known terrestrial rocks older than 3.6 b. y. by saying that the _same_ deluge of large meteorites that melted the Moon 3.6 b. y. ago also melted the Earth at that time. Being larger than the Moon and thereby able to gravitationally attract a proportionately greater share of the postulated deluge of large meteorites 3.6 b. y. ago, the Earth became _completely_ melted then, whereas _only_ the lunar maria were. Under this concept, both the Moon and the Earth were created about 4.6 b. y. ago but only the Earth was completely melted 3.6 b. y. ago. Unanswered, of course, is the cause and nature of the catastrophe that produced the postulated deluge of large meteorites at that time.

Upon closer scrutiny, it appears that the Moon has been subjected to two _types_ or modes of catastrophe during the early part of its evolution: (1) intense heat and (2) shock. Even closer analysis of actual lunar material reveals just how complex such heat and shock must have been. For example, Carter and acgregor (1970) concluded that the presence of breccias, glasses, glass-spattered surfaces, numerous glass lined craters on rock surfaces, and shocked rock and mineral fragments suggests a complex history of impacts, including indications that both the impacting and impacted objects varied from liquids to solids implying the presence of an impact-produced cloud of gas, liquid, or solid particles. Also, the high iron and nickel content of some lunar mounds or hillocks suggests that the impacting bodies were in part iron or nickel-iron.

Interestingly enough, such lunar catastrophic complexities appear to have had counterparts in certain meteoritic evolution. For example, Baldanza and Pialli (1969) have recently reviewed the main deformational structures of meteorites. They concluded, among other things, that some differences appear to exist between meteoritic deformation due to a relatively sudden dynamic event and others which seem more likely to have been caused by stress action with a long time constant (i. e., one which was, in some way, prolonged over an extended period of time).

Mueller (1969) has suggested that the higher percentage of shock effects in carbonaceous chondrites than in ordinary chondrites indicates that the genetic material of carbonaceous chondrites occupied zones closer to the surface of the parent than did ordinary chondrites. Under Mueller's concept, ureilites were produced by a combination of both shock and heating. Such a combined effect might have been produced by: (1) collisions of two meteorite parent bodies; (2) impact of meteorites on the surface of a parent body; or (3) volcanic activity on a parent body.

Other evidence has been found to support the existence of pre-terrestrial deformations in meteorites. In fact, Axon (1969) identified three distinct types of such pre-terrestrial deformation: (1) mild deformation within a meteorite parent body, before kamacite begins to precipitate, which may produce macroscopic twins subsequently decorated by kamacite; (2) stresses from volume changes during precipitation of kamacite, which may produce transformation twins on a microscopic scale that are similar to partially annealed Neumann lines and are decorated by later precipitation of rhabdite, accompanied by a later generation of fresh Neumann lines; and (3) more-violent deformations which occur when parent bodies are fragments and when collisions occur between fragments in space.

However, there is evidence that such did not occur in the case of all meteorites. For example, although shock pressure appears to have played an important role in the eventual capture in the case of some meteorites, such does not appear to have been true in the case of certain hexahedrites which do not contain evidence of having been shocked even sufficiently to have permitted their escape from the Moon. Jain and Lipshutz (1969) suggested that such "unshocked" hexahedrites might have come from an "unshocked" asteroid.

There is evidence that the reheating observed in chondritic meteorites was produced by shock and that it occurred **before** the meteorites were ejected from their parent bodies (See, e. g., Taylor and Heymann, 1969). Another indication of an early catastrophe within the Earth-Moon system is the fact that there is an extremely large quantity of glass on the Moon. If there had not been an early "event" to dehydrate the Moon, the glass would have reverted to a crystalline structure. What was the nature of this early event which drove off not only the water but also almost all other volatiles from the lunar surface?

Evolutionary Hypotheses for the Earth-Moon System.

Even a brief description of the Earth-Moon system reveals that it is a highly complex system and that it has had a very unusual evolutionary history. For example, there is paradoxical evidence supporting both a "hot" Moon and a "cold" Moon. More specifically, it appears that at least part of the Moon began as a hot body that cooled early in its history and has remained cool ever since.

To account for this and other seemingly paradoxical evidence numerous hypothesis have been proposed for the evolution of the Earth-Moon system. Some of these are general, all-inclusive hypotheses covering the entire Earth-Moon system during the past 4.6 b. y. Some cover only certain special or complex problems over a limited period of time. Others cover only one relatively small but important aspect or a single complex event in the history of the Earth-Moon system.

Most evolutionary hypotheses for the Earth-Moon system are directed toward one or more of the following problem areas: (1) that the lunar farside is significantly different from the nearside; (2) that the Moon contains an unusually large number of craters of all sizes, up to several hundred km in diameter; (3) that the bulk of the lunar minerals originated at about the same time as did the Earth; (4) that the environments in which the lunar and terrestrial minerals evolved were somewhat radically different; (5) that the environmental differences appear to have included variations in pressure, temperature, and stress gradients to which the rocks were subjected during their formation; and (6) that the overall density of the Moon is roughly the same as that of the Earth's upper mantle. Any successful hypothesis would, of course, have to be directed toward *all* of these areas.

It now appears that the Moon became an Earth satellite through one of three basic mechanisms or procedures: (1) independent formation in the vicinity of the Earth, probably through accretion from smaller bodies; (2) capture at a time when the Moon, an interloper from some other part of the solar system, approached the Earth under conditions satisfactory to capture; or (3) by some sort of fission or separation from the Earth. Each of these ideas is examined briefly in the following sections.

The Accretion Hypothesis.

Many astronomers, physicists, geologists, and mathematician (as well as a newer group of scientists having an interdisciplinary familiarity with _all_ these areas) prefer to postulate that the Moon is a permanent part of our system and that it developed syngenetically with the Earth. The exact details by which such cogenetic evolution of the Earth and Moon occurred are not normally specified in presently extant hypotheses. Rather, it is postulated simply in general terms that the Moon accreted, by some means, from a cloud of cold solar dust and gas (See, e. g., Urey, 1952; Alfvén, 1954).

Since dust clouds are usually assumed to be homogeneous and of cosmic abundances, the above hypothesis leaves the low density (3.3) of the Moon unexplained. Several ingeneous schemes have sought to circumvent this problem by suggesting that cold "welding" of metallic dust particles would cause primary accretion of dense metallic core material, leaving the fringes of the cloud enriched in silicates.

Even if such schemes are accepted, there are other problems involved in all _simple_ accretion schemes. For example, two simply accreting bodies, such as the Earth and Moon, would have had to maintain a very sensitive balance between their size and orbital energies during the _entire_ accretionary process. As the Earth grew more massive and of greater density, it had two reasons for exerting ever greater gravitational pull on

88

the Moon. As a consequence, the slower-growing, less-dense Moon must have, by some means, constantly managed to gain momentum at precisely the exact rate to keep it independent and to prevent it from falling either into the Earth (if the momentum-increase rate was too _slow_) or into a solar orbit (if the momentum-increase rate was too fast).

The Capture Hypothesis.

The hypothesis that the Moon was once an interloper from another part of the solar system but was captured by the Earth is based on the Moon's low density and tilted orbit. However, its density (3.3) is similar only to the outer parts of the asteroidal belt, and it does not actually match the density of any of the stony meteorites of presumed asteroidal origin. This problem is discussed in greater detail in Chapter 19 (Asteroids, Meteorites, and Tektites).

Another problem posed by the capture hypothesis concerns the Moon's mass which is about 97% greater than that of all the present asteroids combined. Therefore, if the Moon originated in the asteroidal belt, it was by far the largest of such bodies.

Also, the capture hypothesis requires that the relatively massive primordial Moon was, by some means, ejected into a highly eccentric orbit, then captured by the Earth under a low-probability mode, and then perturbed again into a nearly circular but tilted orbit. The mechanics of such a catastrophic event, involving, as it does, the temporal and spatial juxtaposition of several low-probability operating modes, are very difficult to visualize as having occurred in the general mechanical and geometric orderliness of the solar system. Most proponents of the simple capture hypothesis now agree that the Moon was moving in an orbit similar to that of the Earth when captured.

One of the strongest advocates of lunar capture is Urey

ee, e. g., Urey, 1952, 1956, 1959). He feels that the bulk
geophysical evidence supports a "cold" primordial Moon, and
has devised a hypothesis that the Moon is a primitive object
th a bulk composition similar to that of the condensed solar
bula (Urey, 1956). This composition would account for the
nsity and the low internal heat of the present Moon. Urey,
erefore, regards the Moon as being about 4.6 b. y. old with
ts present topography explainable as the result of impact. In
he maria, such impacts caused melting at shallow depths. Under
is concept, the Moon was formed near the Earth, thereby simpli-
ying the mechanics and geometry of capture by the Earth.

Although recently-acquired chemical evidence indicates that
he Moon is neither solar in composition nor wholly primordial
and undifferentiated, there are many elements of Urey's hypothesis
that make it attractive (e. g., that the Moon is very old and
that it formed near the Earth).

From basic considerations of mechanics and geometry, it
follows that, if the Earth had suddenly captured the Moon, the
results in ocean tides, as well as in tectonic manifestations
in the solid Earth, should have been such that there would be
some record in present-day geologic structure. In fact, several
pieces of evidence have been found to support such. For example,
some observers feel that Precambrian stromatolites, 1 to 2 b. y.
old, indicate that ocean tides then were of very large amplitudes,
suggesting the possibility that the Moon was orbiting the Earth
at closer range at that time (See, e. g., Alfvén, 1965; Gerstenkorn,

1967a, 1967b; Goldreich, 1966; Kaula, 1964; MacDonald, 1964, 1967; Singer, 1968). These observers then extrapolate backward from the Precambrian to postulate that the Earth's crust, which does not have any rocks of known age older than about 3.6 b. y., must have undergone a catastrophic melting and outgassing that destroyed all evidence of earlier rocks, a rock's age being reckoned from when it was last in a molten state. According to these observers, the Earth captured the Moon at that time.

Still other observers have studied the global distribution patterns, together with the ages, of anorthosites (See, e. g., Buddington, 1939; Wood et al., 1970). Their studies revealed that all known terrestrial anorthosites are about 1.4 ± 0.3 b. y. old. Also, when known anorthosite outcrops are plotted on a "pre-drift" map, they lose their random distribution and fall on two great-circular belts, one in Laurasia (northern hemisphere pre-drift configuration) and one in Gondwanaland (southern hemisphere pre-drift configuration). From such, they concluded that anorthosites are deep-seated rocks from the inner part of the tectonosphere which were moved to positions near the surface by some unique catastrophe that occurred about 1.4 b. y. ago. They postulate that the "unique catastrophe" was the capture of the Moon by the Earth at that time. The hypothesis was given further quantitative support by MacDonald (1964), whose mathematical calculations showed that the closest approach of the Moon to the Earth occurred 1.78 b. y. ago.

Alfvén (1965) has suggested that the Moon was originally captured in a __retrograde__ orbit. This brought it closer to the Earth until it reached the Roche limit, at 2.89 Earth radii, where tidal forces exceeded the Moon's gravity. As a consequence, the Moon lost much sand, rocks, and larger blocks which were literally ripped from the Moon's surface by the tidal pull of the Earth. This, according to Alfvén's hypothesis, accounts for part or even all of the sialic crust of the Earth. After losing such a large mass of material to the Earth, the angular momentum of the Earth-Moon system was reversed. This caused the Moon to assume a direct, or prograde, orbit and to recede to the vicinity of the present lunar position. As it receded, it was bombarded and cratered by the remnants of the catastrophic debris which was not captured by the Earth during the encounter. Most observers feel that if this event occurred, as hypothesized by Alfvén, then it must have occurred more than about 3.6 b. y. ago when the Earth's crust and the Moon's maria appear to have crystallized. Nevertheless, many observers feel that Alfvén's hypothesis for deriving the Earth's crust from the Moon is an interesting counterpoint to the older hypothesis which postulated that the Moon was derived from the Earth.

The Fission Hypothesis.

Since the overall composition and density of the Moon are roughly the same as those of the Earth's tectonosphere, several evolutionary hypotheses for the Earth-Moon system have considered the possibility that the Moon may be genetically related to the Earth's primordial tectonosphere. The most well-known of these is the "fission" hypothesis, which postulates that the Moon was torn out of the side of the Earth and that the Pacific Ocean is a scar from that catastrophic event.

The origin of the Moon by fission from the Earth was first proposed by George Darwin, second son of Charles Darwin. Darwin (1898) argued that the originally large, protoplanetary Earth rotated with a period of only four hours; that this set up resonance between its free oscillations and the solar tides; that it developed an enormous bulge; and that, as a consequence, it became unstable and finally threw off one or more fragments which formed the Moon.

Later, other scientists suggested that the Pacific Ocean basin is the scar resulting from this catastrophe and that the remaining sialic crust on the Earth ruptured and the continental blocks drifted apart (See, e. g., Moore, 1963, p. 15 ff.). The subject is discussed in more detail in Chapter 15 (Continents and Oceans).

The basic fission hypothesis accounts for the Moon's low density, provided it is assumed that fission occurred after the

Earth's core had already been formed, since the Moon's density and that of the Earth's upper mantle are almost identical. However, the fission hypothesis has very few other good features.

Among the disadvantages of the fission hypothesis are several arising from dynamical problems involved in producing fission of a large body such as the primordial Earth, followed by the ejection of one or more large fragments in the absence of a definable energy source external to the hypothetically fissioning Earth. Also, enormous amounts of energy must be dissipated in such a postulated system without heating the Earth to temperatures that would have driven off even more of its volatile elements, contrary to observational evidence.

One version of the fission hypothesis, proposed by O'Keefe (1970 and earlier), involves splitting the protoplanetary Earth into a contact-binary system (similar to that observed in the case of certain stellar systems). Under O'Keefe's hypothesis, there would be a subsequent loss of much matter and angular momentum to form the primordial Moon. High temperatures would cause the depletion of alkalis and volatiles from both bodies. Since the recent Apollo results have shown that maria rocks are severely depleted in alkalis, many observers feel that O'Keefe's hypothesis bears re-examination, particularly since the latest results show that, in many ways, the Moon appears to be more like the Earth than like chondrites. This question is discussed in another context in Chapter 19 (Asteroids, Meteorites, and Tektites)

Specialized Hypotheses.

In addition to the above described traditional hypotheses, other general hypotheses regarding lunar evolution have been proposed (See, e. g., Baldwin, 1963, pp. 293-313; Markov, 1962; Moore, 1963, pp. 13-24; Alfvén, 1954; Firsoff, 1962; Schmidt, 1958; Sandner, 1965; Alfvén and Arrhenius, 1969). However, even when the best features of all general hypotheses are consolidated into a composite or conglomerate hypothesis, a myriad of questions remains unanswered.

This situation has prompted certain scientists to formulate specialized hypotheses to cover only certain aspects of the Moon's history over a limited span of time, or to explore a single event or problem area in the evolution of the Earth-Moon system. For example, Kaula (1969) used the results of density analyses to conclude that the Moon's history is closely related to some evolutionary process which had certain features of both fission and capture. Cameron (1970) has investigated the consequences of nebular collapse upon the evolution of the Earth-Moon system, particularly in connection with attempts to account for the depletion of iron and the more-volatile elements. Singer (1970) considered the possibility that a modification of the "tidal" hypothesis might be used to increase the inherent probability of the basic lunar capture hypothesis as well as providing heat for the differentiation and loss of volatiles in the primordial Moon. O'Keefe (1970) proposed a postsiderophile fission process for the evolution of the Moon, primarily as a means of accounting for the basic differences in lunar and terrestrial rocks. Wood et al. (1970) proposed that the lunar breccias were formed, fragmented, and then reconstituted over a _long_ repetitious process _unlike_ any known on the Earth. Ringwood (1970) concluded that compositional differences between lunar and terrestrial rocks are difficult to explain on the basis of "traditional" hypotheses and that "new" hypotheses must be found to replace the traditional ones.

The Dual Primeval Planet Hypothesis.

One of the consequences of the author's dual primeval planet hypothesis is a primordial Earth-Moon system composed basically of: (1) the primordial Earth (a body of 5400-km radius, differentiated, and fractured into octants); and (2) six octants of Earth Prime (a similar body which has separated into octants as explained in Chapter 1). This primordial Earth-Moon system is shown schematically in Fig. 2-11-1. In this two-dimensional diagram, half of the octants are hidden behind those in view; i. e., _two_ octants of Earth Prime are at each of the locations marked A, B, C, and D. The two at D assume hyperbolic orbits and are lost to the local system.

The post-primordial Earth-Moon system evolved from the above system as a result of collisions involving the six octants of Earth Prime (shown two at each of the positions marked A, B, and C in Fig. 2-11-1) as they orbited the primordial Earth. A subsequent configuration is shown in Fig. 2-11-2.

Both the present Moon and Earth evolved from the above system. The details of how the Earth evolved are contained in subsequent chapters. Suffice it here to say merely that fragments of Earth Prime play an important role in increasing the size of the primordial Earth from a radius of about 5400 km to its present size. Five octants of a sphere of 5400-km radius, when accreted onto a sphere of similar size, will increase the radius to about the present radius of the Earth.

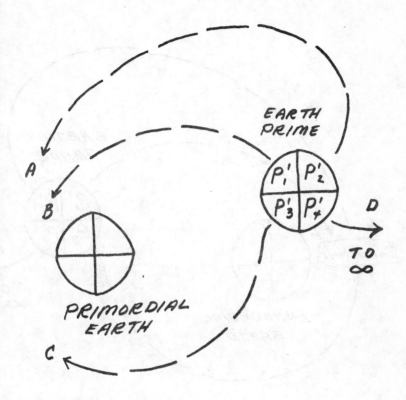

Figure 2-11-1. Schematic representation of the basic elements of the primordial Earth-Moon system. In this two-dimensional schematic diagram, half of the octants are hidden behind those in view; i. e., <u>two</u> octants of Earth Prime are at each location marked A, B, C, and D. The 2 at D assume hyperbolic orbits and are lost to the local system. (Schematic; not to scale).

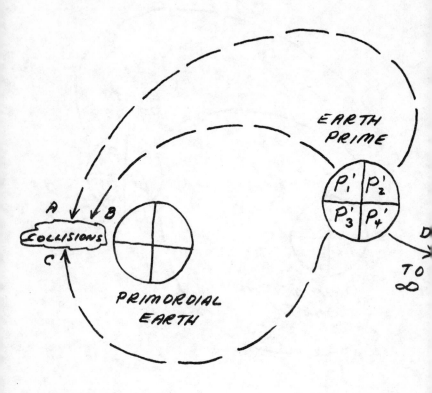

Figure 2-11-2. Schematic representation of the collisions involving the "lunar" octant and 5 "terrestrial" octants of Earth
Prime. Two octants of Earth Prime are at each location marked
A, B, C, and D; those at D assume hyperbolic orbits and are lost
to the local system. (Schematic; not to scale).

100

Depending upon the degree of secondary collisions and fragmentation assumed, two basic types of Moon model may be evolved from the primordial Earth-Moon system shown schematically in Fig. 2-11-2: (1) the "octantal-fragment" Moon model; and (2) the "twice-accreted" Moon model.

A schematic representation of the evolution of the "octantal-fragment" Moon from Earth Prime is shown in Fig. 2-11-3.

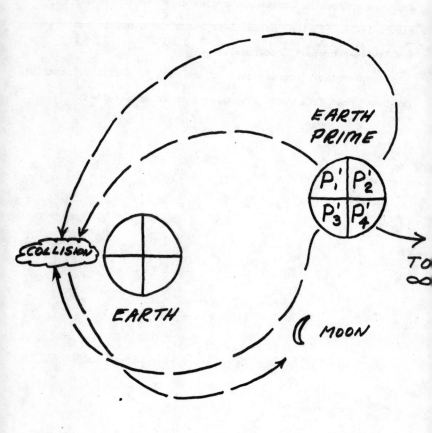

Figure 2-11-3. The evolution of an "octantal-fragment" Moon from Earth Prime in accordance with the dual primeval planet hypothesis. (Schematic; not to scale).

102

The Octantal-Fragment Moon Model.

In the octantal-fragment Moon model, as soon as the lunar octant separated from the quasi-stable environment of the fragmentizing Earth Prime, it was subjected to an extremely rapidly changing environment, since it suddenly found itself no longer confined by the high temperature and pressure of the interior of Earth Prime. This is particularly important in connection with its erstwhile interior molten part (i. e., the vertex of the lunar octant), which was exposed, almost instantaneously, to a rapidly-cooling, highly-reducing environment.

In addition to the expected flow and subsequent solidification of the molten part of the lunar octant, it is assumed that there would be severe buffeting from subsequent collisions with other parts of Earth Prime as they orbited the primordial Earth. Such buffeting would tend to remove the other three "horns" from the lunar octant. In the octantal-fragment Moon model, this becomes the present Moon after additional buffeting, further readjustment of liquid and plastic parts of the lunar octant, and such gravitational sliding, fracturing, and smoothing as might be expected in such a system in that environment. An idea of the small amount of such buffeting, flowing, sliding, and other smoothing required is suggested by the fact that the center of gravity of a homogeneous spherical octant is only 13% removed from that of a sphere of the same volume. For a differentiated body such as Earth Prime, it would be even less.

When the resulting octantal-fragment (or "modified-octant") Moon is analyzed, it is found that some of its salient features may be listed:

    a. Basic triaxial shape.

    b. Non-uniform density distribution.

    c. Mean density about the same as that of the Earth's upper mantle, i. e., about 3.3.

    d. An internal structure not layered like the Earth's but more suggestive of the Earth's tectonosphere.

    e. An age of about 4.6 b. y.

    f. Preponderance of maria on the nearside.

    g. Relatively few maria on the farside.

    h. A somewhat different basic structure on the farside than on the nearside.

    i. A small, unusual magnetic field, if any at all, i. e., such as would be remanent in a modified octant of a previously magnetized planet of 5400-km radius after severe buffeting.

    j. Distinctive fracture and seismic patterns.

    k. Lunar minerals suggestive of a rapidly-cooling, highly-reducing environment.

The Twice-Accreted Moon Model.

In the twice-accreted Moon model, the present Moon evolved from the equivalent of the lighter, unmolten, outer portions of one of the primordial octants of Earth Prime in two basic steps: (1) a more-complete fragmentation through repetitive collisions of the octants of Earth Prime; and (2) a re-accretion of some of this material to form the present Moon.

In the last analysis, the primary difference between the octantal-fragment Moon and the twice-accreted Moon is simply a matter of the degree of primordial repetitive collisions and fragmentations which occurred in the case of that portion of Earth Prime which later became the Moon.

It is not necessary that we consider the myriad degrees of repetitive collisions and fragmentations possibly undergone by the applicable portions of Earth Prime before the Moon was formed therefrom. For the purpose of the level of analysis being pursued in this chapter, it is necessary only that we consider the type and degree of repetitive collisions and frag- mentations that would produce bodies similar to those visualized by Urey (1959) prior to re-accretion to form the present Moon.

Higher-order, multipy-fragmentized and manifoldly accreted modes for lunar evolution differ from the modified-octant or octantal-fragment mode primarily in two salient respects: (1) the equivalent portion of Earth Prime was more highly fragmen- tized prior to re-accretion in the case of lunar evolution by the multiply-fragmentized and manifoldly-accreted modes; and

(2) lunar evolution by the multiply-fragmentized and manifoldly-accreted modes provides a Moon composed _entirely_ of secondary accretions, whereas lunar evolution by the modified-octant or octantal-fragment mode basically provides merely a _mechanical modification_ of _one_ octant of Earth Prime with some accretion thereto.

Neither the details of such differences nor the specific characteristics of the manifoldly-accreted Moon need be considered here. As far as the evolution of the Earth's tectonosphere is concerned, it is inconsequential whether the octantal-fragment mode of a higher-order mode, such as the twice-accreted mode, is used to produce the present Moon. In either case, the Moon that evolves is basically identical. The same may be said of its genetic relationship to the material composing the Earth's tectonosphere.

REFERENCES

Albee, A. L., Burnett, D. S., Chodos, A. A., Eugster, O. J., Huneke, J. C., Papanastassiou, D. A., Podosek, F. A., Russ, G. P., Sanz, H. G., Tera, F., and Wasserburg, G. J., 1970. Ages, irradiation history, and chemical composition of lunar rocks from the Sea of Tranquillity. Science, 167: 463-466.

Alfvén, H., 1954. On the Origin of the Solar System. Clarendon, Oxford, 191 pp.

Alfvén, H., 1965. Origin of the Moon. Science, 148: 476.

Alfvén, H. and Arrhenius, G., 1969. Two alternatives for the history of the Moon. Science, 165: 11-17.

Axon, H. J., 1969. Pre-terrestrial deformation effects in iron meteorites. In P. M. Millman (Editor), Meteorite Research. Springer-Verlag, New York, pp. 796-805.

Baldanzi, B. and Pialli, G., 1969. Dynamically deformed structures in some meteorites. In P. M. Millman (Editor), Meteorite Research. Springer-Verlag, New York, pp. 806-825.

Baldwin, R. B., 1963. The Measure of the Moon. Univ. of Chicago Press, Chicago, 488 pp.

Barabashov, N. P., Mikhailov, A. A., and Lipskiy, Yu. N. (Editors), 1961. Atlas of the Other Side of the Moon. Pergamon, Oxford, 141 pp.

Buddington, A. F., 1939. Adirondack igneous rocks and their metamorphism. Geol. Soc. Am. Mem. 7, 354 pp.

Cameron, A. G. W., 1970. Formation of the Earth-Moon system. Trans. Am. Geophys. Union, 51: 350.

Carter, J. L. and MacGregor, I. D., 1970. Chemistry and morphology of lunar sample surface features. Trans. Am. Geophys. Union, 51: 345.

Darwin, G. H., 1898. The Tides. Houghton, New York, 151 pp.

Doell, R. R., Grommé, C. S., Thorpe, A. N., and Senftle, F. E., 1970. Magnetic studies of lunar samples. Science, 167: 695-697.

Douglas, J. A. V., Dence, M. R., Plant, A. G., and Traill,

R. J., 1970. Mineralogy and deformation in some lunar samples. _Science_, 167: 594-597.

Firsoff, V. A., 1962. _Strange World of the Moon_. Science Editions, New York, 226 pp.

Fredriksson, K., Nelsen, J., Melson, W. J., Henderson, E. P., and Anderson, C. A., 1970. Lunar glasses and micro-breccias: properties and origin. _Science_, 167: 664-666.

Gast, P. W. and Hubbard, N. J., 1970. Abundance of alkali metals, alkaline and rare earths, and strontium-87 / strontium-86 ratios in lunar samples. _Science_, 167: 485-487.

Gerstenkorn, H., 1967a. The importance of tidal friction in the early history of the Moon. _Proc. Roy. Soc._, _Ser. A_, 296: 293.

Gerstenkorn, H., 1967b. On the controversy over the effect of tidal friction upon the history of the Earth-Moon system. _Icarus_, 7: 160.

Goldreich, P., 1966. History of the lunar orbit. _Revs._

Geophys., 4: 411.

Helsley, C. E., 1970a. Evidence for an ancient lunar magnetic field. Trans. Am. Geophys. Union, 51: 348.

Helsley, C. E., 1970b. Magnetic properties of lunar dust and rock samples. Science, 167: 693-695.

Hurley, P. M. and Pinson, W. H., Jr., 1970. Rubidium-strontium relations in Tranquillity Base samples. Science, 167: 473-474.

Jain, A. V. and Lipschutz, M. E., 1969. Shock histories of hexahedrites and Ga-Ge group III octahedrites. In P. M. Millman (Editor), Meteorite Research. Springer-Verlag, New York, pp. 826-837.

Kaula, W. M., 1964. Tidal dissipation by solid friction and the resulting orbital evolution. Revs. Geophys., 2: 661

Kaula, W. M., 1969. A tectonic classification of the main features of the Earth's gravitational field. J. Geophys. Res., 74: 4807-4826.

Keays, R. R., Ganapathy, R., Laul, J. C., Anders, E., Herzog, G. F., and Jeffery, P. M., 1970. Trace elements and radioactivity in lunar rocks: implications for meteorite infall, solar-wind flux, and formation of the Moon. Science, 167: 490-493.

Kosofsky, L. J. and El-Baz, F., 1970. The Moon as Viewed by Lunar Orbiter. NASA, Washington, 152 pp.

LSPET (Lunar Sample Preliminary Examination Team), 1970. Preliminary examination of lunar samples from Apollo 12. Science, 167: 1325-1339.

MacDonald, G. J. F., 1964. Tidal friction. Revs. Geophys., 2: 467.

MacDonald, G. J. F., 1967. Evidence from the surface configuration of the Moon on its dynamical evolution. Proc. Roy. Soc., Ser. A, 296: 298.

Markov, A. V., 1962. The Moon: A Russian View. Univ. of Chicago Press, Chicago, 391 pp.

Michael, W. H., Tolson, R. H., and Gapcynski, 1966. Lunar

orbiter: tracking data indicate properties of Moon's gravitational field. Science, 153: 1102-1103.

Moore, P., 1963. A Survey of the Moon. Norton, New York, 333 pp.

Mueller, G., 1969. Genetical interrelations between ureilite and carbonaceous chondrites. In P. M. Hillman (Editor), Meteorite Research. Springer-Verlag, New York, pp. 535-537.

Muller, P. M. and Sjogren, W. L., 1968. Mascons: lunar mass concentrations. Science, 161, 680-684.

O'Hara, M. J., Biggar, G. M., and Richardson, S. W., 1970. Experimental petrology of lunar material: the nature of mascons, seas, and the lunar interior. Science, 167: 605-607.

O'Keefe, J. A., 1970. Geochemical evidence for the origin of the Moon. Trans. Am. Geophys. Union, 51: 350.

Philpotts, J. A. and Schnetzler, C. C., 1970. Potassium, rubidium, strontium, barium, and rare-earth concentrations in lunar rocks and separated phases. Science, 167: 493-495.

Reynolds, J. H., Hohenberg, C. M., Lewis, R. S., Davis, P. K., and Kaiser, W. A., 1970. Isotopic analysis of rare gases from stepwise heating of lunar fines and rocks. Science, 167: 545-548.

Ringwood, A. E., 1970. Origin of the Moon. Trans. Am. Geophys. Union, 51: 346.

Ringwood, A. E. and Essene, E., 1970. Petrogenesis of lunar basalts and the internal constitution and origin of the Moon. Science, 167: 607-610.

Ronca, L. B., 1965. A geological model for Mare Humorum. Icarus, 4: 390-395.

Ronca, L. B., 1966. Structure of the crater Alphonsus. Nature, 209: 182.

Ronca, L. B., 1967. Minor lunar tectonics. In: S. K. Runcorn (Editor), Mantles of the Earth and Terrestrial Planets. Interscience, New York, pp. 473-479.

Sandner, W., 1965. Satellites of the Solar System. American

113

Elsevier, New York, 151 pp.

Schmidt, O. J., 1958. _Theory of the Earth's Origin_. Foreign Languages Publishing House, Moscow, 139 pp.

Schreiber, E. and Anderson, O. L., 1970. Properties and composition of lunar materials: Earth analogies. _Science_, 168: 1579-1580.

Short, N. M., 1970. Thickness of impact crater ejecta on the lunar surface. _Trans. Am. Geophys. Union_, 51: 346.

Singer, S. F., 1968. The Origin of the Moon and geophysical consequences. _Geophys. J., R. A. S._, 15: 205.

Singer, S. F., 1970. The early history of the Earth-Moon system. _Trans. Am. Geophys. Union_, 51: 350.

Taylor, G. J. and Heymann, D., 1969. Shock, reheating, and the gas retention ages of chondrites. _Earth and Planetary Sci. Letters_, 7: 151-161.

Ulbrich, M. C., 1970. Chemical individuality of lunar, meteoritic, and terrestrial silicate rocks. _Science_, 168:

1375-1376.

Urey, H. C., 1952. The Planets, Their Origin, and Development. Yale, New Haven, 178 pp.

Urey, H. C., 1956. Diamonds, meteorites, and the origin of the solar system. Astrophys. J., 124: 625.

Urey, H. C., 1959. Primary and secondary objects. J. Geophys. Res., 64: 1721-1737.

Vrebalovioh, T. and Jaffe, L., 1967. Summary of results, Surveyor III, Preliminary Report. NASA SP-146, 1-2.

Weill, D. F., McCallum, I. S., Bottinga, Y., Drake, M. J., and McKay, G. A., 1970. Petrology of a fine-grained igneous rock from the Sea of Tranquillity. Science, 167: 635-638.

Whipple, F. L., 1967. Meteoritic environment of the Moon. Proc. Roy. Soc., London, Ser. A, 296: 304-315.

Wood, J. A., Dickey, J. S., Marvin, U. B., and Powell, B. N., 1970. Lunar anorthosites. Science, 167: 602-604.

Chapter 3

A TECTONOSPHERIC EARTH MODEL

The Basic 5400-km Primordial Earth.

A tectonospheric Earth model may be defined as one which describes the Earth's evolution in terms of the mechanics and geometry of the tectonosphere. According to the dual primeval planet hypothesis, it is composed essentially of a 5400-km primordial Earth onto which 5 octants of Earth Prime have been accreted, where, simply stated, the basic 5400-km primordial Earth is a differentiated body which has been fractured into octants as explained in Chapter 1. Such a body is shown schematically in Fig. 3-2-1.

The center, C, represents a solid core, with an undefined radius assumed to lie somewhere within the range 0 to several hundred km. The hatched spherical shell, MC, represents a molten core. The 8 octants of the primordial mantle surround the molten core and are marked S. The preferential flow of heat, and possibly of volatiles, outward from the fracture belts, F, causes igneous activity at I.

The primordial Earth has a volume such that, when 5/8 of an equivalent volume is added, the resulting body has a radius of about 6371 km, the radius of the actual Earth. This gives the primordial Earth a radius of about 5400 km, thus:

1 5/8 V' = V, where V' and V are the volumes of the

116

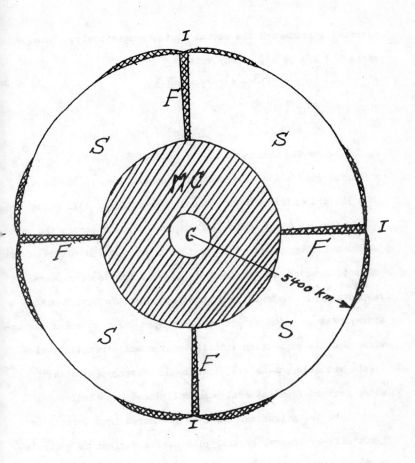

Figure 3-2-1. The basic 5400-km primordial Earth according to the dual primeval planet hypothesis. (Schematic; not to scale).

primordial Earth and the actual Earth, respectively. Then, since $V = 4/3$ pi R^3, we may write

$$1.625 \ (R')^3 = R^3, \text{ from which}$$
$$R' = (1.625)^{-1/3} \ R$$
$$R' = (1.625)^{-1/3} \ (6371 \text{ km}) \doteq 5400 \text{ km}.$$

This is the radius of the primordial Earth, as well as of Earth Prime and of the subtectonosphere used in the model.

The thickness of the tectonosphere is about 6371 minus 5400, equals 971, or about 1000 km, roughly. Although the value 5400km and 1000 km, are used in this model for the radius of the subtectonosphere and for the thickness of the tectonosphere, respectively, it should be pointed out that these values are approximations which depend somewhat upon the particular assumptions made in connection with the assignment of initial values to the variable parameters of the basic tectonospheric Earth model derived from the dual primeval planet hypothesis.

Since the primordial Earth is fractured into octants by 3 mutually—orthogonal central planes, its surface is encircled by 3 mutually—orthogonal great—circular fracture belts, each having a length of 2 pi 5400 = 34,000 km, roughly, or about 3 x 34,000 = 102,00 km total for the 3 belts.

The Geometrico-Mechanical Behavior of the Basic 5400-km Primordial Earth.

In order to analyze the geometrico-mechanical behavior of the basic 5400-km primordial Earth, it will be assumed that the basic behavior of the primordial Earth is such that it tends to equilibrate itself to a state of minimum energy. The driving mechanism of the system may be expressed as a function of the disequilibration energy inherent in its initial state. In simplest terms, it consists essentially of the resultant of 2 factors: (1) the potential energy inherent in the geogenetically disequilibrated shape of the basic 5400-km primordial Earth; and (2) the selectively-channeled energy from the preferential flow of heat, and possibly of volatiles, outward from the three mutually-orthogonal belts which are "active" along the tensile portions of the surficial traces of the 3 planes separating the primordial Earth into 8 octantal subtectonospheric blocks.

When the driving mechanism of the basic model is applied to the primordial Earth, the behavior will be dependent upon certain geometrico-mechanical constraints inherent in the model. The simpler of these constraints may be summarized in terms of several types of equilibration, or tendency to assume a status of minimum energy:

Rotary Equilibration. Fig. 3-3-1 shows schematically the expected basic behavior of the original 5400-km primordial Earth in response to the fundamental driving mechanism of the

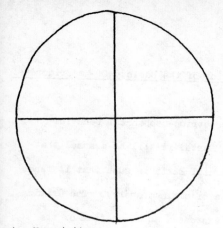

A. No rotation.

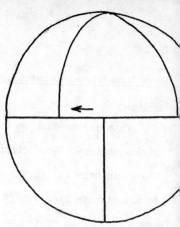

B. Rotation along the geophysic[al]
equatorial plane.

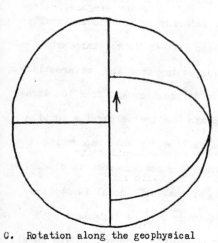

C. Rotation along the geophysical
prime meridian plane.

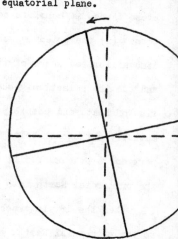

D. Rotation along the geophysic[al]
orthogonal meridian plane.

Figure 3-3-1. The expected basic behavior of the original 5400-[km]
primordial Earth in response to the fundamental driving mechanis[m]
of the tectonospheric Earth model from the standpoint of rotary
equilibration. (Schematic; not to scale).

model from the standpoint of rotary equilibration. Basically there are three degrees of internal rotational freedom, constrained by the configuration of the model to act along the three mutually-orthogonal planes along which the octantal fracturing occurred. It is noticed that, once rotational motion has begun in any one of the 3 mutually-orthogonal fracture planes, gross rotational motion in the other two planes is restricted until a net rotation of n pi/2 shall have transpired in the first plane, where n = 0, 1, 2, 3 - - -.

Radial Translatory Equilibration. Fig. 3-3-2 shows schematically the expected basic behavior of the original 5400-km primordial Earth in response to the fundamental driving mechanism of the model from the standpoint of radial translatory equilibration. The spherical octants are free to move radially with respect to the center of the model. From the principle of hydrostatic adjustment applied on a global basis, about half of the octants are subsided, at any one time, while the others are elevated with respect to the basic 5400-km radius. The magnitude of the radial motion visualized here is of the order of cm/yr with maximum amplitudes of the order of ten km.

Transverse Translatory Equilibration. Fig. 3-3-3 shows schematically the expected basic behavior of the original 5400-km primordial Earth in response to the fundamental driving mechanism of the model from the standpoint of transverse translatory equilibration. Within the constraints of the model,

121

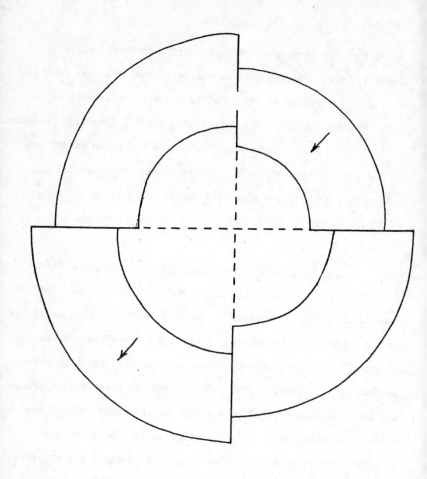

Figure 3-3-2. The expected basic behavior of the original 5400-km primordial Earth in response to the fundamental driving mechanism of the tectonospheric Earth model from the standpoint of radial translatory equilibration. (Schematic; not to scale).

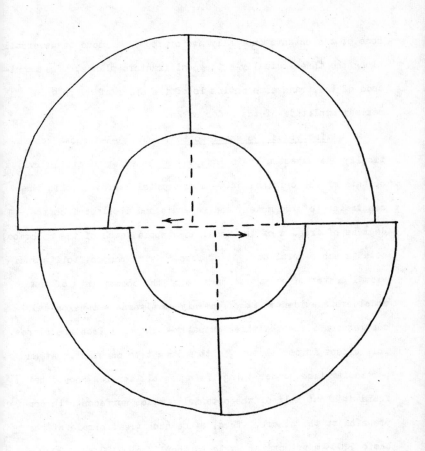

Figure 3-3-3. The expected basic behavior of the original 5400-km primordial Earth in response to the fundamental driving mechanism of the tectonospheric Earth model from the standpoint of transverse translatory equilibration. (Schematic; not to scale).

some of the octants (normally two or four) may move transversally along the three mutually-orthogonal fracture planes. The magnitude of the transverse motion is of the order of cm/yr with maximum amplitude of the order of km.

Individual Octantal Motion. Fig. 3-3-4 shows schematically the expected basic independent behavior of the individual octants of the original 5400-km primordial Earth. Within the constraints of the model, and separate and distinct from the degrees of freedom of the model as a whole, each of the spherical octants has several degrees of freedom with respect to its individual center of gravity. Because of the constraints of the model, this evidences itself mainly as alternate tension and compression along the three mutually-orthogonal fracture planes. This effect is oscillatory, with a long time constant. At any particular time, other things being equal, tension should be found at about half of the octantal contact surfaces with compression at the others. Thus, since each great circle of the basic 5400-km primordial Earth is about 2 pi 5400 = 34,000 km long, or 3 x 34,000 = 102,000 km total for all 3 great circles, then about $\frac{1}{2}$ x 102,000 = 51,000 km of the surficial fracture pattern of the primordial Earth should be in tension while an equal amount is in compression (or neutral). Since the model is continually in the process of equilibrating itself, the locations of the specific segments which are in tension will tend to drift continually along the 102,000-km length of fracture

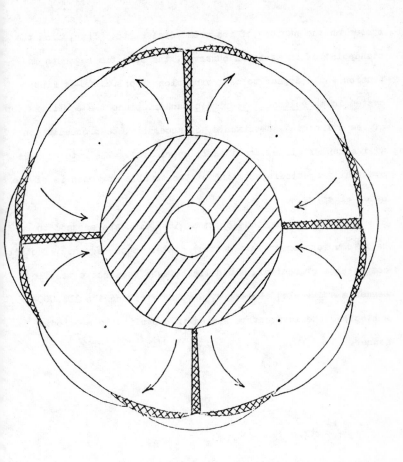

Figure 3-3-4. The expected basic _independent_ behavior of the
individual octants of the original 5400-km primordial Earth
according to the fundamental tectonospheric Earth model.
(Schematic; not to scale).

125

system on the surface of the primordial Earth. Thus, from the standpoint of an external observer, the specific segments in tension would appear to be moving along the 102,000-km linear system in a _cyclically aperiodic_ manner, since the equilibration of the primordial Earth would, of necessity, be proceeding in a like manner (i. e., cyclically aperiodic). The _rate_ of such cyclically aperiodic shifting of the tensile segments is of the order of cm/yr.

Segments not in tension at a given time are assumed to be in compression or in a neutral state. Such segments, whether compressive or neutral, would lie between successive tensile segments and would, therefore, also shift along the 102,000-km subtectonospheric fracture system in a cyclically aperiodic manner.

Post-Primordial Tectonospheric Accretion.

In evolving the post-primordial Earth from the basic 5400-km primordial Earth, three basic evolutionary modes may be assumed, depending upon the degree to which the 5 terrestrial octants of Earth Prime are fragmentized prior to accretion onto the primordial Earth. These are: (1) an octantal-fragment mode; (2) a multiple-fragment mode; and (3) a composite-fragment mode.

The Octantal-Fragment Accretionary Mode.

In the octantal-fragment mode, the 5 octants of Earth Prime are not completely fragmented prior to accretion onto the basic 5400-km primordial Earth. In such case, parts of each of the 5 octants of Earth Prime might still exist today as unmodified portions of the Earth's present tectonosphere, perhaps even as subsurficial portions of presently-existing continental shields. This possibility is considered briefly in Chapter 7 (The Orogenic-Cratonic Structure of the Continents).

The post-primordial Earth, in the case of the octantal-fragment mode, evolves essentially from a combination of two processes: (1) the accretion of 5 octants of Earth Prime onto the basic 5400-km primordial Earth; and (2) the differentiation, metamorphism, reworking, ingeous activity, and other results from the action of the driving mechanism of the basic model during the past 4.6 b. y.

It may be recalled that the driving mechanism of the basic model consists (and has consisted since primordial times) essentially of the resultant of two factors: (1) the potential energy inherent in the geogenetically-disequilibrated shape of the basic 5400-km primordial Earth; and (2) the selectively-channeled energy from the preferential flow of heat, and possibly of volatiles, outward from the 3 mutually-orthogonal "active" belts formed by the basic octantal fracture pattern of the 5400-km primordial Earth.

Here, as in subsequent discussions and analyses regarding the evolution of the Earth's tectonosphere during the past 4.6 b. y., it is necessary to recall that, in a homogeneous sphere of 6371-km radius, the inertia of the inner 5400-km part normally exceeds that of the outer 971-km "shell" by at least 60%. Thus, for most purposes, it may be assumed that the disequilibrating inertia of the inner 5400-km part of the model exceeds (and has always exceeded) the equilibrating inertia of the outer 971-km part by a factor of at least 60%. For a non-homogeneous, differentiated body such as the actual Earth, this inertial factor would, of course, normally by even greater than 60%. The significance of this figure will become more obvious when some of the mechanics and geometry of the model are discussed in connection with some of the observed behavioral patterns of the actual Earth, such as in Chapter 8 (The Earth's Deep Seismicity), Chapter 17 (Continental Drift and Polar Wandering), and Chapter 18 (Plate Tectonics).

The Multiple-Fragment Accretionary Mode.

The multiple-fragment mode differs from the octantal-fragment mode primarily in the degree to which the 5 "terrestrial" octants of Earth Prime were fragmentized prior to accretion onto the basic 5400-km primordial Earth. For example, if the degree of pre-accretion fragmentation was fairly complete, then it is not very likely that any of the actual surficial rocks of the present continental shields would now be identifiable with any unmodified portions of the original octants of Earth Prime. This is discussed in greater detail in Chapter 7 (The Orogenic-Cratonic Structure of the Continents) and Chapter 10 (Intrusive and Extrusive Activity).

The Composite-Fragment Accretionary Mode.

The composite-fragment mode may be considered as an intermediate mode lying between the octantal-fragment mode and the multiple-fragment mode. Depending upon the specific degree of fragmentation assumed for the octants of Earth Prime prior to accretion onto the basic 5400-km primordial Earth, this mode occupies any one of a myriad of intermediate positions within the entire spectrum, or envelope, bounded by the octantal-fragment and multiple-fragment modes as extremes.

By extending the limits of the mean, the composite-fragment mode may be made to include one of both of the extremes, thereby collapsing the three modes into two or one, respectively. In this book, the three modes will be treated as separate and distinct modes. In the last analysis, any salient differences in the Earth models evolved by the 3 modes are virtual rather than real; but, by handling the 3 modes separately, the analysis is more complete and perhaps easier to follow. For this reason, subsequent chapters, including Chapter 6 (The Tectonosphere) will assume that there are actually 3 modes rather than a single mode spread across an entire spectrum, including the end points, of the envelope defined by the two extreme modes. The efficacy of this analytical approach will become more obvious in Chapter 7 (The Orogenic-Cratonic Structure of the Continents) and Chapter 15 Continents and Oceans), where present continental structures are analyzed in terms of the most likely size of the post-

130

primordial subcratonic blocks which existed within the Earth's tectonosphere 4.6 and 3.6 b. y. ago, as well as during other critical times in the orogenic—cratonic evolution of the Earth's tectonosphere.

The Post-Primordial Earth.

The post-primordial Earth, as derived from the dual primeval planet model, consists essentially of two parts: (1) an actively-equilibrating, or mobile, 5400-km primordial Earth, differentiated, and fractured basically into octants by three mutually-orthogonal central planes: and (2) an overlying tectonosphere, consisting essentially of the remnants of 5 fragmented octants of Earth Prime accreted onto the 5400-km primordial Earth to a depth of about 1000 km, as modified continually by the driving mechanism of the basic model. Since the primordial Earth loses its identity by becoming incorporated into the post-primordial Earth, it will henceforth be referred to as the subtectonosphere, or the sub-tectonospheric portion of the model.

Fig. 3-4-1 is a schematic sketch of the post-primordial Earth, showing the earliest effects of both igneous and accretionary activity within the tectonosphere. Accretionary activity is indicated schematically at A; igneous activity at I. The octantal blocks of the subtectonosphere are marked S.

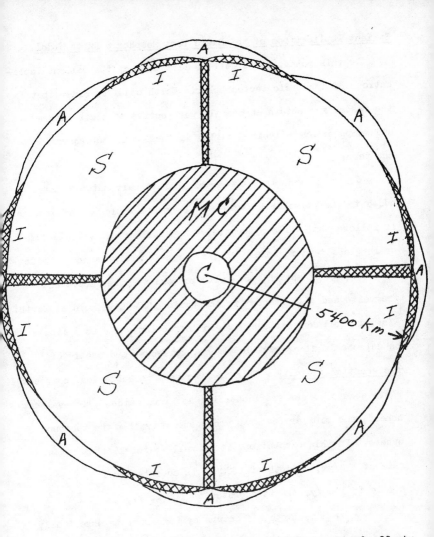

Figure 3-4-1. Post-primordial Earth, showing the combined effects of primeval igneous and accretionary activities. Accretionary activity is indicated by A; igneous activity by I. The octantal blocks of the subtectonosphere are marked S. MC is the molten core. C is a (probable) small central core. (Schematic; not to scale).

Salient Implications of the Basic Tectonospheric Earth Model.

At this point, it is well to list some of the salient implications of the basic tectonospheric Earth model in order that the reader may obtain an overall perspective of their nature and scope prior to their detailed discussion in subsequent chapters:

a. Since the 1000-km of accretionary material was added to the primordial Earth over a considerable period of time, it follows that various accreted fragments, either individually or in global and regional "shells", would have retained considerable freedom of motion, both with respect to other accreted fragments and with respect to the basic 5400-km primordial Earth. Also, it follows that such freedom of motion would most likely be detectable along various planes, surfaces, and shells of preferential rheidity. This is discussed in subsequent chapters where some of these are identified as "wave guides","low-velocity zones", and similar features and phenomena within the tectonosphere. In this connection, it is well to recall that, at a rate of 1 m/yr, a depth of 1000 km is accreted in 1 m. y.; at a rate to 1 cm/yr, it is accreted in 100 m. y.

b. The Earth's fracture system consists essentially of the projections,through the tectonosphere, of the basic primordial fracture system of the subtectonosphere. Since the tectonosphere is not homogeneous, such projections are not normally simple, radial projections but somewhat more complex.

134

Other things being equal, the "probable error" in locating the surficial position of an "energy source" projected from a point on the subtectonosphere through a vertical thickness of 1000 km of average tectonospheric material is roughly 1000 x pi/4, or about 785 km. The significance of this "probable error" figure, as well as the causes for, and magnitudes of, variations in its value are discussed in Chapter 6 (The Tectonosphere) when "cones of activity" are considered in connection with the mechanics and geometry of the evolution of the Earth's tectonosphere during the past 4.6 b. y.

c. The Earth's basic deep-fracture system is continuous and worldwide in both the inner (subtectonospheric) and outer (tectonospheric) parts. The latter is basically a consequence of the former, within the constraints of the driving mechanism and the octantally-fractured geometry and mechanics. From this, it appears that the Earth's deep trenches should form a global pattern identifiable with the 3 mutually-orthogonal great-circular belts of the subtectonosphere.

d. "Active" regions (new islands, new mountains, seismic ridges, new volcanoes, etc.) lie _roughly_ above the basic fracture planes of the inner, subtectonospheric part. Due to the effects of planes, surfaces, zones, and shells of preferential rheidity and of heat flow, these "active" regions would not, of course, be expected to lie _directly_ above the subtectonospheric fracture system.

e. "Inactive" or "fossil" regions (older mountains, older islands, older volcanic belts, seamounts, aseismic ridges, etc.) are identifiable with _former_ positions of projections from the subtectonospheric fracture system. To an external observer, these fossil, or inactive, regions appear to have "drifted", in time, with respect to the presently active regions. Normally, such "drifting" involves both a translation and a rotation, where neither is necessarily linear with respect to time. Also, it might be more accurate to say that the subtectonospheric fracture system, rather than the surficial manifestations thereof, has "drifted", or, perhaps even more accurately, that _both_ the surficial manifestations and the basic subtectonospheric fracture system have drifted, depending upon whether one wishes to assume a fixed or mobile datum during the analysis. In either case, of course, the _results_ of the analysis are the same. Also, according to the basic model, the "fossil" features are spatially parallel to, and temporally derivable from, the "active", or present, features by backward extrapolation procedures. Needless to say, such spatial and temporal relationships are _not_ necessarily _linear_ functions of time. This is discussed in greater detail in applicable subsequent chapters, such as Chapter 7 (The Orogenic-Cratonic Structure of the Continents), Chapter 8 (The Earth's Deep Seismicity), and Chapter 9 (Global Patterns of Terrestrial Heat Flow and Other Geothermal Activity).

f. Other things being equal, the _age_ of any particular

surficial manifestation (such as a mountain range, a crustal rift-ridge system, a volcanic belt, etc.) may be approximated roughly by measuring the gross distance by which the particular surficial manifestation has "drifted" from the present position of its genetic subtectonospheric fracture plane. Thus, it is not too surprising that the Appalachian-Caledonian (Upper Paleozoic) orogeny is geographically removed by about 30° from the surficial trace of the nearest subtectonospheric fracture plane, a drift of 1 cm/yr being roughly equivalent to about 30° per 300 m. y. This is discussed in Chapter 7 (The Orogenic-Cratonic Structure of the Continents), Chapter 8 (The Earth's Deep Seismicity), Chapter 9 (Global Patterns of Terrestrial Heat Flow and Other Geothermal Activity), Chapter 10 (Intrusive and Extrusive Activity), Chapter 11 (Morphology of the Earth), Chapter 12 (Mountain Building), Chapter 13 (The Earth's Gravity Field), Chapter 16 (Sea-Floor Spreading), Chapter 17 (Continental Drift and Polar Wandering), Chapter 18 (Plate Tectonics), and Chapter 20 (The Integrated Earth and Its Future).

g. The oldest geological features (or surficial manifestations), such as the Precambrian shields, presently occupy positions identifiable with the basic fracture system of the subtectonosphere. Specifically, they are either: (1) identifiable with the centers of the surficial octants of the subtectonosphere, if stable; or (2) identifiable with the edges of the surficial octants of the subtectonosphere, if they are unstable, active, or in the process of rifting. An example of the latter might be the South American-African "shield" 200 m. y. ago, which has since been rifted and separated as shown schematically in Fig. 3-8-1. A more-detailed discussion of this is contained in Chapter 17 (Continental Drift and Polar Wandering).

h. Likewise, the _newest_ surficial features would be expected to be located _near_ the surficial traces of the 3 mutually-orthogonal subtectonospheric fracture planes projected to the surface of the Earth.

i. One of the several "unbalanced" conditions existing in the post-primordial Earth resulted from the fact that portions of the Earth's surface must, at various times during the history of the post-primordial Earth, have become squarely juxtaposed above the surficial manifestations of the tectonospheric fracture system. Among the consequences of such a situation would be sea-floor spreading and continental drift, which are described in Chapters 16 and 17, respectively.

j. The stress release and other equilibrating effects from the potential energy inherent in the geogenetically-disequilibrated shape of the subtectonosphere are the basic cause of the Earth's deep seismicity. This is discussed in Chapter 8 (The Earth's Deep Seismicity).

k. "Mantle convection" may be motivated by the preferential flow of heat along the pattern provided by the subtectonospheric fracture system. Under the concept of the model, mantle convection is a "surficial manifestation" of the driving mechanism of the basic model. Thus, unlike most other hypotheses, the dual primeval planet hypothesis does _not_ invoke mantle con-

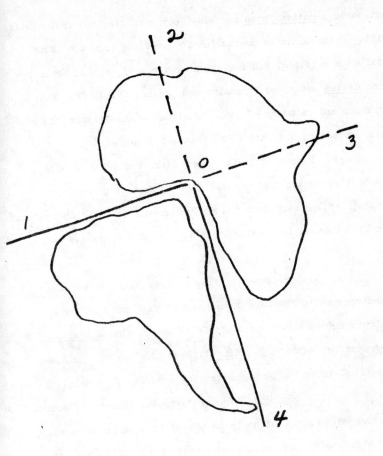

Figure 3-8-1. Two mobile belts of the tectonospheric Earth model severing Africa from South America during the Mesozoic. Mobile belt segments 1-0 and 4-0 were "tensile" during the Mesozoic; segments 2-0 and 3-0 were compressive or neutral.

139

vection as a __motive__ __force__ for sea-floor spreading and continental drift. Rather, mantle convection is derivable from the basic model as a surficial manifestation of the __same__ driving mechanism which __drives__ sea-floor spreading and continental drift. The mechanics and geometry of such are discussed in Chapter 16 (Sea-Floor Spreading) and Chapter 18 (Plate Tectonics).

l. Igneous activity, both intrusive and extrusive, is motivated by the preferential flow of heat (and possibly of volatiles) outward from the subtectonospheric fracture system. This is discussed in Chapter 10 (Intrusive and Extrusive Activity).

m. The preferential flow of heat outward from the subtectonospheric fracture system may be employed to furnish the basic motive power and a stable framework for __maintaining__ a geomagnetic dynamo with a __long__ time constant. This is discussed in Chapter 14 (Geomagnetism and Polarity Reversals).

n. The roughly antipodal juxtaposition of the oceans with respect to the continents is basically a consequence of two factors: (1) the radial translatory motion of the subtectonospheric octantal blocks with respect to the center of the model; and (2) hydrostatic adjustment, or isostasy, on a global scale over a long period of time. This is discussed in Chapter 15 (Oceans and Continents).

o. If viewed on a global scale, features such as fault lines, rift-ridges, mountain belts, seamount chains, etc., should normally tend to display roughly parallel surficial patterns, intersected by transverse features. As a result, other things being equal, they should tend to form surficial manifestations suggestive of pinnate, grid, and arcuate patterns. These reflect both translation and rotation between the subtectonospheric blocks and the surface of the Earth, most probably along planes, surfaces, shells, and zones of preferential rheidity within the tectonosphere. This is discussed in Chapters 6 (The Tectonosphere) and 18 (Plate Tectonics).

The above are a few of the "first order" implications of the basic tectonospheric Earth model. The details of these and other implications are discussed in appropriate subsequent chapters. "Second order" implications of the model are beyond the scope of this book. However, a brief discussion of the nature and extent of these "second order" implications is included in Chapter 20 (The Integrated Earth and Its Future).

Global Surficial Manifestations Predictable from the Basic Tectonospheric Earth Model.

One of the postulates of the basic tectonospheric Earth model is that almost all surficial features and phenomena of the Earth, both past and present, are fundamental consequences of a single driving, or causal, mechanism which has resided within the Earth since the formation of the primordial Earth, about 4.6 b. y. ago. The details of exactly how such surficial manifestations evolved as a consequence of the behavior of the Earth's tectonosphere during the past 4.6 b. y. will be discussed in subsequent chapters. However, it may be well to list a few of them here, in order that the reader may have an appreciation for the general nature and scope of those surficial manifestations which are tectonospherically-induced according to the tectonospheric Earth model: global and regional tectonism and orogenesis; sea-floor spreading; global and regional heat-flow patterns; the Earth's magnetic field and its variations, including cyclically aperiodic reversals; the Earth's gravity field; volcanic activity; mountain building; the development of continental shields, mobile belts, fracture systems, rifts, and ridges; continental splitting, separation, and drifting; earthquakes and global and regional seismicity patterns; gravity anomalies and global and regional geoidal patterns; heat-flow anomalies and global and regional thermal patterns; electrical conductivity anomalies; etc.

Before proceeding, it may be well to examine one of these surficial manifestation in an attempt to provide a tentative answer to the compound question: Is there any observational evidence that the Earth might, in fact, be representable by the tectonospheric Earth model derived from the dual primeval planet hypothesis; and, if so, how might the present positions of the 8 hypothetical subtectonospheric blocks be identified or located?

To answer this compound questions, consideration may be given to a preliminary analysis of any one of several sets of applicable observational data regarding the gross surficial characteristics of the Earth. For example, one such set of data is that showing the locations of the active volcanoes. Since volcanoes appear to be thermally induced and since they are surficial manifestations of tectonospheric activity of some sort, they might be expected, according to the model, to lie along global belts defined roughly by the surficial traces of projections from the 3 mutually-orthogonal great-circular belts forming the subtectonospheric fracture system.

Therefore, as a first approximation to the location of the surficial manifestations of the subtectonospheric fracture system, we may perform a 3-dimensional least-squares analysis of the distribution of the Earth's active volcanoes. Allowing for the probable error expected from the 1000-km projection from the subtectonosphere through heterogeneous material, we should

143

expect almost all of the Earth's active volcanoes to lie within a surficial distance of about 1000 x pi/4 ≐ 785 km from the _radial_ projection of the subtectonospheric fracture pattern.

When the suggested analysis is done, it is found (Tatsch, 1964) that approximately 93% of the active volcanoes fall within the belts predicted by the model. Specifically, they appear to form segments of 3 great-circular belts intersecting at the following 6 points, which may be designated tentatively as the basic "tectonospheric points" of the _present_ surficial manifestations of the basic tectonospheric Earth model: 55°N, 165°W (Aleutians); 55°S, 15°E (Bouvet); 5°S, 85°W (Galapagos); 30°N, 5°E (Gibraltar); 5°N, 95°E (Bengal); and 30°S, 175°W (Kermadecs).

Since the probable error of the projection from the subtectonosphere to the surface of the Earth is about 785/111 ≐ 7°, no purpose is served in attempting to locate the coordinates of the tectonospheric points more accurately than about 5° in either latitude or longitude. The geographic names (Aleutians, etc.) shown above in parentheses are intended for _associative_ identification of the _general_ areas of the points, rather than as _exact_ _geographical_ locations of the points. The associative names will normally be used in subsequent discussions as a matter of convenience; but it should be understood that these points are _projections_ of 6 points which form a pattern on the surface of the _subtectonosphere_ 1000 km _beneath_ the surface of the Earth.

For the sake of simplicity, the first two points (Aleutians

and Bouvet) may be referred to as the "tectonospheric poles" of the present surficial manifestations of the basic tectonospheric Earth model; the other 4 points (Galapagos, Gibraltar, Bengal, and Kermadecs) may be considered as defining roughly the "tectonospheric equator" of the present surficial manifestations of the basic tectonospheric Earth model.

Fig. 3-9-1 shows the above tentative "tectonospheric coordinate system" of the present surficial manifestations of the basic tectonospheric model superimposed upon a Mollweide projection of the salient geographical features of the actual Earth. In the figure, the 3 mutually-orthogonal great circles (ACDFA, ABDEA, and CBFEC) are the radial projections of 3 similar great circles defining the subtectonospheric fracture system. Each of the intersections at A, B, C, D, E, and F is an orthogonal intersection, and each of the 8 triangles (CBA, CDB, CED, CAE, FEA, FDE, FBD, and FBA) is an equilateral spherical triangle (i. e., with all sides and angles equal to 90°). On the surface of the tectonosphere, 90° equals about 10,000 km. On the surface of the subtectonosphere, 90° equals about 8500 km.

The great circle ABDEA, also marked QQQQ in Fig. 3-9-1, is a projection of the subtectonospheric equator and defines roughly the tectonospheric equator of the present surficial manifestations of the basic tectonospheric Earth model. The great circle CBFEC, also marked PPPP, is a projection of the subtectonospheric prime meridian and defines roughly the tec-

145

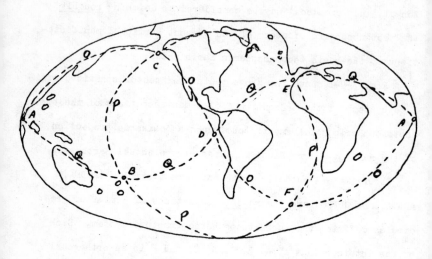

Figure 3-9-1. Tentative "tectonospheric coordinate system"
of the present surficial manifestations of the basic tectonospheric Earth model superimposed upon a Mollweide projection
of the salient geographical features of the actual Earth. See
text for explanation of symbols.

tonospheric prime meridian of the present surficial manifestations of the basic tectonospheric Earth model. Similarly, the great circle ACDFA, also marked 0000, is a projection of the subtectonospheric orthogonal meridian.

The tectonospheric coordinate system is particularly useful as a base for performing harmonic analyses of the Earth's surficial manifestations. According to the dual primeval planet hypothesis, the Earth's surficial manifestations are tectonospherically induced and, therefore, harmonic with respect to the _tectonospheric_ coordinate system, rather than with respect to the _geographic_ coordinate system normally used for harmonic analyses when using other models. This coordinate transformation is applicable to analyses in some subsequent chapters, including: 6, The Tectonosphere; 7, The Orogenic-Cratonic Structure of the Earth; 8, The Earth's Deep Seismicity; 9, Global Patterns of Terrestrial Heat Flow and Other Geothermal Activity; 10, Intrusive and Extrusive Activity; 13, The Earth's Gravity Field; 14, Geomagnetism and Polarity Reversals; 15, Continents and Oceans; 16, Sea-Floor Spreading; 17, Continental Drift and Polar Wandering; and 18, Plate Tectonics.

REFERENCES:

Tatsch, J. H., 1964. Distribution of active volcanoes: summary of preliminary results of three-dimensional least-squares analysis. Geol. Soc. Am. Bull., 75: 751-752.

147

Chapter 4

THE EARTH AS A PLANET

This chapter proposes to give the reader a global overview or perspective of the Earth's internal behavior and the various manifestations thereof which are observable or measurable on the surface of the Earth. It is hoped, thereby, that the reader will gain a fuller appreciation for the mechanics and geometry of the Earth's internal behavior and its possible relationships to the surficial manifestations thereof, both spatially and temporally, during the past 4.6 b. y.

Armed with sophisticated sensors, instruments, and computing devices, a modern scientist can measure and otherwise observe many surficial features, phenomena, and manifestations of the Earth's internal behavior. He is not always sure just what causal relationships exist between what he observes and what is happening inside the Earth. He can, for example, see the crust rifting at certain places, and, when this occurs, he calls it "sea-floor spreading"; but he has not yet been able to completely understand just _what_ sea-floor spreading _is_ or what _causes_ it. Nor does he yet understand precisely why sea-floor spreading occurs at certain places and not at others.

In a similar manner, when he sees lithospheric plates moving about on the surface of the Earth, he computes their instantaneous centers of rotation and calls this phenomenon "plate tectonics", but he has not been able to completely comprehend precisely _what_ plate tectonics _is_ or what _causes_ it. Nor does he yet understand exactly _why_ the lithospheric plates form into pieces of certain sizes and shapes rather than others. He suspects that the _cause_ lies deep below, but he is not sure just _where_, nor what the causal mechanism is or how it operates to break the lithosphere into a few large plates and many small subsidiary and peripheral plates. The break-up of the lithosphere is obviously a surficial manifestation of the Earth's internal behavior. But _what_ precisely is the nature of the Earth's internal behavior which causes this and other surficial manifestations?

For ease of presentation, three aspects of the global surficial features and phenomena will be considered: (1) a global view of the Earth's _present_ surficial features: (2) the most probable surficial features of the proto-Earth _4.6 b. y. ago_; and (3) the evolution of the Earth's present surficial features from those of the proto-Earth during the past 4.6 b. y. Since the evolutionary period appears to have been interrupted _3.6 b. y. ago_, the period will be considered in two parts: 4.6 b. y. to 3.6 b. y. ago, and 3.6 b. y. to the present.

149

Orogenic-Cratonic Structure of the Continents.

Scientists have long suspected that the orogenic-cratonic structure of all continents is more-or-less the same and that, in the simplest analysis, continents appear to have grown by accretion onto a central shield or nucleus. Exactly what caused them to follow such a pattern of quasi-organized accretionary growth is not known, of course. Besides, most modern hypotheses do not extend back as far as the 3.6 b. y. during which observational evidence supports such growth-from-the-center evolution for the continents.

In viewing the orogenic-cratonic structure of the continents on a global scale, it appears: (a) that orogenic belts are long and rectilinear, i. e., the accretionary growth may occur along segments of great circles, which unlike small circles plot on the surface of the globe as rectilinear features; (b) that the continents are old and appear to have formed by accretion from the center outward, beginning at least 3.6 b. y. ago; (c) that the oldest orogenic belts are nearest the centers of the continental nuclei, but that farther from the centers there may be a certain amount of overprinting by orogenic belts of different ages; (d) that generally speaking orogenic belts appear to have been active on a worldwide basis in a cyclically aperiodic manner during at least the past 3.6 b. y.; (e) that concentration of radioactive elements in continental rocks suggests chemical differentiation of the mantle, or an equivalent process; and (f) that the possibility of such differentiation suggests that

the entire continental areas may be of ultimate mantle origin
and that the primordial Earth might have been completely devoid
of a crust or continents as we know them today.

The above indicated oroginic-cratonic structure is fairly
well preserved in certain continents (See, e. g., Douglas, 1969;
Wynne-Edwards, 1969; Drake, 1970; Martin, 1969; Hurley, 1969).
In general, what is seen is an old Precambrian shield successively
surrounded by long rectilinear orogenic or geosynclinal belts,
often broken or otherwise modified by overprinting by subsequent
orogenies, with the last three orogenies fairly well preserved
in at least some continents; e. g., the Appalachian, Innuitian,
and Cordilleran orogens of the Canadian shield. Unanswered
are questions as to why the orogenic belts are rectilinear and
of different ages within any one continent but appear to extend
worldwide, with their activity having occurred in a cyclically
aperiodic manner for at least the past 3.6 b. y., and with concentrations of this activity having occurred at about 0.9 b. y.
intervals. Also not understood is why the spatial-temporal
relationships within all orogenic belts are remarkably similar,
regardless of where found or to which period they belong.

Considered as a surficial manifestation of the Earth's
internal behavior, the orogenic-cratonic structure of the continents suggests a remarkably orderly behavior over an extremely
long period of time, perhaps as long as 3.6 b. y., or even 4.6
b. y.

The Earth's Seismicity and Earthquakes.

It is universally recognized that earthquakes are the result of fracturing occurring at various depths of the crust and upper mantle to depths of almost 800 km. Many of the faults associated with earthquakes do not extend to the Earth's surface, and the epicentral areas of even severe earthquakes provide no evidence of the subsurface faults or fractures. In such cases, the only surficial manifestation of the Earth's internal behavior is the seismic shock itself.

Other studies indicate that causal relations are not the same in regions of different geological structure and history, such as young folded regions, young platforms, platforms of any age involved in intensive tectonic movements during recent times, and rift zones within continents (See, e. g., Drake, 1970; McBirney, 1969; Barazangi and Dorman, 1969; Petrushevsky, 1969; McGarr, 1969; Molnar and Olvier, 1969; Oliver et al., 1969; Raitt et al., 1969; Balakina et al., 1969).

When viewed on a global scale, in terms of ordinary analyses, neither the Earth's seismicity nor the distribution of earthquakes appear to form an orderly pattern. However, by using special techniques, much progress has been made in recent years to develop the specifications of an unambiguous worldwide earthquake mechanism. Many observers report evidence indicating that the pressure axis is perpendicular to the trend of the continental seismic belts and that the tension axis is perpendicular to mid-oceanic belts, when analyzed in terms of the earthquake energy-release mechanism (See, e. g., Balakina et al., 1969). However, such evidence tends to be somewhat misleading because some of the so-called "mid-ocean" belts lie within continental areas (e. g., in North America, Central America, Africa, Iceland, etc.). It does suggest, however, that the earthquake energy-release mechanism is truly global and independent of present positions of continents and oceans. Thus, considered as surficial manifestations of the Earth's internal behavior, earthquakes suggest that the Earth's internal behavior is controlled by a global mechanism completely independent of continental margins, island arcs, and similar features, except insofar as they also may be surficial features of the same mechanism.

Heat Flow and Other Geothermal Activity.

A detailed description of the Earth's thermal history is, of necessity, a rough guess, because the distribution of heat sources within the Earth is not known with any certainty. However, the following comments may be made in connection with surficial manifestations of the Earth's internal thermal behavior on a global scale: (a) thermal Earth models based on spherical symmetry are not consistent with the assumption that radioactive elements are more highly concentrated in continental than in oceanic crust; (b) the distribution of radioactive material below the M discontinuity is usually assumed to be somewhat different under oceans than under continents; (c) in a body the size and composition of the Earth, thermal diffusion is so slow that little heat could have escaped from depths greater than about 750 km during the past 4.6 b. y. (Tikhonev et al., 1969; Stacey, 1969), unless it has escaped along paths of preferential heat flow; (d) it is often assumed that "for any reasonable distribution" of radioactivity in an oceanic Earth model, the actual effective thermal conductivity at depth must be substantially greater than the average value of thermal conductivity of igneous rocks; otherwise the entire mantle would be molten; (e) thermal conduction is usually presumed to be confined to the upper 700 km; and (f) there is evidence that the thermal conductivity of the mantle maintains a temperature profile in equilibrium with internal heat generation (See, e. g., Stacey, 1969).

154

Although the outward flow of heat through the crust is the only directly measurable surficial manifestation of the Earth's internal thermal behavior, certain tentative conclusions may be drawn from present-day analyses made in connection with heat-flow data: (a) there is a general correlation of heat flow with geology, the higher heat flows being associated with the more recent orogeny; (b) in specific regions, such as the mid-Atlantic ridge, for example, heat flow varies considerably, such as is also true for volcanic areas, of course; (c) high heat flow appears to correlate with low gravity (Lee and Mac-Donald, 1963); (d) continental and oceanic heat flow averages are roughly the same; (e) the features of the geoid do not appear to correlate with present locations of continents and oceans; (f) the oceanic mantle is neither hotter nor colder, on average, than the continental mantle; (g) the equality of average outward flow of heat from continents and oceans may be the result of a dynamic balance in a convecting (or otherwise well-behaved) upper mantle (Elsasser, 1967); (h) although radioactivity is approximately the same beneath oceans and continents, there appear to be some differences within the upper mantle (Polyak and Smirnov, 1968); (i) although the global correlation between heat flow and gravity is not understood, the lack of a global spatial correlation may be the result of a temporal lag between the Earth's internal thermal behavior and the internal behavior producing changes in gravity, perhaps amounting to a temporal lag of millions of years; (j) a more definite correlation at

155

the present time may exist between heat flow and age of basement rock (See, e. g., Lubimova, 1969), bacause heat flow values appear very uniform in all shield areas (0.92 ± 0.38) and in all Meso-Cenozoic orogenic areas (1.92 ± 0.49), the general correlation being that heat flow decreases with increase in age of basement; (k) preliminary analyses show that heat flow data correlate with other geophenomena and features provided the low "signal" of the system (i. e., the paucity of heat-flow data) can be offset by decreasing the "noise" of the system (such as by transformation from geographical coordinates to geotectonospheric coordinates, as suggested in Chapter 3); and (l) temporal variations in heat-flow rates are significantly different for regions of tectonogenesis of dissimilar ages if the deepest levels of the upper mantle are assumed to be responsible for the formation of the tectonic zones and if the tectonogenesis of dissimilar age is characterized by various intensities in the inflow of sialic material from the mantle (Lubimova, 1969).

Thus, viewed on a global scale, it appears that the Earth's internal behavior reveals itself through thermal manifestations on the surface; but the exact correlation between the internal behavior and the surficial thermal manifestations is not clearly understood, perhaps because of the paucity of global heat-flow data. If a generalization may be made, it is that regions of modern orogenesis correlate with high heat-flow rates.

156

Intrusive and Extrusive Activity.

In considering the surficial manifestations of the Earth's internal behavior from the standpoint of intrusive and extrusive activity on a global scale, it is necessary to consider all intrusive and extrusive activity which appears to represent mantle material or which appears to be produced by deep-seated tectonics. This includes volcanic activity, the creation of basaltic traps both in continental and oceanic areas, mantle upwelling at crustal rifts, the emplacement of batholitic, laccolitic, and other intrusive bodies during at least the past 3.6 b. y.

What, if any, global patterns are formed by the Earth's intrusive and extrusive activity? To answer this question, it would be well to consider the following observations regarding the Earth's intrusive and extrusive activity: (a) it is usually agreed that the ultramafic rocks contained in layered, stratiform, and other intrusions involving gabbro and diabase, together with accumulations or concentrations of mafic minerals, were formed from mantle-derived basaltic magma (See, e. g., Wyllie, 1969); (b) all other ultramafic rocks appear to be associated with major tectonic features of the Earth's crust with distributions controlled by deep-seated tectonics with linear trends; (c) Alpinotype ultramafic rocks are presently found along deformed mountain chains and island arcs, usually with associated gabbro or basic volcanic rocks; (d) the occurrence of serpentinites peridotites along mid-ocean ridges suggests a third type of ultramafic belt, but exactly how it may be related to the other two is not known; (e) although the petrogenesis is extremely complicated in some cases by possible "coupling" between mantle-derived and crustal materials, there are reasons to believe that all ultramafic rocks are derivable ultimately from a mantle source; (f) most observers feel that the petrogenesis of Alpine intrusions in orogenic regions is complicated by the fact that it normally involves several destructive metamorphic episodes but that Alpine intrusions represent parts of a solid, or partially-fused, mantle that flowed into or through the crust along an unstable, linear orogenic belt; (g) vast composite sheets of Mediterranean and Himalayan areas appear to represent voluminous extrusions of mantle material breaking through the floor of the Tethyan sea floor to depths of 8 and 10 km (See, e. g., Maxwell, 1968); (h) genetic links between oceanic ultramafic rocks and Alpine ultramafic belts are suggested by the observation that Puerto Rican serpentinites represent uplifted oceanic crust composed of altered mantle material exposed at the surface, as well as by the observation that a "spreading" ocean floor could cause tectonic incorporation of the serpentinites of the oceanic crust into the underlying sediments of the continental rise, thus producing ultramafic rocks of the Alpine type upon metamorphism of the sedimentary pile (See, e. g., Wyllie, 1969); (i) similarities between the basic pillow lavas and peridoties which appear to characterize the mid-ocean ridges (Maxwell, 1968) and the extrusion of mantle material in the ophiolitic complexes of the Alpine mountain belts (i. e., those formed during the past 200 m. y.) suggest that mid-ocean ridges may be the loci of ultramafic belts just as extensive as the Alpine untramafic belts (See, e. g., Wyllie, 1969).

From the above and related analyses (See, e. g., Drake, 1969, 1970; Simonen, 1969; Drake and Kosminskaya, 1969; Green, 1969), it may be seen that, viewed on a global scale, intrusive and extrusive activity of both the mid-ocean-ridge type and the Alpine-ultramafic-belt type may be considered to be surficial manifestations of a single global mechanism, characterized by cyclically aperiodic activity along long, linear belts of deep-seated control. Closer scrutiny reveals that intrusive and extrusive activity forms a complete temporal-spatial spectrum within each cycle of orogenic activity whether of the mid-ocean-ridge type or of the Alpine-ultramafic-belt type (Chapter 10).

Mountains and Geosynclines.

The relationship that exists between mountains and geosynclines is not completely understood. It appears, however, that there may be an antithetical, Phoenix-like association between them and that the general nature of such might become more obvious if mountains and geosynclines are analyzed simultaneously as surficial manifestations of the Earth's internal behavior on a global scale during the past 3.6 b. y. The most obvious feature common to both mountains and geosynclines is their linearity when viewed on a global scale. Also, both appear to date far back into the history of the Earth, perhaps back at least 3.6 b. y. Following are other significantly interesting data regarding mountains and geosynclines: (a) a geosyncline appears to be an elongated downwarp in the Earth's crust, the bottom apparently subsiding deeply beneath accumulating sediments; (b) the Alleghany mountain range evolved from a long-continued linear subsidence, followed by an accumulation of sediments during the entire Paleozoic (a span of 300 m. y.), completed by great breakings, faultings, and foldings of strata along with other results of deep-seated disturbance; (c) since it appears that the deposition of sediments was in shallow water, some observers (See, e. g., Hall, 1857) feel that the geosynclinal depression was caused by the weight of the accumulating sediments, but others (See, e. g., Dana, 1873) see evidence suggesting that the accumulation of sediments was a consequence, rather than a cause, of the subsidence; (d) some, but not all, mountains and geosynclines appear to form along the edges of continents; some appear to form entirely within continental areas; some seem to form partly in and partly out; and, if crustal rifting is to be considered as the earliest stage of the mountain-building cycle, then some may form completely within oceanic areas; (e) particularly when the paired-belt concept of miogeosynclines (non-volcanic) and eugeosynclies (volcanic) is used (See, e. g., Stille, 1940), the identification of geosynclines along continental margins becomes confusing to the circumspect analyst and often meaningless because the separation between the mio and eu element of the geosynclinal pair is observed to be more than a thousand km in many cases; (f) there are indications that many eugeosynclines (e. g., the American Cordilleran and Appalachian) were analogous to present island arcs and that sediments were derived from active volcanoes within the sedimentary basins associated therewith (See, e. g., Kay, 1951); (g) there is evidence that both present and past geosynclines form in an orderly manner in all parts of the globe, suggesting the possibility of orderly control on a global scale (See, e. g., Drake et al., 1959; Dietz, 1963; Dietz and Holden, 1966); (h) there is evidence that much of the "fill" for North American geosynclines was derived from "borderlands" situated on the oceanic side of the continental margins (See, e. G., Hsu, 1965); (i) the structure and development of the Alpine geosyncline seem to have been similar to those of the Indonesian arcs, consisting of a complex system of troughs and ridges throughout most of its history (See, e. g., Aubouin, 1965); (j) when geosynclines are considered to form along all continental margins, there are five types of geosynclines: Atlantic Andean, island—arc, Japan Sea, and Mediterranean (Mitchell and Reading, 1969); (k) on a global scale, these 5 types of geosynclines may be regrouped into 2 basic types: those occurring near a mobile belt (the Andean, island—arc, Japan Sea, and Mediterranean types) and those not occurring near a mobile belt (the Atlantic type); (l) in the first of the regrouped types, there is both associated orogeny and differential plate movement, whereas, in the second, there is neither.

From the above and other analyses (See, e. g., Argand, 1916; Kuenen, 1967, Laubscher, 1969; Schuchert, 1923; Sylvester-Bradley, 1968), it appears that when geosynclines and mountains are analyzed simultaneously on a global scale, and as "active" features (i. e., excluding Atlantic type geosynclines), their origin and evolution appear to be controlled by a single global mechanism deep within the Earth.

The Earth's Gravity Field and the Shape of the Geoid.

If the Earth's gravity field and the shape of the geoid are surficial manifestations of the internal behavior of the Earth during the past 4.6 b. y., then they should have certain characteristics correlatable with the Earth's internal behavior on a global scale during that period of time. Under ideal conditions of isostatic equilibrium, all crustal columns exert equal pressure at some depth, and the column extending above sea level has a mass exactly equal to that of the compensating mass at depth. Much has been written in this area (See, e. g., Heiskanen and Vening Meinesz, 1958; Heiskanen and Moritz, 1967; Caputo, 1967; Woollard, 1969a, 1969b, 1969c). Although the Earth does not behave as an ideal isostatic system, much can be gained from a brief analysis of the nature and extent to which the Earth's internal behavior departs from that of an ideal isostatic system when viewed on a global scale and over a considerable period of time.

Specifically germane to an analysis of the Earth's surficial manifestations in this connection are the following observations related to the isostatic behavior of the Earth: (a) that significant changes (25 to 50 mgal) in isostatic anomalies are to be expected where there are changes in crustal composition (horizontal and vertical heterogeneities); (b) that there is a strong correlation between gravity values and topographic relief, changes in geology, and areas of change in crustal composition and thick-

163

ness, particularly when the areas are narrow; (c) that there appears to be a slight elevation dependence, governed more by the wavelength (width) of a particular topographic feature than by its amplitude (elevation); (d) that major changes appear to result from variations in crustal and upper mantle parameters, including deep mass anomalies; (e) that, in cases of major changes, sea-level intercept values may vary by as much as $\pm$ 50 mgal; (f) that, generally speaking, gravity anomaly gradients exceed normal values in the case of graben-type surficial features and are subnormal in the case of horst-type surficial features.

Thus, viewed on a global scale, it appears that the Earth's gravity field and the shape of the geoid may reflect the Earth's internal behavior during the past 4.6 b. y. provided appropriate spatial and temporal corrections are made for the fact that the Earth is a dynamic body and that its surficial features may lag behind its deep internal behavior. For example, corrections must be made for the fact that ideal conditions of isostatic equilibrium exist if, and only if, the unbalanced force acting on a vertical column is zero; in all other cases the column tends to move in a direction parallel to that of the unbalanced force vector. It is extremely difficult to visualize a case of absolute ideal isostatic equilibrium within the real dynamic Earth, and actual gravity surveys appear to confirm this. However, approximate isostatic balance of subglobal or regional features would be expected, and such is observed, but not in agreement with present continental positions. The independence of geoidal and continental features compels the inference that geoidal features are due either to: (1) density differences deep within the mantle (deeper than the so-called "low-velocity" layer in the upper mantle, by virtue of which isostatic balance might be feasible): or (2) density differences maintained by convection, or by a similar mechanism, in which case they are more apt to lie in the upper mantle. There is no reason why they cannot be a combination of both (1) and (2), that is, with rheid surfaces, shells, and/or planes at more than one depth within the tectonosphere (See Chapter 6).

On the basis of present evidence, it appears that the Earth's gravity field and the shape of the geoid are surficial manifestations of the internal behavior of the Earth during the past 4.6 b. y. However, it appears that mass redistributions lag behind the forces causing the Earth's internal behavior. The shape of the geoid therefore should correlate with the tectonospheric driving mechanism rather than with the surficial manifestations of the Earth's internal behavior.

Geomagnetism and Polarity Reversals.

If geomagnetism and polarity reversals are surficial manifestations of the Earth's internal behavior, then the following
should be applicable in connection therewith: (1) the geomagnetic dynamo has evolved and is driven by the same causal mechanism
which has produced the Earth's internal behavior during the past
4.6 b.y.; (2) the geomagnetic field correlates, now and in the
past, with global characteristics of the Earth's driving mechanism; and (3) cyclically aperiodic polarity reversals are definable in terms of the geometry and mechanics of the Earth's
driving mechanism.

Most geomagnetic dynamos fall into one of two categories:
(1) convective geomagnetic dynamos which derive their driving
force from convective processes deep within the Earth; and (2)
precessional dynamos which derive their driving force from the
Earth's precession.

Theoretically, a convective dynamo could have evolved from
an adiabatic temperature gradient in the outer core, resulting
from the latent heat of a progressively solidifying inner core
(See, e. g., Verhoogen, 1969). However, such a mechanism would
not have maintained the dynamo for more than about 3 b. y. For
this and other reasons, a precessional dynamo might be preferred
(See, e. g., Malkus, 1963). There is a possibility that the
Earth's actual dynamo is neither convective nor precessional.

In order to determine the true nature of the geomagnetic

166

dynamo, it is necessary to examine the Earth's thermal history during the past 4.6 b. y. For example, if the Earth's core is in a state of thermal convection, then it is cooling at a rate of about $100^{\circ}C$/b. y. In the case of the precessional dynamo, nothing need be said about deep cooling, although lunar precession implies a decrease in the internal heat generated by the precessional torques. Thus, the heat flux from the Earth's core need not be appreciably different for the two basic types of dynamo because a "stirred" _outer_ core implies an adiabatic temperature gradient against conduction in both types.

Present-day observations show that the dipole moment of the Earth is changing by about 1/20 of 1% per year, indicating a time constant of only about _2000 years_, from which it follows that the geomagnetic field cannot be simply a decaying vestige of the primordial Earth but must be continuously motivated by some means.

The motive force for the geomagnetic dynamo poses no problem if the Earth's rotation enters into the system (See, e. g., Bullard, 1949). For example, since the core is considered to be ellipsoidal rather than purely spherical, it may not follow _mantle_ precession exactly, with the result that internal motions must occur and that they might be more than adequate to drive a geomagnetic dynamo. However, the issue is still in doubt because the dynamo currents appear to be of _thermo-electric_ origin rather than purely thermal. Also, the dominance of the

dipole field and its axial character (at least over periods
of tens of thousands of years) suggest that the Earth's rotation
(or some other axisymmetric characteristic) exercises a controlling
influence, leading to the conclusion that the motive force for
the geomagnetic dynamo might be a linear combination of several
forces.

Geomagnetic polarity reversals appear to occur with greater
frequency during periods of rapid polar wander and/or continental
drift (See, e. g., Irving, 1966). This is a particularly intriguing
observation, because it suggests that the configuration of the
mantle might have a direct influence upon the core motions which
are thought to be responsible for the geomagnetic field. In an
attempt to resolve this paradox, several conclusions may be
drawn. First, it is difficult to explain the correlation between frequency of polarity reversals and rate of polar wander
and/or continental drift on the basis of ordinary convective
dynamo processes, but it appears to be somewhat relatable to
the precessional dynamo, since the process of polar wander would
appear to cause the axis of symmetry of the core to depart
somewhat from the axis of rotation of the Earth and, thereby,
alter the differential precessional torques acting on the core
and mantle. What seems completely unresolved, however, is that
the geomagnetic dynamo with a computed time constant of only
about 10^3 years could be sensitive to axial changes on a time
scale of 10^6 years, or 3 orders of magnitude greater.

Electrical conduction in the lower mantle, such as would be required in a thermo–electrical dynamo, may result from the semiconducting properties of silicates at elevated temperatures. Other observations of interest include the following: (1) the crust is less conductive than sedimentary rocks; (2) conductivity is increased by the amount _and_ salinity of interstitial water; (3) the mantle shows a _rapid increase_ in conductivity with depth; (4) upper–mantle conductivities amy vary by two orders of magnitude within _small_ horizontal distances (e. g., the upper mantle conductivity under Arizona is 50 to 100 times greater than that under New Mexico); and (5) conductivities are _higher_ in regions of high heat flow rates and low seismic velocities (See, e. g., Madden and Swift, 1969).

It is generally concluded that the _nondipole_ components of the geomagnetic field reflect some characteristics of _deep-seated_ features, but the details are not clearly understood (See, e. g., Stacey, 1969). This arises partly from the fact that spherical harmonic analyses of the geomegnetic field do _not_ produce _individual_ harmonic terms identifiable with features of the field except perhaps for the _dipole_ terms. Other analyses suggest that the field may be due to _several_ deep-seated dipoles (See, e. g., Bullard, 1956). It appears that 8 non–central radial dipoles reasonably represent the field when considered on a global basis, with the dipoles at about 0.25 Earth radii, or possibly farther, from the center (See, e. g., Lowes and

Runcorn, 1951; Alldredge and Hurwitz, 1964).

In any analysis of geomagnetic polarity reversals, the above observations about the geomagnetic and geoelectric fields must be supplemented by considerations regarding the applicable global frames of reference, as well as the geometrical and mechanic environment in which the fields exist. If one had measurable inducing fields with a broad continuum of frequencies and length scales to probe the entire Earth (or at least certain selected portions of it), it might be possible to achieve a satisfactory solution to the problem of the actual distribution pattern of electrical conductivity within the mantle. Unfortunately, such is not practical, and it is possible to observe only the natural magnetic variations, which might be considered to reflect the geoelectric field. However, these are usable at only relatively few, discrete frequencies, in a narrow bandwidth, with the result that the conductivity solution is subject to large systematic errors. The situation is further aggravated by the following unknowns: (1) the effects of the oceans and sedimentary regions and pockets; (2) lack of knowledge regarding the size and nature of the so-called "low-conductivity layer"; and (3) the role of heat flow.

Symmetry and other geometrical considerations seem to enter into the origin, evolution, and present characteristics of the geomagnetic field. For example, the observation that the geo-magnetic field is roughly symmetrical about the Earth's axis

170

of rotation has at least two important implications: (1) it has an important bearing on paleomagnetic analyses, and (2) it poses important constraints on hypotheses regarding the origin and perpetuation of the Earth's magnetic field (See, e. g., Stacey, 1969).

Within the geological time scale, the angular variation (inclination) between the geomagnetic axis and the Earth's rotational axis appears as a transient. Consequently, there is little reason to search for asymmetric elements in the Earth or in geomagnetic dynamo mechanisms hypothesized to explain the evolution and present behavior of the geomagnetic field. Although the principles of symmetry are normally applied intuitively to physical cause-and-effect situations, a more explicit approach should be used in cases such as this where cause-and-effect relations are closely intermingled (and even, on occasion, possibly interchanged). (See, e. g., Elsasser, 1966).

As far as can be determined, the relevant contributing causes of the geomagnetic field are simply the rotation of the Earth (which, for all practical purposes, is axisymmetric) and various motions, temperature gradients, etc., in the core, all of which, apart from the effect of rotation, are assumed by most observers to be spherisymmetric. Thus, if the geomagnetic field were to deviate in a consistent (asymmetric) manner with respect to the axis of rotation, then the situation would be that of a field with lower symmetry than that attributable to a combination

171

of the known contributing causes. This would require us to search for some other (currently unknown) contributing cause of lower symmetry. However, since there is no consistent inclination of the dipole field, there is no need for such search. Thus, it is reasonable to assume that the geomagnetic axis is constrained to transient variations about the Earth's rotational axis as an axis of symmetry, but the nature of the causal mechanism for such constrained control is not known.

Evidence for the reality of geomagnetic polarity reversals on a global scale may be summarized as follows: (1) there is a marked correlation between polarity reversals observed on different continents and within different rocks on a single continent in the case of both igneous and sedimentary complexes; (2) sediments which were baked by contact with subsequently-emplaced igneous rocks have acquired thermo-remanent magnetism of the same polarity as the igneous rocks in almost all cases; and (3) the actual process of reversal has been traced both in rapid sequences of lava and in deep-sea cores (See, e. g., Stacey, 1969).

Fluctuations in the polarity of the geomagnetic field are known only for the past 4.5 m. y. (i. e., during the past one-tenth of one percent of the Earth's existence) but enough events are involved to permit certain conclusions to be drawn on the basis of what has happened during this short period: (1) reversals appear to have occurred cyclically, but in an aperiodic

172

manner, on an average of once every 200,000 yr.; (2) each reversal process lasts about 5000 yr (i. e., about 1.5% of the cycle);
(3) the magnitude of the global field strength is reduced by a
factor of 4 or 5 during the reversal process, suggesting that
the nondipole field and possibly an equatorial component of the
dipole field may remain effective during the actual reversal
process; (4) the total number of magnetized rocks is about equally
divided between normal and reversed polarities; (5) the reversals
are complete 180° reversals (rather than a succession of increments totaling 180°); (6) the basic causes contributing to the
geomagnetic field appear to be axisymmetric without polar symmetry, i. e., there is no distinction between the opposite axial
directions, either polarity being eqully probable; (7) during
certain periods, such as the Permian, the geomagnetic field
apparently had a constant polarity (i. e., without detectable
reversals) for tens of millions of years; and (8) reversals
appear to have occurred more frequently during periods of rapid
polar wander and/or continental drift, thereby implying that
mantle configuration may have a direct influence upon those
core motions that are responsible for the geomagnetic field
(Irving, 1966).

An analysis of the above indicates that the following tentative conclusions may be drawn: (1) that the geomagnetic dynamo
has evolved, and is presently driven, by the same global causal
mechanism that has produced the Earth's internal behavior during

173

the past 4.6 b. y.; (2) that the geomagnetic field correlates, now and in the past, with global characteristics of the Earth's driving mechanism; and (3) that cyclically aperiodic polarity reversals are definable in terms of the geometry and mechanics of the Earth's driving mechanism. On this basis, it appears that, when viewed on a global scale, geomagnetism and polarity reversals are surficial manifestations of the Earth's internal behavior and that such has been the case for at least part of the past 4.6 b. y. (Chapter 14).

Continents and Oceans.

In order to determine whether continents and oceans are surficial manifestations of the Earth's internal behavior, it is necessary to (1) analyze the interactions now occuring between continents and the floors of the oceans, and (2) to consider the types of interaction that might have occurred during the past 4.6 b. y., with particular attention to global and regional interactions that might be interpreted as having deep-seated causal connections.

It is generally accepted that there are two basic types of continental margin differing quite significantly from each other. From geographical associations, representative members of these are known as the Atlantic and the Pacific type margins (See, e. g., Drake, 1969).

The Atlantic type has stable unmodified blocks on the landward side and no evidence of post-Paleozoic activity. Ocean-to-continental crustal dimensions are rather abrupt and may, in places, be characterized by deep sedimentation. Atlantic type margins do not normally contain deep trenches, although a trench-like feature may be found beneath the sediments of the associated continental rise. In many cases, Paleozoic and earlier features are truncated by the continental margin.

The Pacific type, on the other hand, has young tectonic belts paralleling the shores. Shallow-focus earthquakes are very common, and intermediate and deep-focus earthquakes are abundant, particularly along <u>certain</u> segments of the Pacific type margin. Volcanic activity occurs, or has occurred historically, along the entire margin. Many portions of the margins contain <u>deep</u> trenches. Ocean-to-continental transition is much more <u>variable</u> than in the Atlantic type, and sedimentary accumulations do <u>not</u> appear to be great.

Analyzed on a global scale, it appears that the <u>differences</u> between Atlantic and Pacific type margins may be a function of <u>position</u> with respect to an <u>active</u> portion of the Earth's driving mechanism, as previously discussed in connection with geosynclines.

A deeper analysis leads to the question as to whether the above categories may be applicable to the oceans and continents <u>themselves</u>, as well as to their margins. To answer this question, it is necessary to analyze not only the oceans and continents as actual surficial manifestations but also the <u>understructures</u> beneath them. Also, it is essential that such be done over a <u>considerable</u> period of time, because it is well known that continents and oceans have <u>not</u> always been where they are today.

As long as the continents are exposed, erosion tends to lower sea level, which action disturbs the isostatic balance and renews uplift. If sea level is lowered, erosion is hastened by relative uplift; but, if sea level is raised, the flooding of continents produces no such restoring effect. The elevation of most continents averages less than a few hundred meters above sea level; the most-elevated, the Eurasian, averages only about

176

a km. Several processes tend to raise sea level, including the deposition of the products of erosion in the oceans, the release of new sea water, and the addition of fresh lava. The effect of these processes is to reduce the area of the ocean basins and to deepen the ocean. Such being the case, it is surprising that most of the continents were not long ago submerged by shallow seas.

The above suggests that there must be restoring forces tending to increase the heights of the continents (or, more generally stated, to increase the differences in "heights" of continents and oceans). Such restoring forces are not in evidence on the surface and, being invisible, are the subject of speculation. This is particularly true regarding cyclically aperiodic uplift of continental areas, which seems to have taken place on 3 scales, i. e., involving areas having diameters of thousands, hundreds, and tens of km, or perhaps a complete spectrum of diameters ranging from km to thousands of km.

For our purpose, the important point is that most of them are Tertiary in age and that they represent uplift of land over very extensive areas. Some continents (e. g., Africa and Antarctica) have been subjected to recent uplift; others (e. g., Australia) have not. In other regions of the globe, only portions of continents have undergone uplift, e. g., the Colorado Plateau and the Basin and Range Province of North America.

Alternating with uplift, there have been periods of sinking

177

and marine invasion. For example, the central portion of North America has been widely invaded by the ocean, roughly once every 100 m. y., since at least the Precambrian, alternating with periods of uplift and withdrawal of the sea. These sequences are temporally independent of similar activity in other parts of the Earth. In fact, they are temporally distinct from those of the Appalachian and Cordilleran belts on the same continent.

Viewed on a global scale, it appears that most areas, whether oceanic or continental, have suffered as many as 100 cyclically aperiodic episodes of uplift during the past 3.6 b. y. Consequently, present continental areas have not always been continental, nor have present oceanic areas always been oceanic. From this, it appears that oceans and continents may be surficial manifestations of the Earth's internal behavior to the extent that the Earth's driving mechanism appears to have given the Earth's surface an ever-changing temporal and spatial panorama of continents and oceans in all parts of the globe during at least the past 3.6 b. y.

Crustal Rifting and Sea-Floor Spreading.

To determine whether crustal rifting and sea-floor spreading are surficial manifestations of the Earth's internal behavior, it is necessary to understand the exact nature and possible interrelationships of these two phenomena. Basically, when viewed on a global scale and over an extended period of time, it appears that crustal rifting occurs on certain parts of the globe when surficial manifestations of the internal behavior appear in the form of crustal "tension". Since crustal rifting of this type was originally thought to occur only in the oceans, it was called "sea-floor spreading". There is evidence now that oceanic crustal rifting sometimes extends into continental areas (e. g., North America, Central America, Iceland, Africa). In spite of this, the original term of sea-floor spreading is often used even when referring to the type of crustal rifting that extends into continental areas. Such being the case, this book will use the term "sea-floor spreading" in those cases where current usage dictates. However, the author prefers the more general term, "crustal rifting", for both oceanic and continental rifting of this type.

According to the sea-floor spreading hypothesis, new oceanic lithosphere is formed more-or-less continuously at certain mid-ocean ridges and then spreads outward at rates of a few cm/yr (Vine and Matthews, 1963; Heirtzler et al., 1968; Le Pichon, 1968, 1969). Many geological observations indicate that sea-

179

floor spreading, as defined above, has been occurring for at least the past 100 m. y. This is based on evidence that most ocean floors are no older than that. Since continents are at least an order of magnitude older than the oceans, this suggests that a process similar to the above might have created the continents during the past 3.6 b. y.

Wilson (1965) reported evidence that the ages of fossil volcanoes (seamounts) increase with their distance from ridges, suggesting that these fossil volcanoes were once active volcanoes near the ridges. However, in deriving the locations of fossil volcanoes from those of active volcanoes, it is well to realize that sea-floor spreading probably does not occur at a linear rate nor at the same rate throughout the entire surface of the Earth at any given time. Most evidence suggests a cyclically aperiodic rate.

Since sea-floor spreading is commonly attributed to some type of mantle convection as a "driving force", Lee and MacDonald (1963) attempted to obtain evidence of a convective pattern by comparing spherical harmonic analyses of gravitational potential and surface heat flow, on the assumption that a correlation between geoidal lows and high heat flow would represent rising limbs in the hypothesized convection pattern. Their results were not conclusive, primarily because the amount of data available was too small for the particular type of correlation attempted. Subsequent, similar analyses have not clarified the nature of the suspected relationship between sea-floor spreading and mantle convection, other than to suggest that <u>both</u> sea-floor spreading <u>and</u> the hypothesized mantle convection might be surficial manifestations of the Earth's internal behavior.

Summarizing, it may be said that crustal rifting (and sea-floor spreading as a subsidiary phenomenon of crustal rifting) appears to be a surficial manifestation of the Earth's internal behavior. Furthermore, such crustal rifting appears to occur when and wherever the Earth's internal behavior causes crustal tension.

Plate Tectonics and Related Matters.

If plate tectonics is a surficial manifestation of the Earth's internal behavior, then the behavior of the plates should be correlatable with certain characteristics of the Earth's internal behavior on a global scale and over a period of considerable time.

Basic to the hypothesis of plate tectonics is the concept of a rigid upper layer of the Earth (lithosphere), composed of the crust and uppermost portion of the mantle. The lithosphere is assumed to have considerable strength, to be roughly 100 km thick, and to rest or float upon a second layer, the asthenosphere, which is assumed to have practically no strength and to extend from the base of the lithosphere to a depth of several hundred km below. The floating lithosphere is divided into vast plates or blocks which are bounded by ocean ridges, by certain faults, and by the great system of quasi-arcuate structures such as the circum-Pacific system. The plates spread apart at the ridges, grind against each other at the faults, and are underthrust at the island arcs and similar structures (See, e. g., Drake, 1970).

The system contains 6 major plates (and some minor ones). The major ones are: (1) the American Plate including North and South America and the western half of the Atlantic floor; (2) the African Plate, including Africa, the eastern half of the South Atlantic floor, and the Indian Ocean floor out to the

182

Mid-Indian Ridge; (3) the Eurasian Plate, including Europe and Asia except for India and Arabia; (4) the Pacific Plate, containing both the North and South Pacific; (5) the Australian Plate, including Australia plus some less well-defined ocean areas; and (6) the Antarctic Plate, including Antarctica plus some less well-defined ocean areas.

To cover the entire globe, the above 6 major plates are supplemented by various minor plates, such as the Caribbean, Cocos, Nacza, Philippine, India, Arabian, and Aleutians plates. These plates, both major and minor, move and supposedly have been moving during at least part of the past 3.6 b. y. over the surface of the Earth. They have been carrying with them the continents and/or oceanic sediments and have been spreading apart at the ocean ridges where hot basaltic material from the mantle rises and joins the trailing edge.

Major collisions between plates have created, over at least most of geologic time, some of the Earth's major topographical features. Minor collisions cause earthquakes. If the velocity of collision is less than 6 cm/yr, the plates tend to buckle and fold, creating ranges of new mountains, such as the Himalayas (See, e. g., Le Pichon, 1968). If the velocity of collision is more than 6 cm/yr, one plate thrusts itself under the other and descends into the asthenosphere at an angle of about 45°. Upon reaching the mantle, it is hypothesized that the leading edge of the underthrusting plate is progressively reconverted into mantle material. The underthrusting plate creates the deep, linear trenches that are observable in locations such as Tonga, the Kuriles, Japan, the Marianas, the Peru-Chile coast, Puerto Rico, New Zealand, etc.

It is further hypothesized that the impact of two colliding plates produces the islands and volcanoes, and that the impact accounts for major earthquakes observed in the vicinity of the trench-and-island-arc systems. Collisions at the edge of a continent produce long quasi-linear mountain ranges, such as the American cordilleras, which stretch from Alaska to Tierra del Fuego, a distance of about 15,000 km, on the assumption that some of the ocean-bottom sediments on the underthrusting plate are scraped off and piled up along the edge of the underthrust continents. The heat of the scraping supposedly causes volcanoes to develop in the piled-up sediments (See, e. g., Dewey and Bird, 1970; Isacks et al., 1968; Le Pichon, 1968; McKenzie and Parker, 1967; Morgan, 1968).

The origin and evolution of the forces that supposedly have driven these huge plates over the surface of the Earth, during at least part of the past 3.6 b. y., have not been defined in any detail in current hypotheses. The same is true of the geometrico-mechanical framework which has controlled these forces during all this time.

What initiated the global driving mechanism that has evolved, either syngenetically with the Earth, 4.6 b. y., ago, or at some time since then, say 3.6 b. y. ago? Once initiated, what has

184

sustained this fairly well organized driving mechanism during the past 4.6 or 3.6 b. y.? To answer these questions, consideration must be given to the development of the forces that appear to have been both long-lived and quasi-organized in a manner sufficient to perform this function in the internal behavior of the Earth and the creation of its kaleidoscopic surficial features during at least part of the past 3.6 b. y.

One possible driving mechanism could have been the convection within the mantle created by differences in temperature under oceans and continents. Once started, the convection could then slowly move like a conveyor belt, splitting the plates apart to form a ridge and then carry them along on the huge conveyor-belt system. But, granted that forces were available to have started the system, say 4.6 or 3.6 b. y. ago, what would have kept it going? If it started, not 4.6 or 3.6 b. y. ago, but at some time since then, say 1 b. y. or 200 m. y. ago, what made it start then, rather than at some other time? Does this mean that perhaps it started syngenetically with the Earth 4.6 b. y. ago and that it has been going on ever since?

Is there an alternative to the above system for splitting and moving these plates? For example, are gravitational forces, acting on the downthrusting lithospheric plate, exerting a pull sufficient to drag it downward at an angle of about 45° to depths of at least 700 or 800 km? Or, perhaps, does the underthrusting plate cool the mantle to produce the convection currents that

185

move the plates?

The geometry and mechanics involved in these and other questions are examined in Chapter 18 (Plate Tectonics). Suffice it here to say that plate tectonics appears to be a surficial manifestation of the Earth's internal behavior on a global scale during at least the past 3.6 b.y. but that the geometry and mechanics of the Earth beneath the plates, to a depth of at least 1000 km, must be more completely defined and analyzed than they are now.

Further analyses are also necessary to explain surficial manifestations such as: (1) seismic activity under Nevada; (2) the creation of the Rocky Mountains and certain other mountains of the Earth; (3) the existence of a complete spectrum of magma and associated rocks which appear to follow certain global distribution patterns (such as is suggested by the "Andesite Line" or by basaltic flows of voluminous proportions). Answers are also required for certain related questions, including: (1) how the geomagnetic dynamo was initiated and maintained, over a period of at least several b. y., and why, if the plates are moving in a more-or-less random manner, it has managed to reverse its polarity in a cyclically aperiodic manner; (2) the nature and origin of the numerous rectilinear orogenic-tectonic features observed on the surface of the Earth and what mechanico-geometrical configuration causes them to be rectilinear (since Le Pichon's plates, by-and-large, do not appear to have

186

rectilinear edges); (3) <u>why</u> the leading edge of a downthrusting plate descends at an angle of about 45° (rather than vertically or at some other angle).

Continental Drift and Polar Wandering.

If portions of the Earth's surface are undergoing a certain amount of drifting and if, as a result of this or in addition to it, the Earth's pole of rotation is wandering, then these effects probably fall within the category of surficial manifestations of the Earth's internal behavior. What mechanism within the Earth's interior causes continental drift and polar wandering? Is this something relatively new, or has it been going on for the past 4.6 b. y.? If the latter, what has managed to move the continents over the surface of the Earth, during all these years? If this mechanism started recently, say 200 m. y. ago, what caused it to start then, rather than 300 m. y. ago, or 900 m. y, or 2.9 b. y. ago? Is continental drift proceeding in a strictly random manner, or is it following a quasi-organized pattern? Is pre-drift continental splitting (e. g., Africa-South America) induced by random forces, or are they quasi-linear as would be expected if they were induced by an organized, non-random global driving mechanism operating over a long period of time?

The hypothesis of continental drift, which proposes that continent-size blocks of the Earth's crust have retained their form but have moved relative to one another, is today fairly well established on the basis of global paleomagnetism analyses, supplemented by regional and global studies of other types. Although the earlier concepts of continental drift are associated

188

with Du Toit, Taylor, and Wegener, the modern approach is usually
associated with Runcorn (1962), Bullard (1964), Blackett (Blackett
et al., 1965), and Isacks (Isacks et al., 1968).

General concepts and observations supporting continental
drift and/or polar wandering may be summarized as follows: (1)
the southern continents, Africa, Australia, South America, India,
and Antarctica appear to have been grouped around the latter
during the Jurassic (about 150 m. y. ago); (2) the polar-wander
curves for Africa, South America, and Australia follow paths
which are almost identical for the 3 continents and have the
shape of a large Z in which the diagonal is about 90°, or about
10,000 km, or a linear trace roughly one-quarter of the Earth's
circumference; (3) polar-wander curves suggest continental drift
rates of 90° in 500 m. y., or about 2 cm/yr; (4) since other
evidence suggests that large-scale tectonic processes occur
at a few cm/yr, polar wander of this magnitude may be globally
associated with at least some of the large-scale tectonic processes; (5) stresses supported by the _lower_ mantle are _too large_
to permit lower-mantle convection of a few cm/yr, equivalent
to mantle creep rates of about 10^{-15} to 10^{-16} sec^{-1}; (6) both
polar wandering and continental drift appear to be caused by
the _same_ general process of mass distribution as may be envisioned
by a lower-mantle adjustment or equilibration of the equatorial
bulge to a changing axis of rotation; (7) there could have been
two separate super-continents (Laurasia and Gondwanaland) or only

189

one (Pangaea), but this concept may be too simple and temporally restrictive to the extent that it must be supplanted by one in which the super-continents have repeatedly broken up and rejoined in different global surficial configurations; (8) polar-wander and continental-drift episodes are _not_ necessarily steady-state processes but appear to be cyclically aperiodic processes in which relatively long quiescent periods are separated by short drift and/or wander episodes; (9) all polar movement relative to Australia, and extending over the 300 m. y. period from upper Silurian to mid-Cretaceous, occurred within a 20 m. y. episode during the Carboniferous (about 300 m. y. ago) (Irving, 1966); (10) some correlation exists between drift episodes and strong orogenic-tectonic activity occurring within certain continents during episodes of drift and/or polar wander; (11) the ages of rocks are grouped in a manner suggesting a more-or-less cyclical sequence of tectonically active and quiescent periods, with the cycles occurring every 250 to 500 m. y.

Present astronomical determinations of _longitude_ are too imprecise to indicate relative movements of a few cm/yr; however, _latitude_ measurements made with reliable precision indicate a currently steady movement of the pole toward about 70°W at a rate of 10 cm/yr (Markowitz et al., 1964; Markowitz and Guinot, 1968). Francheteau and associates (Francheteau et al., 1969) have found evidence suggesting that the northwestern and northeastern Pacific have _not_ always been rigidly connected. From this, they concluded that the present Pacific Plate (Le Pichon, 1968) is really a composite of at least 2 distinct plates and that volcanic aseismic ridges within the Pacific Ocean may be the fossil boundaries of these plates now coalesced into a single plate. By plotting available geochronological data and zones of isochronous tectonic and igneous activity on a pre-drift reconstruction of Laurasia, Hurley (1969) has shown that the configurations of the age provinces and the prevailing trends of orogenic zones suggest that there were successively younger belts of tectonism around ancient super-continents. After a review of observational evidence bearing on continental drift and polar wandering, Lyustikh (1969) concluded that present simple models must be extended and specialized to more-complex realistic models that take into account real-world observations regarding the behavior of the Earth, its structure, and the inhomogeneities and anomalies of its interior, with particular attention to the upper 1000 km portion, i. e., the tectonosphere.

From the above and other analyses (See, e. g., Allan et al., 1967; Anderle and Smith, 1969; Anderson, 1967, 1969; Barazangi and Dorman, 1969; Beloussov, 1969a, 1969b; Brune and Dorman, 1963; Bullard et al., 1965; Cox et al., 1964; Crawford and Wilson, 1969; Dietz et al., 1969; Ewing and Ewing, 1967; Goldreich and Toomre, 1969; Gough, 1970; Griggs, 1939; Grommé et al., 1967; Hurley et al., 1967; Isacks et al., 1968; Johnson, 1967; Knopoff, 1964, 1967a, 1967b, 1969; Knopoff and Shapiro, 1969; Le Pichon,

1968; Maxwell, 1969; McKenzie and Parker, 1967; Heservey, 1969; Morgan, 1968; Oliver and Isacks, 1967; Oliver et al., 1969; Salop and Scheinmann, 1969; Tozer, 1965; Verhoogen, 1965; Vine, 1966, 1969a, 1969b; Vine and Matthews, 1963; Wegener, 1915; and Wilson, 1965), it may be concluded that continental drift and polar wander are surficial manifestations of the Earth's internal behavior, during possibly the entire age of the Earth, but that the geometry and mechanics of the Earth's interior and the nature of its driving mechanism remain to be defined.

The Earth's Most Probable Surficial Features 4.6 b. y. Ago.

In analyzing the present surficial features that might have been caused by possible internal behavior patterns of the Earth during at least part of the past 4.6 b. y., it is well to consider briefly the possible surficial configuration of the Earth 4.6 b. y. ago and at various critical times since then.

The configuration of the primordial Earth's most probable surficial features depends largely upon which hypotheses are adopted to explore the evolution of the Earth from the solar nebula (Chapter 1) and to explain the evolution of the Earth-Moon system. If it is assumed that the Earth accreted from the solar nebula and that the Moon formed in the vicinity of (but not part of) the Earth, then the Earth's surface would probably have been devoid of any features other than those that would be genetic to an accretionary process. Thus, the Earth's surface would have presented a more-or-less homogeneous aspect 4.6 b. y. ago, both from the standpoint of topography (smooth) and composition (the same) over the entire surface.

If, on the other hand, it is assumed that the Earth's surficial composition and/or topography were not homogeneous 4.6 b. y. ago, then such features most probably would not have survived until today without internal or external rejuvenation, at least on one occasion since then. For the purpose of this analysis, therefore, it may be assumed that the Earth's surface

4.6 b. y. ago was devoid of any features other than those that would have been genetic to an accretionary process of planetary formation from the solar nebula at that time.

The Earth's Most Probable Surficial Features 3.6 b. y. Ago.

If it is assumed that the Earth had no appreciable surficial features 4.6 b. y. ago, _when_ were the surficial features _first_ acquired and _what_ caused them? Were they acquired through _external_ or _internal_ behavior? If external, _what_ was it? If internal, why did it occur _when_ it did rather than sometime earlier or later?

If the surficial features resulted from _external_ sources during the period 4.6 b. y. ago to 3.6 b. y. ago, such features would not _now_ be active but would resemble the Precambrian shields and would have an age of about 3.6 b. y.

If the Earth acquired surficial features from _internal_ sources, during the period 4.6 b. y. to 3.6 b. y. ago, such features would not _now_ be active _unless_ the same causative internal behavior has been operating since 3.6 b. y. ago, in which case it would be reasonable to assume that the internal mechanism has been operating more-or-less continuously during the past 4.6 b. y. (since nothing _internal_ to the Earth would have been available to activate it 3.6 b. y. ago).

In any case, it appears that the Precambrian shields existed 3.6 b. y. ago, but not necessarily in their present forms and locations, and that they have existed in one or more locations and forms, without having been severely heated or shocked, since then.

194

The Evolution of the Earth's Present Surficial Features.

Regardless of whether it is assumed that the Earth's surficial features of 3.6 b. y. ago were externally or internally induced, we must consider certain questions about what has happened to the Earth's surface since then. Have there been significant externally induced catastrophes that might have altered the Earth's surficial features during the past 3.6 b. y.? If so, what were they, when did they occur, and what caused them? If no such external influences have operated during the past 3.6 b. y., is it reasonable to assume that all modifications to the Earth's surface since then were induced by the Earth's internal behavior?

To answer these and related questions about the evolution of the Earth's present surficial features, global-scale analyses must be made regarding both the present and past internal behavior of the Earth. Several international scientific bodies have been devoted to such analyses. In discussing the results of the International Upper Mantle Commission and the purpose of the Interunion Commission on Geodynamics, Drake (1970) has summarized many of the problems involved in analyzing the Earth's internal behavior and the surficial manifestations thereof. For the interested reader, Drake's paper constitutes an excellent summary of the results of the International Upper Mantle Commission and the aspirations of the Interunion Commission of Geodynamics.

Other authors have made similar analyses but from different viewpoints or in lesser inclusive regions (See, e. g., Delany and Smith, 1969; Hart, 1969; Knopoff et al., 1968; Stacey, 1969; Takeuchi et al., 1967; Wynne-Edwards, 1969). Earlier analyses of the Earth's internal behavior and related surficial manifestations have been summarized in certain standard texts on the Earth (See, e. g., Aubouin, 1965; Beloussov, 1962; Cailleux, 1969; Cook and Gaskell, 1961; De Sitter, 1956; Dunbar, 1966; Garland, 1966; Gaskell, 1967; Gutenberg, 1951; Holmes, 1965; Jacobs, 1963; Jacobs et al., 1959; Jeffreys, 1962; King, 1962; Lee, 1965; Macdonald and Kuno, 1962; Markowitz and Guinot, 1968; Matsumoto, 1967; Orlin, 1966; Poldervaart, 1955; Rikitake, 1966; Rittmann, 1962; Runcorn, 1962; Scheidegger, 1963; Shimer, 1968; Steinhart and Smith, 1966; Stevenson, 1962: Vening Meinesz, 1964).

When the Earth is considered as a global entity, the results of the above analyses indicate that most past and present geophenomena are manifestations of the Earth's internal behavior during the past 4.6 b. y., but that the geometry and mechanics of the causative internal behavior and the nature of the ultimate driving mechanism reamin to the defined. It is the author's hypothesis that the tectonospheric Earth model derived from the dual primeval planet hypothesis provides answers for the undefined geometry and mechanics and for the nature of the driving mechanism. One purpose of this book is to explain these in detail and to examine them alongside other hypotheses regarding the evolution of the Earth's tectonosphere.

196

REFERENCES

Allan, D. W., Thompson, W. B., and Weiss, N. O., 1967. Convection in the Earth's mantle. In: S. K. Runcorn (Editor), Mantles of the Earth and Terrestrial Planets. Interscience, New York, pp. 507-512.

Alldredge, L. R. and Hurwitz, L., 1964. Radial dipoles as the sources of the Earth's main magnetic field. J. Geophys. Res., 69: 2631-2640.

Anderle, R. J. and Smith, S. J., 1968. Observation of 27th and 28th order gravity coefficients based on Doppler observations. J. Astronaut. Sci., 15: 1-4.

Anderson, D. L., 1967. Phase change in the upper mantle. Science, 157: 1165-1163.

Anderson, D. L., 1969. Geophysical evidence on the petrology of the mantle. Geol. Soc. Am. Abstracts with Programs, 7: 2.

Argand, E., 1916. Sur l'arc des Alpes occidentales. Eclogae Geologicae Helvetiae, 16: 179-182.

Aubouin, J., 1965. Geosynclines. Elsevier, Amsterdam, 335 pp.

Balakina, L. M., Misharina, L. I., Shirokova, E. I., Vveden-skaya, A. V., 1969. The field of electric stresses associated with earthquakes. In: P. J. Hart (Editor), The Earth's Crust and Upper Mantle. Am. Geophys. Union, Washington, pp. 166-171.

Barazangi, M. and Dorman, J., 1969. World seismicity map of ESSA Coast and Geodetic Survey epicenter data for 1961-1967. Bull. Seis. Soc. Am., 59: 369-380.

Beloussov, V. V., 1962. Basic Problems in Geotectonics. McGraw-Hill, New York, 816 pp.

Beloussov, V. V., 1969a. Continental rifts. In: P. J. Hart (Editor), The Earth's Crust and Upper Mantle. Am. Geophys. Union, Washington, pp. 539-544.

Beloussov, V. V., 1969b. Interrelations between the Earth's crust and upper mantle. In: P. J. Hart (Editor), The Earth's Crust and Upper Mantle. Am. Geophys. Union, Washington, pp. 698-712.

Blackett, P. M. S., Bullard, E. C., and Runcorn, S. K.,

(Editors), 1965. A symposium on continental drift. Phil. Trans. Roy. Soc., Ser. A., 258: 1.

Brune, J. N. and Dorman, J., 1963. Seismic waves and Earth structure in the Canadian Shield. Bull. Seis. Soc. Am., 53: 167–210.

Bullard, E. C., 1949. The magnetic field within the Earth. Proc. Roy. Soc. A, 197: 433.

Bullard, E. C., 1956. Edmund Halley (1656–1742). Endeavour, 15: 189.

Bullard, E. C., 1964. Continental drift. Quart. J. Geol. Soc. Lond., 120: 1.

Bullard, E. C., Everett, J. E., and Smith, A. S., 1965. The fit of the continents around the Atlantic. Phil. Trans. Roy. Soc. Lond., Ser. A, 258: 41–51.

Cailleux, A., 1968. Anatomy of the Earth. McGraw-Hill, New York, 255 pp.

Caputo, M., 1967. The Gravity Field of the Earth. Academic New York, 202 pp.

Cook, A. H. and Gaskell, T. F. (Editors), 1961. The Earth

<u>Today</u>. Roy. Astron. Soc., London, 404 pp.

Cox, A., Doell, R. R., and Dalrymple, G. B., 1964. Reversals of the Earth's magnetic field. <u>Science</u>, 144: 1537-1543.

Crawford, A. R. and Wilson, J. T., 1969. Continental drift in the Arctic and Indian oceans and before the Mesozoic era. <u>Geol</u>. <u>Soc</u>. <u>Am</u>. <u>Abstracts</u> <u>with</u> <u>Programs</u>, 7: 41.

Dana, J. D., 1873. On some results of the Earth's contraction from cooling, including a discussion of the origin of mountains and the nature of the Earth's interior. <u>Am</u>. <u>J</u>. <u>Sci</u>., Ser. 3, 5: 423-443; 6: 6-14, 104-115, 161-172.

Delany, P. and Smith, C. H., (Editors), 1969. <u>Deep-Seated</u> <u>Foundations</u> <u>of</u> <u>Geological</u> <u>Phenomena</u>. Special issue <u>Tectonophysics</u>, 7: 359-610.

De Sitter, L. U., 1956. <u>Structural</u> <u>Geology</u>. McGraw-Hill, London, 552 pp.

Dewey, J. F. and Bird, J. M., 1970. Mountain belts and the new global tectonics. <u>J</u>. <u>Geophys</u>. <u>Res</u>., 75: 2625-2647.

Dietz, R., 1963. Collapsing continental rises: an actualistic concept of geosynclines and mountain building. J. Geol., 71: 314-333.

Dietz, R. and Holden, J. C., 1966. Miogeosynclines in space and time. J. Geol., 74: 566-583.

Dietz, R. S., Holden, J. C., and Sproll, W. P., 1969. Geotectonic evolution and subsidence of Bahama Platform. Geol. Soc. Am. Abstracts with Programs, 7: 43.

Douglas, R. J. W., 1969. Orogeny, basement, and geological map of Canada. Geol. Assoc. Can. Spec. Paper 5, pp. 1-6.

Drake, C. L., 1969. Continental margins. In: P. J. Hart (Editor), The Earth's Crust and Upper Mantle. Am. Geophys. Union, Washington, pp. 549-556.

Drake, C. L., 1970. A long-range program of solid Earth studies. Trans. Am. Geophys. Union, 51: 152-159.

Drake, C. L. and Kosminskaya, I. P., 1969. The transition from continental to oceanic crust. Tectonophysics, 7: 363-384.

Drake, C. L., Ewing, M., and Sutton, G. H., 1959. Continenta[l]
margins and geosynclines: the east coast of North America north
of Cape Hatteras. In: L. H. Ahrens et al. (Editors), _Physics
and Chemistry of the Earth_, Vol. 3. Pergamon, New York, pp.
110-198.

Dunbar, C. O., 1966. _The Earth_. World, New York, 252 pp.

Elsasser, W. M., 1966. Thermal structure of the upper
mantle. In: P. M. Hurley (Editor), _Advances in Earth Sciences_.
MIT Press, Cambridge, pp. 461-471.

Elsasser, W. M., 1967. Interpretation of heat flow equality.
J. Geophys. Res., 72: 4768-4770.

Ewing, J. and Ewing, M., 1967. Sediment distribution on
the mid-ocean ridges with respect to the spreading of the sea
floor. _Science_, 156: 1590-1592.

Francheteau, J., Harrison, C. G. A., Sclater, J. S., and
Richards, M. L., 1969. Magnetization of Pacific seamounts: a
preliminary polar curve for the northeastern Pacific. _Trans._

Am. Geophys. Union, 50: 634.

Garland, G. D. (Editor), 1966. Continental Drift. Univ. of Toronto Press, Toronto, 140 pp.

Gaskell, T. F. (Editor), 1967. The Earth's Mantle. Academic, London, 509 pp.

Gastil, G., 1960. The distribution of mineral dates in time and space. Am. J. Sci., 258: 1.

Goldreich, P. and Toomre, A., 1969. Some remarks on polar wandering. J. Geophys. Res., 74: 2555-2567.

Gough, D. I., 1970. Did an ice cap break Gondwanaland? J. Geophys. Res., 75: 4475-4477.

Green, D. H., 1969. The origin of basaltic and nephelinitic magmas in the Earth's mantle. Tectonophysics, 7: 409-422.

Griggs, D., 1939. A theory of mountain building. Am. J. Sci., 237: 611-650.

Gromme, C. S., Merrill, R. T., and Verhoogen, J., 1967. Paleomagnetism of Jurassic and Cretaceous plutonic rocks in the

Sierra Nevada, California, and its significance for polar wandering and continental drift. _J. Geophys. Res._, 72: 5661-5684.

Gutenberg, B., 1951. _Internal Constitution of the Earth_. Dover, New York, 439 pp.

Hall, J., 1857. Direction of the currents of deposition and source of the materials of the older Paleozoic rocks. _Can. Naturalist and Geologist_, 2: 284-286.

Hart, P. J. (Editor), 1969. _The Earth's Crust and Upper Mantle: Structure, Dynamic Processes, and Their Relation to Deep-Seated Geological Processes_. Am. Geophys. Union, Washington, 735 pp.

Heiskanen, W. A. and Moritz, H., 1967. _Physical Geodesy_. Freeman, San Francisco, 364 pp.

Heiskanen, W. A. and Vening Meinesz, F. A., 1958. _The Earth and Its Gravity Field_. McGraw-Hill, New York, 470 pp.

Heirtzler, J. R., Dickson, G. O., Herron, E. M., Pitman, W. C., and Le Pichon, X., 1968. Marine magnetic anomalies,

geomagnetic field reversals, and motions of the ocean floor
and continents. J. Geophys. Res., 73: 2119-2136.

Holmes, A., 1965. Principles of Physical Geology. Ronald,
New York, 1288 pp.

Hsu, K. J., 1965. Collapsing continental rises: a discussion. J. Geophys. Res., 73: 897-900.

Hurley, P. M., 1969. Some observations on the geological
history of Laurasia. Geol. Soc. Am. Abstracts with Programs,
7: 112.

Hurley, P. M., de Almeida, F. F. M., Melcher, G. C., Cordani,
U. G., Rand, J. R., Kawashita, K., Vandoros, P., Pinson, W. H.,
and Fairbairn, W. H., 1967. Test of continental drift by comparison of radiometric ages. Science, 157: 495-500.

Irving, E., 1966. Paleomagnetism of some carboniferous
rocks from New South Wales and its relation to geological events.
J. Geophys. Res., 71: 6025-6051.

Isacks, B., Oliver, J., and Sykes, L. R., 1968. Seismology

and the new global tectonics. _J. Geophys. Res._, 73: 5855-5899.

Jacobs, J. A., 1963. _The Earth's Core and Geomagnetism._
Pergamon, Oxford, 137 pp.

Jacobs, J. A., Russell, R. D., and Wilson, J. T., 1959.
Physics and Geology. McGraw-Hill, New York, 424 pp.

Jeffreys, H., 1962. _The Earth: Its Origin, History, and
Physical Constitution._ Cambridge Univ. Press, Cambridge, 438 pp.

Johnson, L. R., 1967. Array measurements of velocities in
the upper mantle. _J. Geophys. Res._, 72: 6309-6325.

Kay, M., 1951. North American geosynclines. _Geol. Soc.
Am. Mem._ 48, 143 pp.

King, L. C., 1962. _The Morphology of the Earth._ Hafner,
New York, 699 pp.

Knopoff, L., 1964. The convection current hypothesis.
Rev. Geophys., 2: 89-122.

Knopoff, L., 1967a. Thermal convection in the Earth's
mantle. In: T. F. Gaskell (Editor), _The Earth's Mantle._ Academic,

London, pp. 171-196.

Knopoff, L., 1967b. A convection in the upper mantle. *Geophys*. *J*., 14: 341-346.

Knopoff, L., 1969. Continental drift and convection. In: P. J. Hart (Editor), *The Earth's Crust and Upper Mantle*. Am. Geophys. Union, Washington, pp. 683-689.

Knopoff, L. and Shapiro, J. N., 1969. Comments on the interrelationships between Grüneisen's parameter and shock of isothermal equation of state. *J*. *Geophys*. *Res*., 74: 1439-1450.

Knopoff, L., Drake, C. L., and Hart, P. J. (Editors), 1968. *The Crust and Upper Mantle of the Pacific Area*. Am. Geophys. Union, Washington, 522 pp.

Kuenen, P. H., 1967. Geosynclinal sedimentation. *Geologische Rundschau*, 56: 1-19.

Laubscher, H., 1969. Mountain building. *Tectonophysics*, 7: 551-563.

Lee, W. H. K. (Editor), 1965. *Terrestrial Heat Flow*. Am.

Geophys. Union, Washington, 276 pp.

Lee, W. H. K. and MacDonald, G. J. F., 1963. The global variation of terrestrial heat flow. J. Geophys. Res., 68: 6481-6492.

Le Pichon, X., 1968. Sea floor spreading and continental drift. J. Geophys. Res., 73: 3661-3697.

Le Pichon, X., 1969. Models and structure of oceanic crust. Tectonophysics, 7: 385-401.

Lowes, F. J. and Runcorn, S. K., 1951. The analysis of geomagnetic secular variation. Phil. Trans. Roy. Soc., Ser. A, 243: 525-535.

Lubimova, E. A., 1969. Thermal history of Earth. In: P. J. Hart (Editor), The Earth's Crust and Upper Mantle. Am. Geophys. Union, Washington, pp. 63-77.

Lyustikh, E. N., 1969. Problem of convection in the Earth's mantle. In: P. J. Hart (Editor), The Earth's Crust and Upper Mantle. Am. Geophys. Union, Washington, pp. 639-692.

Macdonald, G. A. and Kuno, H. (Editors), 1962. The Crust of the Pacific Basin. Am. Geophys. Union, Washington, 195 pp.

Madden, T. R. and Swift, C. M., Jr., 1969. Magnetotelluric studies of the electrical conductivity structure of the crust and upper mantle. In: P. J. Hart (Editor), The Earth's Crust and Upper Mantle. Am. Geophys. Union, Washington, pp. 469-479.

Markowitz, W. and Guinot, B. (Editors), 1968. Continental Drift, Secular Motion of the Pole, and Rotation of the Earth. Reidel, Dordrecht, 107 pp.

Markowitz, W., Stoyko, N., and Fedorov, E. P., 1964. Longitude and latitude. In: H. Odishaw (Editor), Research in Geophysics, Vol. 2. MIT Press, Cambridge, pp. 149-162.

Martin, H., 1969. Problems of age relations and structure in some metamorphic belts of southern Africa. Geol. Assoc. Can. Spec. Paper 5, pp. 17-26.

Matsumoto, T. (Editor), 1967. Age and Nature of the Circum-Pacific Orogenesis. Special issue Tectonophysics, 4: 317-623.

Maxwell, J. C., 1968. Continental drift and a dynamic Earth. _Am. Sci._, 56: 35-51.

Maxwell, A. E., 1969. Recent deep sea drilling results from the South Atlantic. _Trans. Am. Geophys. Union_, 50: 113.

McBirney, A. R., 1969. Andesitic and rhyolitic volcanism of orogenic belts. In: P. J. Hart (Editor), _The Earth's Crust and Upper Mantle_. Am. Geophys. Union, Washington, pp. 501-507.

McGarr, A., 1969. Amplitude variations of Rayleigh waves: propagation across a continental margin. _Bull. Seis. Soc. Am._, 59: 1281-1305.

McKenzie, D. P. and Parker, R. L., 1967. The north Pacific: an example of tectonics on a sphere. _Nature_, 216: 1276-1280.

Meservey, R., 1969. Topological inconsistency of continental drift on the present-sized Earth. _Science_, 166: 609-611.

Mitchell, A. H. and Reading, H. G., 1969. Continental margins, geosynclines, and ocean floor spreading. _J. Geol._, 77: 629-646.

Molnar, P. and Oliver, J., 1969. Lateral variations of
attenuation in the upper mantle and discontinuities in the lithosphere. *J. Geophys. Res.*, 74: 2648–2682.

Morgan, W. J., 1968. Rises, trenches, great faults, and
crustal blocks. *J. Geophys. Res.*, 73: 1959–1982.

Oliver, J. and Isacks, B., 1967. Deep earthquake zones,
anomalous structures in the upper mantle, and the lithosphere.
J. Geophys. Res., 72: 4259–4275.

Oliver, J., Sykes, L., and Isacks, B., 1969. Seismology
and the new global tectonics. *Tectonophysics*, 7: 527–541.

Orlin, H. (Editor), 1966. *Gravity Anomalies: Unsurveyed
Areas.* Am. Geophys. Union, Washington, 142 pp.

Petrushevsky, B. A., 1969. Earthquakes and tectonics.
In: P. J. Hart (Editor), *The Earth's Crust and Upper Mantle.*
Am. Geophys. Union, Washington, pp. 279–282.

Polvervaart, A. (Editor), 1955. *Crust of the Earth.* Geol.
Soc. Am., New York, 761 pp.

Polyak, B. G. and Smirnov, Ya. B., 1966. Heat flow on continents. *Dokl. Akad. Nauk SSSR*, 168: 170-172.

Raitt, R. W., Shor, G. G., Francis, T. J. G., and Morris, G. B., 1969. Anisotropy of the Pacific upper mantle. *J. Geophys. Res.*, 74: 3095-3109.

Rikitake, T., 1966. *Electromagnetism of the Earth's Interior*. Elsevier, Amsterdam, 308 pp.

Rittmann, A., 1962. *Volcanoes and Their Activity*. Wiley, New York, 305 pp.

Runcorn, S. K. (Editor), 1962. *Continental Drift*. Academic, New York, 338 pp.

Salop, L. I. and Scheinmann, Yu. M., 1969. Tectonic history and structures of platforms and shields. *Tectonophysics*, 7: 565-597.

Scheidegger, A. E., 1963. *Principles of Geodynamics*. Academic, New York, 362 pp.

Schuchert, C., 1923. Sites and natures of North American

geosynclines. Geol. Soc. Am. Bull., 34: 151-260.

Shimer, J. A., 1968. This Changing Earth. Harper & Row, New York, 233 pp.

Sinonen, A., 1969. Batholiths and their orogenic setting. In: P. J. Hart (Editor), The Earth's Crust and Upper Mantle. Am. Geophys. Union, Washington, pp. 483-489

Stacey, F. D., 1969. Physics of the Earth. Wiley, New York, 324 pp.

Steinhart, J. S. and Smith, T. J. (Editors), 1966. The Earth beneath the Continents. Am. Geophys. Union, Washington, 663 pp.

Stevenson, J. S. (Editor), 1962. The Tectonics of the Canadian Shield. Univ. of Toronto, Toronto, 180 pp.

Stille, H., 1940. Einführung in den Bau Nordamerikas. Bontraeger Verlagsbuchhandlung, Berlin, 717 pp.

Sylvester-Bradley, P. C., 1968. Tethys, the lost ocean.

Takeuchi, H., Uyeda, S., and Kanamori, H., 1967. Debate about the Earth. Freeman Cooper, San Francisco, 253 pp.

Tikhonov, A. N., Lubimova, Ye. A., and Vsalov, V. K., 1969. Heat flow from the Earth's interior depending on inner parameters variations. Bull. Volcanol. 33(1): 261-280.

Tozer, D. C., 1965. Heat transfer and convection currents. Phil. Trans. Roy. Soc. Lond., Ser. A., 258: 252-271.

Vening Meinesz, F. A., 1964. The Earth's Crust and Mantle. Elsevier, Amsterdam, 124 pp.

Verhoogen, J., 1961. Heat balance of the Earth's core. Geophys. J., Roy. Astr. Soc., 4: 276.

Verhoogen, J., 1965. Phase change in convection in the Earth's mantle. Phil. Trans. Roy. Soc. Lond., Ser. A., 258: 276-283.

Vine, F. J., 1969a. Spreading of the ocean floor: New evidence. Science, 154: 1405-1415.

Vine, F. J., 1969b. Sea-floor spreading and continental drift. Geol. Soc. Am. Abstracts with Programs, 7: 231-232.

Vine, F. J. and Matthews, D. H., 1963. Magnetic

anomalies over ocean ridges. Nature, 199: 947-949.

Wegener, A., 1915. Die Entstehung der Kontinente und Ozeane. Sammlung Vieweg, Braunschweig, 94 pp.

Wilson, J. T., 1965. Convection currents and continental drift: evidence from ocean islands suggesting movement in the Earth. Phil. Trans. Roy. Soc., Ser. A., 258: 145-155.

Woollard, G. P., 1969a. Tectonic activity in North America as indicated by earthquakes. In: P. J. Hart (Editor), The Earth's Crust and Upper Mantle. Am. Geophys. Union, Washington, pp. 125-133.

Woollard, G. P., 1969b. Standardization of gravity measurements. Ibid., pp. 283-293.

Woollard, G. P., 1969c. Regional variations in gravity. Ibid., pp. 320-341.

Wyllie, P. J., 1969. The origin of ultramafic and ultrabasic rocks. Tectonophysics, 7: 437-455.

Wynne-Edwards, H. R. (Editor), 1969. Age Relations in High-Grade Metamorphic Terrains. Geol. Assoc. Can., Toronto, 228 pp.

Chapter 5

THE CRUST OF THE EARTH

The crust of the Earth is the outer portion of the tectono-sphere. The crust is separated from the rest of the tectonosphere by the M Discontinuity, the crust-mantle interface at which there is a marked increase in seismic velocities. This book considers the crust as a global entity in order to emphasize the concept of tectonic behavior as a controlled global system operating over a long period of time.

The crust has an average thickness of 35 km in continental areas; about 5 km, in oceanic areas. In both areas, the crust is characterized by a layered plate-block structure with both continuous and piecewise-continuous boundaries at all depths.

In a sense, the crust presents a panorama of the Earth's internal behavior. When viewed on a global scale, and considered together with the ages of the various rocks, the crust reveals much about the temporal and spatial variations in the evolution of the Earth's tectonosphere during the past 3.6 b. y. Definite differences exist among various sections of the Earth's crust. Continental and oceanic platforms differ from folded structures. Recently-folded structures differ from ancient eroded folded structures. Land and marginal seas differ from large oceans. Where once there was an ocean there is now land and vice versa.

216

All these ever-changing crustal types emphasize the variations in the Earth's internal behavior over a long period of time. The most strongly pronounced examples of such differences are the variations in the depth of the M Discontinuity as well as the relationship of the thickness of the sedimentary cover to the thickness of the consolidated crust. The roots of the mountains of the Asiatic Pacific coast, for example, are shallower than the roots of the mountains in the inland areas of Asia. There also appears to be a general decrease in the thickness of the crust from the inland regions of Asia (65 to 70 km) toward the marginal areas.

Other marked crustal differences appear in island arcs, in the shields of the eastern European platform (e. g., Ukraine and Voronezh massifs), and in the Mediterranean folded areas. Considered independently, these crustal types suggest different behavior at different depths of the tectonosphere. Similar crustal types considered collectively suggest a global control over tectonospheric behavior.

Some of the greatest crustal thicknesses of the Earth are found in the Carpathians, the Kopetdags, and certain systems of the Mediterranean orogenic belt. Kosminskaya and others (Kosminskaya et al., 1969) found that in these deep systems the maximum activity, as revealed by maximum crustal thickness, is not concentrated along the axis of the orogenic belt but

217

rather toward the foredeep. These structures, when viewed in a dynamic spatio-temporal frame, appear as a system of wave-like dislocations of folded syneclises created by the mobility of the line of maximum activity as the mountains grew always toward the foredeep. In these and similar systems, the depth of the M Discontinuity decreases with the shield depth, suggesting a close relationship between the behavior of the crust above and the mantle beneath the M Discontinuity.

In discussing the Earth's crust, it is well to keep in mind that at least 10% of the Earth does not have a crustal layer in the ordinary sense. Two examples of this absence of crustal structure are: (1) areas such as the Ivrea zone of Italy, where the mantle is "exposed"; and (2) in certain tectonomagmatically active zones where magmatic mantle material is currently upwelling to the surface, e. g., certain mid-ocean ridges, volcanoes, etc.

To understand such anomalous crustal areas as these and others it is necessary to look closer at the structure, composition, and behavior of the crust.

The Structure of the Earth's Crust.

The Earth's crust, when viewed on the basis of physical parameters, displays unique structure. The primary physical parameters that characterize crustal structure are stability, thickness, seismic velocity, heat-flow rate, and gross composition. Brune's (1969) ten types of crustal structure may be used as the basis for evaluating the Earth's crust in terms of tectonospheric behavior during the past 3.6 b.y. Brune's basic crustal types may be summarized:

a. Shield Crustal Type. A very stable crustal structure averaging 35 km thickness; $P_n = 8.3$ km/sec; heat flow = 0.7 to 0.8 HFU; and a composition of exposed batholithic rocks of Precambrian age having little or no sedimentary cover.

b. Mid-Continental Crustal Type. A stable crustal structure averaging 38 km thickness; $P_n = 8.2$ km/sec; heat flow = 0.8 to 1.2 HFU; and a composition of a moderate thickness of post-Precambrian sediments.

c. Basin-Range Crustal Type. A very unstable crustal structure averaging 30 km thickness; $P_n = 7.8$ km/sec; heat flow = 1.7 to 2.5 HFU; and characterized by recent normal faulting, volcanism, and intrusion of high mean elevation. Since no sharp boundary of $P_n = 8.0$ km/sec is evident in this type of structure, it is not certain that the layer with $P_n = 7.8$ corresponds (in a strict geophysical sense) with the M Discontinuity found in the adjacent regions.

   d. <u>Alpine Crustal Type</u>. A very unstable crustal structure averaging 55 km thickness; P_n = 8.0 km/sec; heat flow = 0.7 to 2.0 HFU (highly variable); and characterized by rapid recent uplifts and relatively recent intrusions of high mean elevation. The formation of mountains during rapid uplift was preceded by crustal downwarping, accumulation of large thicknesses of sediments, and formation of batholithic rocks. There may be one or more stages of uplift between periods of quiescence in the cyclically aperiodic evolution of the Alpine crustal type. Bouguer gravity anomalies are low, reflecting isostatic compensation. Heat-flow rates are inversely proportional to the age of the mountains. The upper mantle in Alpine regions has a low-velocity channel with S-wave velocities as low as 4.3 km/sec at depths of about 150 km. The fundemental mode Rayleigh-wave group velocities at periods of 40 sec are lower for Alpine crustal types than for any other crustal type, a result of both the thick crust and the pronounced low-velocity channel in the upper mantle.

   e. <u>Great Island-Arc Crustal Type</u>. Very unstable crustal structure averaging 30 km thickness; P_n = 7.4 to 7.8 km/sec; heat flow = 0.7 to 4.0 HFU (highly variable); and characterized by high volcanism, intense folding, and marked faulting. Although the crustal thickness is roughly the same as that for normal continents, rapid lateral variations occur more in great island arcs (such as Japan and New Zealand) than in larger continents. As in the Basin-Range crustal type, this type evidences no sharp boundary with P_n = 8.0, since the upper mantle velocities are low (7.6 to 7.8). The dispersions for island arcs are uncertain because the crustal structure is so variable and because the narrow dimensions make it difficult to eliminate the effects produced by surrounding regions. Within a given island-arc structure, it is difficult to find paths long enough to permit accurate measurements of group velocities. For phase velocity measurements, it is often necessary to use waves that originate outside the island arc, thereby introducing uncertainties due to mode conversion at the boundaries. The fundamental Rayleigh wave group and phase velocities near periods of 60 sec are much lower than for shields, a result of low upper mantle velocities under great island arcs. However, the observed fundamental Love wave group and phase velocities are not consistent with this result, suggesting relatively high veolcities in the lower crust and upper mantle. In Japan, this discrepancy has been interpreted as indicating strong anisotropy within the crust.

f. <u>Deep-Ocean-Basin Crustal Type</u>. Very stable crustal structure averaging 11 km thickness; P_n = 8.1 to 8.2 km/sec; heat flow = 1.3 HFU; and characterized by very thin sediments overlying basalts, linear magnetic anomlaies, absence of thick

Paleozoic sediments, and no evidence of ordinary orogenic activity. The recent discovery of remarkable linear magnetic anomalies parallelling oceanic ridges (Chapters 14 and 16) has been interpreted as representing sea-floor spreading in the reversing magnetic field of the Earth, implying that the ocean floor is a relatively young feature (Chapter 15). However, there is some evidence that the age of the deep ocean basins is indeterminate (See, for example, Kamen-Kaye, 1970). The approximate oceanic crustal structure consists of 5 km of water overlying less than 1 km of low rigidity sediment (S = 0.5 to 1.0 km/sec) and 5 km of rock (P = 6.4). The M Discontinuity at about 11 km is represented by an abrupt increase in P velocity to 8.1 km/sec and S velocity to 4.0 km/sec. The upper mantle beneath the deep-ocean-basin crustal type has a low-velocity channel with S = 4.5 at a depth of about 100 km. Love and Rayleigh waves between 30 and 60 sec are invariably influenced by components from continental margins, ridges, rises, and other crustal types with lower velocities included within the path.

g. Shallow-Water Mid-Ocean Ridge Crustal Type. These unstable crustal types include the Mid-Atlantic ridge near Iceland, the Easter Island Rise, the Hawaiian Islands, and other large mid-ocean volcanic islands. Thicknesses average 10 km, P_n = 7.4 to 7.6 km/sec, and they are composed of recently active basaltic volcanism with little or no sediments. Heat flows may be very high. Small earthquakes and volcanic eruptions are common. Typical structure of this type has less than 3 km of water overlying an undetermined thickness of basaltic lava flows. The exact crustal thickness is uncertain, but the underlying mantle appears anomalously low in both P and S velocities, i. e., about 7.5 and 4.3, respectively. The Love-wave group velocity is similar to that for mid-ocean basins, since the water layer has no effect. Because of the low mantle velocities, however, the Love-wave group velocities approach 4.3 only at longer periods (30 sec and over). For shallower water depths, the steep portion of the Rayleigh-wave group-velocity curve occurs in the period range of 5 to 9 sec and occurs at progressively longer periods as the water depth increases.

h. <u>Continental Plateau Crustal Type</u>. Not adequately studied to permit analytical definition.

i. <u>Deep-Ocean-Trench Crustal Type</u>. Not adequately studied to permit analytical definition.

j. <u>Transitional and Composite Crustal Types.</u> These crustal structures belong to areas having crustal characteristics embodying a linear combination of two or more of the above crustal types.

The Composition of the Crust.

The most salient characteristic of the crust is the homogeneity of its composition. Many researchers feel that this global homogeneity indicates that the processes of metamorphism in the crust occur in a closed system (Ronov and Yaroshevsky, 1969). A direct comparison of the composition of geosynclinal sedimentary rocks with that of the granitic shell suggests a significant supply of silica and alkalis during the granitization of the sediments.

These two processes (metamorphism of geosynclinal rocks and their subsequent granitization, spatially and temporally related) are, from the geochemical standpoint, the principal hypogenic processes influencing the crust.

The following paragraphs describe how these and related processes influence the composition of the crust, primarily by: (1) the supply of material and energy from the mantle; and (2) the fundamental reworking of the crustal material by surficial processes.

Composition of the Sedimentary Shell.

The total volume of the Earth's sedimentary shell is 9.4 x 10^8 km^3 (considering the consolidation of recent sedimentation) and 9.0 x 10^8 km^3 without volcanic rocks, i. e., about 10% of the crustal volume and 0.1% of the Earth's entire volume.

The average thickness of the sedimentary shell is 1.8 km, or 2.0 km excluding shield areas not covered by sediments. The bulk of the sediments is on the continental areas (5.0 x 10^8 km^3) and on continental margins (1.9 x 10^8 km^3), leaving only 2.5 x 10^8 km^3, or 28%, in the oceanic areas. Since the oceans, surficially, are almost $2\frac{1}{2}$ times as vast as the continents, the sedimentary shell is only about 10% as deep in the oceanic areas as in the continental areas.

On the continents, about 75% of the volume of all sedimentary rocks is found in geosynclinal areas and only 25% in the platforms, their average thickness being about 10 km and 1.8 km, respectively. These depths take into account a reduction, over time, of the sedimentation regions in the geosynclines and corresponding accretions on the platforms.

Clay and shale are the most widespread sedimentary rocks on the continents (42%). Arenaceous, volcanic, and carbonate rocks are approximately equally abundant (20, 19, and 18%, respectively). All other rock types, mainly evaporites, comprise about 1%.

A fundamental peculiarity of sedimentary rocks is the clearly pronounced difference between their composition and the average composition of rocks of the granitic shell, which was the chief source of sedimentary material, at least during the past 1.5 b. y. This difference is reflected in the strongly heightened content of water, carbon dioxide, and organic carbon, as well as sulfur, chlorine, fluorine, boron, and other excess volatiles (Rubey, 1955) in the stratisphere and the hydrosphere. This is an indication of direct release of these excess volatiles from the mantle during the process of its degassing (Vinogradov, 1967; Goldschmidt, 1954; Rubey, 1951, 1955).

The other important peculiarity of the composition of sedimentary rocks is their high calcium content, which is a most enigmatic feature of the geochemistry of the outer shells. Very typical is the displacement of the ratio of potassium to sodium in favor of potassium which is not compensated by sodium excess in the ocean. This leads to some deficiency of sodium in the stratisphere and hydrosphere, taken together, relative to the granitic shell.

These peculiarities are most clearly manifested in platform sediments, since these are products of deep weathering and strongly developed surface differentiation. In contrast, geosynclinal sediments have undergone less extensive alteration (especially sands), their composition approaching that of the parent rocks (Ronov and Yaroshevsky).

226

Composition of the Granitic Shell.

The granitic shell is completely concentrated on the continents, with a volume and mass of approximately 3.6×10^9 km^3 and 9.8×10^{24} g, respectively. Acidic granitoids and metamorphic rocks are the primary rock types of the granitic shell, the basic and ultrabasic rocks comprising less than 15% of the shell's volume. These relations determine its generally acidic composition: a typical high content of silica plus concentrations of alkalis (K greater than Na) and of the majority of rare elements (U, Th, rare earths, Zr, Nb, and others).

The average composition of the granitic shell differs considerably from that of the Neogene sedimentary rocks. This indicates that the sedimentary rocks are derived from two sources: (1) acidic materials from the granitic shell; and (2) basic materials of volcanic eruptions from deep within the tectonosphere. A closer examination of the nature of the sediments comprising the present granitic shell leads to the conclusion that a substantial portion of these sediments was changed by surface and hypogenic processes. Since the most ancient sediments appear to have been the products of weathering of more basic rocks than granites, the assumption is inevitable that the excess of bases and iron had to be moved by weathering processes into the oceans and pelagic sediments (Ronov, 1964). On the other hand, the deficiency of silica and alkalis should

227

have been compensated as a result of the fundamental reworking of sediments by processes of regional metamorphism developed at certain stages in the open system and accompanied by the supply of solutions of silica and alkalis. According to some researchers, this is the essence of the granitization process (Ronov and Yaroshevsky, 1969).

Composition of the Basaltic Shell.

The basaltic shell consists of two parts, the continental and oceanic, which differ in both structure and composition. The continental portion of the basaltic shell is composed of deeply metamorphosed acidic and basic rocks containing considerable magmatic material, the volume of basic rocks approximating half of the total volume and compositionally resembling geosynclinal basalts. This portion of the basaltic shell has considerable thickness and heterogeneous composition.

The oceanic portion of the basaltic shell, on the other hand, is highly homogeneous, being roughly 90% tholeiitic basalts characterized by: (1) a low content of K, Rb, Sr, Ba, P, U, Th, and Zr; and (2) high ratios of K/Rb and Na/K. These characteristics stronly distinguish the oceanic basalts from the continental types. Further characteristics of the oceanic portion of the basaltic shell are: (1) an absence of granitization; (2) a paucity of intensive differentiation and fusion from the mantle.

Much of the oceanic portion of the basaltic shell consists of ultrabasic rocks in rift valleys, medial ranges, and other deep faults (Vinogradov, 1967). Since these ultrabasic rocks appear to be outcrops of mantle material, this implies a possible participation of ultrabasic rocks in the formation of lower horizons of the oceanic crust. By analogy, ultrabasic rocks may also form the base of the continental portion of the basaltic shell (Ronov and Yaroshevsky, 1969). There are, however, no quantitative means for estimating the portion of the basaltic shell produced in this manner.

Any attempt to construct a chemical model of the basaltic shell must take into account: (1) the increase in the velocity of seismic waves, which indicates an increase in density with depth from 2.5 in the sedimentary rocks to 2.7 in the granitic layer and to 2.9 in the basaltic layer (Magnitsky and Zharkov, 1969); and (2) the possibility that an increase in density is the result not only of the compaction of rocks but also of the increase of their basicity. The latter assumption is supported by a significant increase in the amount of the basic rocks (amphibolite) in the deepest of visible zones of the crust in the ancient Lower Archean series (greenstones). The data on the geochemical balance of the elements and isotopes in the Earth's crust also indicate that a large part of the basic (basaltic) material is required as a source of these elements and isotopes in sediments and sea water.

Basaltic magmas have been erupted abundantly on both oceanic and continental regions of the Earth throughout most of the Earth's existence. The most probable immediate source region of basalts is the tectonosphere (Ringwood, 1969).

Composition of the Continental Crust.

The continental crust consists essentially of an upper granitic layer and an underlying basaltic layer. Very little is known about the continental crust deeper than the upper part of the granitic layer. The upper half of the granitic layer consists of granite; the lower half, of igneous and metamorphic rocks, primarily gneiss.

The basaltic layer of the continental crust is identical to the third (main) layer of the oceanic crust. This layer, in most crustal models, comprises the entire lower part of the continental crust. It is separated from the granitic layer by the Conrad discontinuity and extends uninterrupted under the oceans where, because there is no granitic layer, it constitutes the entire consolidated crust.

Closer scrutiny of the continental crust reveals temporal and spatial variations within layers. However, when considered on an overall basis, the basaltic layer of the continental crust is a complex mixture of metamorphic rocks, mostly of high-grade metamorphism, and of basic and ultrabasic igneous rocks from the mantle.

Among the significant anomalies within the continental crust is a transitional layer having seismic velocities ranging from 7.2 to 7.4 km/sec and consisting of a mixture of crustal and mantle material where the latter has probably intruded into the crust.

Composition of the Oceanic Crust.

The oceanic crust, in contrast to the continental crust, is composed of very few rock types: (1) serpentinites, (2) peridotites, and (3) basic igneous rocks that may be fresh, metamorphosed, or weathered.

The oceanic crust probably originated in most regions through a process now known as sea-floor spreading, wherein new oceanic crust is supplied from below at the crests of mid-ocean ridges. In some regions, the oceanic crust is highly fractionated; in other regions, actual upper mantle material appears to be exposed.

The ocean-floor basalts have a highly convergent chemical composition. They are distinct from almost all other basalts and show many primitive characteristics. Closer scrutiny reveals that the chemical variations in the ocean-floor basalts are systematic and apparently related to their mode of origin within the mantle.

The Behavior of the Crust.

The behavior of the crust, when viewed on a global scale and considered together with crustal structure and composition, reveals much about the temporal and spatial variations in the evolution of the Earth's tectonosphere during the past 3.6 b. y. This behavior, for our purposes, may be considered in terms of its horizontal and vertical movements. Horizontal movements, during the past 3.6 b. y., are thought to have involved mass transports through thousands of km; vertical ones, through hundreds of km. At a rate of 1 cm/yr, a crustal plate could circumnavigate the Earth in approximately 4 b. y. At 1 mm/yr, mantle material could rise up from depths of 1000 km in approximately 1 b. y.

Vertical crustal movements of recent origin have been studied more than horizontal ones. Vertical displacement gradients appear to be about an order of magnitude less than horizontal ones. In many cases, there appears to be a relationship between the two types of movement, thereby permitting a distinction between areas of uplift and extension and areas of subsidence and compression. Vertical and horizontal crustal movements may be analyzed as components of 3-dimensional deformations of the Earth's crust (Artyushkov and Mescherikov, 1969).

Major Horizontal Motions of Crustal Material.

Major horizontal motions of crustal material are thought to be of two types: (1) those in which two crustal plates move away from each other; and (2) those in which two crustal plates move transversally with respect to each other along a common boundary. The first type of horizontal motion is usually referred to as continental drift, sea-floor spreading, or crustal rifting; the second type, as horizontal motion along a fault, or shear displacement.

Continental drift, sea-floor spreading, and crustal rifting are covered in later chapters and will not be considered here except insofar as they are related to major horizontal motions along faults.

The magnitudes of actually measurable horizontal displacements of crustal plates are relatively small. The horizontal dislocation along the San Andreas fault, for example, is only about 1 km. The evident displacement along the Alpine fault of New Zealand is even less than one km. However, on the basis of more general geological data, many investigators conclude that within geological time horizontal displacements along these two faults have amounted to hundreds of km (Hart, 1969).

Even greater horizontal displacements are suggested by magnetic anomalies along large latitudinal faults in the eastern Pacific Ocean. Some of these displacements exceed 1000 km, equivalent to 10 surficial degrees or 1/36 of the Earth's circumference. An analysis of other large displacements indicates numerous shear displacements orthogonal to the mid-ocean ridges. The Great Glen Fault of Scotland involves such a shear displacement amounting to over 100 km.

In some cases evidence regarding shear displacements is paradoxical and contradictory. Geological features in numerous places along the San Andreas fault are so similar on opposite sides of the fault that it is difficult to believe that any shear has occurred. On opposite sides of the Great Glen Fault, two granite massifs, offset by some distance, appear to represent two separate parts of a single massif. Since the two parts are quite different petrologically, however, it is doubtful that any shear displacement has actually occurred.

Large horizontal shear displacements of crustal material require an explanation for the disposition of the large volumes of material that should accumulate at the ends of the fault along which the motion occurs. An explanation is required, for example, for the disposition of the crustal material that must have been carried 1200 km shoreward along the Mendocino fault orthogonal to the California coast. Paradoxes of this type led to the concept of "transform faults" in which horizontal displacements are compensated by sinking of crustal material into the mantle on one side and creation of new crust (from mantle material) on the other.

Major Vertical Motions of Crustal Material.

Major vertical motions of crustal material are of two types: (1) upward motions or elevations, and (2) downward motions or subsidences. In almost all parts of the Earth, the crust is either rising or sinking. Rates vary from about zero to centimeters per year but average a few millimeters. Epeirogenic vertical motions usually involve large, more-or-less quadrate regions and occur at extremely slow rates. Orogenic vertical motions, on the other hand, usually involve long, more-or-less rectilinear regions and occur at faster rates than do the epeirogenic motions.

In some areas of the Earth, both subsidence and uplift are occurring within relatively short distances, with differential rates exceeding 1.6 cm/yr within a horizontal distance of 1000 km (Artyushkov and Mescherikov, 1969). Uplift normally is associated with positive structural elements (e. g., arches); subsidence, with negative structural elements (e. g., basins). Analyses of movements in platform areas suggest block structure with maximum differential motion along the boundaries between adjacent blocks.

There are two primary causes for uplift of crustal material: (1) the traditional glacio-isostatic process; and (2) mantle-induced uplift. In some extraglacial regions uplift due to the second cause is as great as that due to the first cause.

Geological and geomorphological data indicate, moreover, that the glacio-isostatic process ceases to be of importance approximately 5 or 6 thousand years after the melting of the glacier involved.

Gradients of recent vertical movements in orogenic belts may be two orders of magnitude greater than those of epeirogenic areas. Harmonic analyses of these various vertical movements indicate that wavelengths of multiples of about 45 km can be distinguished in many regions (Artyushkov and Mescherikov, 1969). This represents a resolution equivalent to the 450th order in the harmonic expansion. The coincidence of similar values of wavelengths for both epeirogenic and orogenic movements suggests that recent vertical movements of the Earth's crust in these orogenically different regions have a common, deep-seated, globally-oriented causal mechanism. Details regarding hypothesized globally-oriented causal mechanisms are described in Chapter 6.

Sharp changes in the sign and intensity of vertical crustal movements detected in the tectonomagmatic belts appear to be associated with earthquakes and related seismic activity. An analysis of applicable seismic data identifies three types of vertical crustal movements in active regions: (1) slow secular movements occuring during the relatively long quiescent periods separating the active episodes in the cyclically aperiodic seismic behavior at any given location; (2) accelerated movements preceding an earthquake and reflecting a deformation of the Earth's crust during the preliminary stage of an incipient earthquake; and (3) quick movements directly related to the earthquake and subsiding with it.

REFERENCES

Artyushkov, E. V. and Mescherikov, Y. A., 1969. Recent movements of the Earth's crust and isostatic compensation. In: P. J. Hart (Editor), The Earth's Crust and Upper Mantle. Am. Geophys. Union, Washington, pp. 379-390.

Brune, J. N., 1969. Surface waves and crustal structure. In: P. J. Hart (Editor), The Earth's Crust and Upper Mantle. Am. Geophys. Union, Washington, pp. 230-242.

Goldschmidt, V. M., 1954. Geochemistry. Clarendon, Oxford, 730 pp.

Hart, P. J. (Editor), 1969. The Earth's Crust and Upper Mantle: Structure, Dynamic Processes, and Their Relation to Deep-Seated Geological Processes. Amer. Geophys. Union, Washington, 735 pp.

Kamen-Kaye, M., 1970. Age of basins. Geotimes, 15(7): 6-8.

Kosminskaya, I. P., Belyaevsky, N. A., and Volvosky, I. S., 1969. Explosion seismology in USSR. In: P. J. Hart (Editor), The Earth's Crust and Upper Mantle. Am. Geophys. Union, Washington, pp. 195-208.

Magnitsky, V. A. and Zharkov, V. N., 1969. Low-velocity layers in the upper mantle. In: P. J. Hart (Editor), The Earth's Crust and Upper Mantle. Am. Geophys. Union, Washington, pp. 664-675.

Ringwood, A. E., 1969. Composition and evolution of the upper mantle. In: P. J. Hart (Editor), The Earth's Crust and Upper Mantle. Am. Geophys. Union, Washington, pp. 1–17.

Ronov, A. B., 1964. General tendencies in evolution of composition of the Earth's crust, ocean, and atmosphere. Geokhimiya, 1964 (8): 715–743.

Ronov, A. B. and Yaroshevsky, A. A., 1969. Chemical composition of the Earth's crust. In: P. J. Hart (Editor), The Earth's Crust and Upper Mantle. Am. Geophys. Union, Washington, pp. 37–57.

Rubey, W. W., 1951. Geologic history of sea water: An attempth to state the problem. Bull. Geol. Soc. Am., 62: 1111–1147.

Rubey, W. W., 1955. Development of the hydrosphere and atmosphere with special reference to probable composition of the early atmosphere. Geol. Soc. Am. Spec. Paper 62: 631–650.

Vinogradov, A. P., 1967. The formation of the ocean. Izv. Akad. Nauk SSSR, Ser. Geol., 1967 (4): 3–9.

Chapter 6

THE TECTONOSPHERE

The tectonosphere is the upper approximately 1000 km of
the Earth. It includes the crust and upper mantle to a depth
of about 1000 km below the surface. In analyzing the geometry
and mechanics germane to tectonospheric evolution, it is well
to consider both _vertical_ and _horizontal_ differentiations, inhomogeneities, and fractures, as well as the basic driving mechanism
that has produced them during the past 4.6 b.y. Only through
such an all-encompassing analysis can one hope to understand
how the behavior patterns of the tectonosphere have evolved
together with their surficial manifestations. This chapter,
therefore, discusses the origin, development, structure, and
status-behavior of the upper 1000 km of the Earth during the
past 4.6 b. y. and from a global viewpoint.

Classical Concepts of the Earth's Crust and Upper Mantle.

In discussing classical concepts of the Earth's crust and
upper mantle, it is well to recall that two recent developments
have made most classical concepts more-or-less obsolete: (1)
that the Earth is not radially symmetric: and (2) that important
lateral inhomogeneities exist in at least the Earth's upper
700 to 800 km (See, e. g., Hart, 1969). However, since most
classical concepts were based on interpretations of surficial

manifestations of the Earth's internal behavior, a brief summary of these concepts will be discussed.

Earliest concepts visualized the crust and mantle as forming concentric shells about the core. In the basic concept, the crust extended to the M discontinuity, which varies from place to place but averages about 35 km under continental areas and about 5 km under oceanic areas. Although the M discontinuity is a _physical_ reality, it does _not_ seem to be important to an analysis of the _dynamic_ behavior of the Earth. Consequently, most modern classical concepts consider the crust and upper mantle together to various depths of 700 to 1000 km. This book considers them to a depth of 1000 km and refers to the crust-plus-upper-mantle combination as the tectonosphere.

One reason for considering the crust and upper mantle as a unit was the discovery, about 100 yr ago, that the Earth behaves as though the upper 100 to 200 km were a strong, fairly rigid plate (or lithosphere) floating on a fluid substructure (or asthenosphere) extending down to a depth of several hundred km.

The terms lithosphere and asthenosphere were introduced by Barrell (1914a, 1914b) when he summarized the evidence for a strong outer layer overlying a weaker layer in the uppermost part of the Earth. Barrell reasoned that the Earth is very strong basically because it provides the support to explain: (1) the surface loads of the major deltas and topography; and (2) the large variations in vertical stress resulting from

differences in density within the crust. Opposed to evidence
for a __strong__ lithosphere, he found equally convincing evidence
for a __weak__ lithosphere. That is, it appeared that there was
approximate isostatic equilibrium of the continents and oceans
when viewed on a global scale, in spite of the great tectonic
and erosional activity (up to 15 km) that has occurred repeatedly
throughout at least the past 3.6 b. y.

Consequently, although the outer lithosphere obviously
possesses almost unbelievable strength, it appears equally obvious
that deeper seated material must be able to flow or otherwise
readjust itself to compensate for the lateral movements observed
in various surficial manifestations. Barrell concluded that
there must be an asthenospheric layer, shell, or zone of weakness
in which material may move laterally, given sufficiently large
stress differences.

Since Barrell's (1914a, 1914b) analysis, very little contradictory evidence has been found; but some studies have tended
to complicate the original concept of the lithosphere. For
example, some studies supported the basic concept of considerable
mobility for relatively small stress differences at great depths
(See, e. g., Bentley, 1964; Gutenberg, 1941; Hamilton et al.,
1956; Harrison, 1955; Haskell, 1937; Hospers, 1965; and Lawson,
1942). However, Barton (Barton et al., 1933) concluded from
depth-to-basement analyses that lithospheric __collapse__ had occurred
under certain surface loads and environments. Also, Walcott

(1968, 1970) reasoned that the great vertical density-stress differences could reflect an absolute flexural rigidity for the lithosphere, particularly if the lithosphere, fractured as it is, were regarded as a family of vertical prisms, because then most differences in vertical stress could be considered as purely local compensation of each prism rather than an entire lithospheric plate. This idea was supported by Crittenden's (1963) earlier analysis.

Popelar (1968) concluded that surface stresses as small as 5 bars would produce movement within the asthenosphere. Most other studies showed this same tendency for isostatic compensation to occur at the local level.

Molnar and Oliver (1969) demonstrated the existence, and determined the pattern, of lateral variations of attenuation in the uppermost mantle on a worldwide basis. Their findings include the following: (1) that S_n, which does not penetrate the low-velocity channel, propagates very efficiently across the stable regions of the Earth, the continental shields, and deep-ocean basins, but very inefficiently when the paths cross the crests of the mid-ocean ridge system or the concave sides of most island arcs; (2) that these observations suggest that attenuation is more pronounced in the uppermost mantle near the ridge crests and the island arcs than in the more stable regions; (3) that, if low attenuation, or high Q, correlates with high strength, the data imply that the uppermost mantle

244

is considerably weaker under the ridge crests and the concave sides of most island arcs than it is elsewhere; (4) that, thus, the uppermost part of the strong outer shell, or lithosphere, in the mantle is discontinuous with gaps in it at the ridge and island arcs; and (5) that the low attenuation for S_n for paths crossing the transform faults, connecting ridge crests, suggests that any gap at the transform fault is very narrow.

As for deeper structure, very few global geometrico-mechanical concepts exist for the structure and behavior of the mantle beneath the asthenosphere. However, it is generally conceded that the lower mantle is a separate subsystem from the upper mantle (See, e. g., Stacey, 1969, p. 234).

Summarizing, most modern concepts hypothesize a very weak quasi-fluid asthenosphere and a laterally-weak but strongly anisotropic lithosphere that, apparently capable of supporting the largest mountains, suggests not a continuous elastic sheet but rather a family or stack of piles, or prisms, each floating independently under its own load and each separated from adjacent stacks by surfaces of preferential rheidity, by virtue of which their independence is assured.

International Upper Mantle Project.

The International Upper Mantle Project (See, e. g., Hart, 1969; Drake, 1970) concluded that significant lateral inhomogeneities within at least the upper 700 km of the Earth may be evidence for the driving mechanism for large-scale motions of the upper portions of the Earth and that these motions, in turn, may provide the process whereby the inhomogeneities are produced. Although the IUMP failed to clearly identify the _exact_ geometrico-mechanical processes involved in the above conclusion, this and other findings left little doubt that the driving mechanism for the Earth's internal behavior is _deep-seated_ and _worldwide_.

Other findings of of the IUMP that are of interest to an analysis of the evolution of the Earth's tectonosphere include the following: (1) that there are reasons to believe the history of the development of the Earth's crust is fundamentally dependent upon processes operating at depths to 1000 km (Beloussov, 1969); (2) that the lateral and vertical variations within the upper mantle are correlatable with the motions of material within the upper mantle and with surficial manifestations of the internal behavior (e. g., earthquakes, volcanism, the building and deformation of mountains, and the differentiation of various rock types); and (3) that an analysis of the above interrelationships will lead to an understanding of the Earth's internal behavior and the surficial manifestations thereof.

Detailed findings of the IUMP are contained in Hart (1969).

246

A good summary is contained in Drake (1970). These are discussed in subsequent chapters when the detailed behavior of the tectonosphere is considered in connection with its evolution. Suffice it here to say that, although the IUMP left many questions unanswered, it did, however, provide answers to an unbelievably large number of questions during its 10 years of existence, from 1960 to 1970. Among these are the following: (1) that, considered separately and on a local basis, the data for continental drift and sea-floor spreading are _not_ convincing, but, when considered on a _global_ basis, they increase the probabilities that large-scale horizontal motions take place and that they have occurred in the past (Knopoff, 1969); (2) that it is as yet undetermined whether the ideas of sea-floor spreading and continental drift can be used in their _present_ forms to explain geologic processes in areas _remote from_ oceanic ridges, trenches, and rift zones (ibid.); (3) that it is undetermined yet whether a process of convection is required to explain the spreading and drifting observations (ibid.; Lyustikh, 1969); (4) that the following details regarding the possibility of mantle convection remain undetermined: whether it can extend over the _entire_ mantle or only part of it, whether it is continuous or intermittent, what the most probable number of convection cells is, whether the cells are arranged in one or several layers, and what causes _changes_ in the convection pattern (Lyustikh, 1969); (5) that simple models of mantle convection must be extended

247

to more complex models embodying more details of the known
structure and inhomogeneities of the actual Earth (ibid.); (6)
that both tensional and compressional tectonic features within
the Earth at a given time have a single driving mechanism (Runcorn, 1969); (7) that regional metamorphism may be set in motion
by the flow of heat from below (Beloussov, 1969; Hehnert, 1969);
(8) that the composition of the consolidated part of the oceanic
crust is unknown (Beloussov, 1969; Vogt et al., 1969; Ronov and
Yaroshevsky, 1969); (9) that almost all of the oldest oceanic
sediments (i. e., those lying on the second layer) belong to
the Cretaceous (Beloussov, 1969; Ewing, 1969); (10) that the
composition of molten basalt probably depends upon the depth
at which the complete separation between basaltic liquid and
crystalline substrata occurred (Beloussov, 1969); (11) that
the low—velocity zone in the upper mantle is significantly related
to deep processes (ibid.); (12) that it is unknown whether geosynclinal deformation is due to horizontal or vertical forces
or to a combination of both (ibid.; Khain and Muratov, 1969);
(13.) that any oceanic material older than the Cretaceous must
lie within the second and/or third layers (Beloussov, 1969);
(14) that the most important elements of oceanic structure are
the mid—ocean ridges (ibid.); (15) that not all mid—ocean ridges
are genetically, compositionally, and structurally alike (ibid.);
(16) that a study of the two basic types of oceanic margins
leaves many unanswered questions (ibid.); (17) that the relationship

between the distribution of geosynclines within different tectonic cycles and the Earth's internal behavior remain to be determined (ibid.); and (18) that a consolidated hypothesis for the behavior of the upper mantle must embody a _joint_ and _balanced_ utilization of both _oceanic_ and _continental_ geological, geophysical, and geochemical data on a _worldwide_ basis and _temporally-oriented_ to include as much of the past 4.6 b. y. as possible.

Interunion Commission on Geodynamics and Related Projects.

Many contradictory hypotheses and the paucity of data upon which they are based indicate that much remains to be done in connection with both data collection and data analysis before we shall have a _unified_ model for the behavior of the upper mantle on a _global_ basis during the past 4.6 b. y. Although the Upper Mantle Project has ended, much of its work is being continued by the Interunion Commission on Geodynamics (Drake, 1970) and related projects of a global nature.

The primary mission of the Interunion Commission on Geodynamics is to examine plate tectonics and related hypotheses bearing on the dynamic behavior of the upper mantle. This includes an analysis of the dynamic processes, both present and past, that have shaped the Earth's surface. Thus, the new program will attempt to gain an understanding of the underlying forces at work producing the _horizontal_ movements of the Earth's surface, such as continental drift and sea-floor spreading. Although much work has been done on _what_ happens at the _surface_, little is known about the _how_ and _why_ of the _internal_ driving mechanism that _causes_ continental drift and sea-floor spreading. In order to explain the horizontal motion, it will be necessary to analyze certain _vertical_ motions, such as the uplifting of the Colorado Plateau of southern Utah and northern Arizona, for example.

Other scheduled projects include: (1) deep drilling in order to determine the composition of various types of crust (and mantle, if possible); (2) systematization of data on the temporal and spatial relationships among tectonic, metamorphic, and magmatic processes, now and in the past, in order to more clearly define their possible relation to a __single__ __global__ driving mechanism; (3) comparison of the deep structure of __different__ tectonic zones; (4) analysis of information regarding the composition and origin of the oceanic second and third crustal layers; (5) the study of similarities and differences between structure of oceanic and continental crust; and (6) the establishment of various regional projects supporting the above global projects.

The Tectonosphere According to the Dual Primeval Planet Hypothesis.

Basically, according to the dual primeval planet hypothesis, the tectonosphere is an accretion, roughly 1000 km deep, that formed onto the 5400-km primordial Earth. The latter existed about 4.6 b. y. ago and is now the Earth's subtectonosphere.

The Tectonospheric Earth Model.

A tectonospheric Earth model, as defined in Chapter 3, is one that describes the Earth's evolution in terms of the mechanics and geometry of the tectonosphere. The tectonosphere is assumed to have accreted onto the primordial Earth in one of three basic evolutionary modes previously described. The evolution and behavior of the tectonosphere will depend largely upon which of the three modes is assumed: (1) octantal-fragment; (2) multiple-fragment; or (3) composite-fragment. For example, in the octantal-fragment mode, parts of each of the 5 terrestrial octants of Earth Prime might still exist today as unmodified portions of the Earth's present tectonosphere, perhaps even as subsurficial portions of presently-existing continental shields. In the case of the multiple-fragment mode, such would not be likely, however, except perhaps as fairly small fragments now coalesced with, or imbedded within, modified material less than 4.6 b. y. old.

The Subtectonosphere.

The subtectonosphere is essentially the 5400-km primordial Earth that now underlies the tectonosphere. It is fractured into octants by 3 mutually-orthogonal planes passing through the center.

Fig. 6-2b-1 is a schematic view of the Earth's subtectonosphere, showing the geometric interrelationships of the 8 subtectonospheric blocks and the subtectonospheric fracture system. Subtectonospheric blocks and points bear the same names as their surficial counterparts lying 1000 km roughly above them on the surface of the Earth. Thus, the subtectonospheric point Aleutians (AL) is 1000 km roughly beneath the Aleutians area on the surface of the Earth. Similarly, the North American subtectonospheric block is 1000 km roughly beneath the North American continent.

Note that the North American subtectonospheric block is larger than the North American continent because the former covers 1/8 of the entire subtectonosphere (and, therefore, underlies 1/8 of the entire Earth's surface, rather than only the North American continent as usually defined). Similar analyses may be made for other continents. In making such analyses, adjustments must be made also for the fact that a 6371-km sphere has a surface area about 1.4 times that of a sphere with a radius 1000 km less.

The surface of each of the 8 subtectonospheric blocks forms

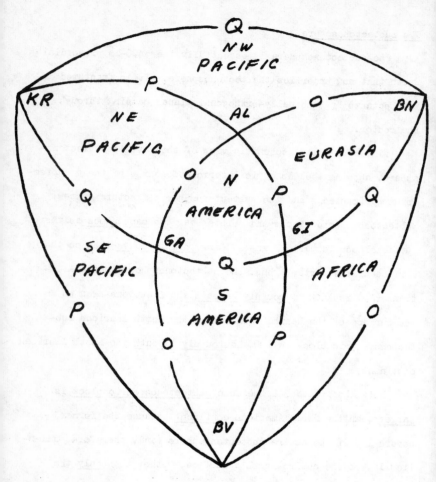

Figure 6-2b-1. Schematic view of the surface of the Earth's subtectonosphere, showing the geometric interrelationships of the 8 subtectonospheric blocks and the subtectonospheric fracture system (SFS). The north pole of the SFS is at AL; the south pole, at BV. The equator of the SFS is marked QQQQ. The prime meridian of the SFS is marked PPPP; the orthogonal meridian, OOOO. See text for details. (Note that the Antarctic subtectonospheric block is on the "reverse", defined by the points KR, BN, and BV).

254

an equilateral spherical triangle with all angles and sides equal to $90°$ (about 8500 km on the surface of the subtectonosphere). The poles of the subtectonospheric fracture system are located at the subtectonospheric points Aleutians (AL) and Bouvet (BV), respectively. The equator of the subtectonospheric fracture system lies along the great circle formed by the subtectonospheric points Galapagos (GA), Gibraltar (GI), Bengal (BN), and Kermadecs (KR). The prime meridian of the subtectonospheric fracture system lies along the great circle formed by the subtectonospheric points Gibraltar (GI), Bouvet (BV), Kermadecs (KR), and Aleutians (AL). The orthogonal meridian of the subtectonospheric fracture system lies along the great circle formed by the subtectonospheric points Bengal (BN), Bouvet (BV), Galapagos (GA), and Aleutians (AL).

Fig. 6-2b-2 presents the same information as Fig. 6-2b-1 but uses a Mollweide projection for superimposing an outline of the continents over the basic elements of the subtectonospheric fracture system.

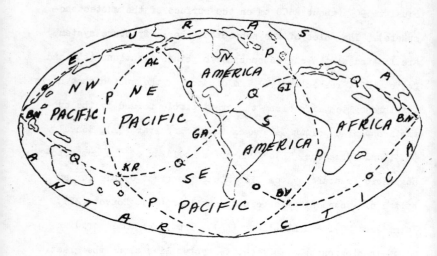

Figure 6-2b-2. Mollweide projection of continents superimposed over basic elements of the subtectonospheric fracture system (SFS). The north pole of the SFS is at AL; the south pole, at BV. The equator of the SFS is marked QQQQ. The prime meridian of the SFS is marked PPPP; the orthogonal meridian, OOOO. See text for details.

The Driving Mechanism for the Tectonospheric Earth Model.

The basic behavior of the primordial Earth (the present subtectonosphere) is assumed to be such that it tends to equilibrate itself to a state of minimum energy. The driving mechanism of the system may, therefore, be expressed as a function of the disequilibration energy inherent in its initial state. In simplest terms, it consists essentially of the resultant of two factors: (1) the potential energy inherent in the geogenetically disequilibrated shape of the basic 5400-km primordial Earth; and (2) the selectively-channeled energy from the preferential flow of heat, and possibly of volatiles, outward from the subtectonospheric fracture system (described in the following section).

The driving mechanism of the tectonospheric Earth model operates within the geometrico-mechanical constraints described in Chapter 3, and these constraints must always be taken into consideration when making analyses using the model. The importance of this will become clearer in subsequent chapters. Suffice it here to say merely that the driving mechanism and the geometrico-mechanical framework in which it operates constitute inseparable parts of an integrated system.

The Subtectonospheric Fracture System.

The subtectonospheric fracture system consists of three mutually-orthogonal planar fractures. Where these intersect the surface of the subtectonosphere, they form 3 mutually-orthogonal great circles. Each circle has a length (or circumference) of 5400 2 x pi = 34,000 km, approximately, or 3 x 34,000 = 102,000 km for the system.

In accordance with the constraints and degrees of freedom of the subtectonosphere (See Chapter 3), approximately half of the 102,000 km subtectonospheric fracture system would be expected to be in tension at any given time, while the remaining 51,000 km would be in compression. Thus, at any point along the 102,000-km system, during the past 4.6 b. y., there would be alternating states of tension and compression, as the model attempts to equilibrate itself (See Fig. 3-3-4 and related text).

Because the above-indicated behavior is oscillatory with a long time constant, the _locations_ of the specific segments of the subtectonospheric fracture system in tension will tend to drift continually (but relatively slowly because the time constant is long) along the 102,000-km system. From the viewpoint of an external observer, the specific segments in tension would appear to be moving along the subtectonospheric fracture system in a _cyclically aperiodic_ manner, because the equilibration of the primordial Earth would, of necessity, be proceeding in a like manner (i. e., cyclically aperiodic). The present

258

average rate of such cyclically aperiodic shifting of these tensile segments is of the order of cm/yr.

Segments of the subtectonospheric fracture system not in
tension are assumed to be in compression or in a neutral state;
but, because the pressure gradients along the system are not
constant, neutral states would be short-lived. Compressed and
neutral segments, being bounded by tensile segments, would also
shift along the 102,000-km subtectonospheric fracture system
in a cyclically aperiodic manner and at an average rate of the
order of cm/yr.

Superimposed upon the effects of the cyclically-aperiodic
"tensile-compressive" behavior of the subtectonopsheric fracture
system would be other types of activity expected from the basic
driving mechanism and constraints of the model (See Chapter 3,
section on the geometrico-mechanical behavior of the basic 5400-km
primordial Earth). Thus, in addition to the cyclically-aperiodic
tensile-compressive behavior, one would expect the subtectonospheric fracture system to display the following types of geometrico-mechanical behavior: (1) rotary equilibration; (2)
radial translatory equilibration; (3) transverse translatory
equilibration; (4) independent octantal motion of the 8 subtectonospheric blocks; and (5) such other behavior as might be
produced in the system by the basic driving mechanism of the
model described in the previous section.

The nature of the specific types of behavior expected along

259

tensile segments of the subtectonospheric fracture system is
somewhat different from that expected along compressed segments.
For example, the size and polarity of the pressure gradient
(i. e., whether increasing or decreasing) at a given time are
instrumental in determining the exact nature of thermal, tectonic, seismic, geochemical, electromagnetic, and other effects
produced both at the surface of the subtectonosphere and, as
surficial manifestations thereof, on the surface of the Earth.
Thus, the nature of magmatic production (i. e., whether intrusive or extrusive; whether acidic, basic, basaltic, ultrabasic,
etc.) may be analyzed in terms of the cyclically-aperiodic behavior
of the subtectonospheric fracture system during the past 4.6 b.y.
(See Chapter 10 for details). Similar analyses may be made
in connection with the global distribution pattern of the Earth's
deep seismicity (See Chapter 8). Other chapters will discuss
the role of subtectonospheric behavior in producing specific
features, phenomena, and manifestations on the surface of the
Earth, including: (1) the nature, composition, and global topography of the crust of the Earth; (2) the orogenic-cratonic
structure and evolution of the continents; (3) global patterns
of terrestrial heat flow and other geothermal activity; (4)
global geomorphology of the Earth during the past 3.6 b. y.;
(5) mountain building and modification; (6) the Earth's gravity
field; (7) geomagnetism and polarity reversals; (8) the relative
juxtaposition and antipodal location of oceans and continents

when viewed on a global scale: (9) sea-floor spreading and other
indications of crustal rifting in both oceanic and continental
areas; (10) continental drift and polar wandering; and (11)
plate tectonics and related phenomena.

Local Cones of Tectonospheric Activity.

Other things being equal, the "probable error" in locating the surficial position of an "energy source" projected from a point on the subtectonosphere, and through a vertical thickness of 1000 km of "average" tectonospheric material, is approximately 1000 x pi/4 ≐ 785 km. Such being the case, "activity" due to a given energy source (existing on the surface of the subtectonosphere) would be expected to occur somewhere within a cone centered on the "energy source" as a vertex and intersecting the surface of the Earth in a circle with a radius of 785 km. "Activity", as used here, includes all direct results from the Earth's driving mechanism (e. g., tectonic, metamorphic, magmatic, etc.), all of which would normally be expected to occur within these cones of activity according to the model. Indirect results of the Earth's driving mechanism (e. g., readjustments resulting from directly-caused activity) may, of course, occur anywhere within the Earth, without regard to where the cones of activity are located at any particular time.

Fig. 6-2e-1 shows a schematic sketch of typical cones of activity resulting from an energy source, D, along the subtectonospheric fracture system. O is the center of the Earth. OX, OY, and OZ are the 3 mutually-orthogonal fracture planes of the subtectonosphere, extended to the surface of the Earth. Cones of activity intercept the Earth's surface in circles, AA. To allow for variations in heterogeneities within the tectonosphere,

262

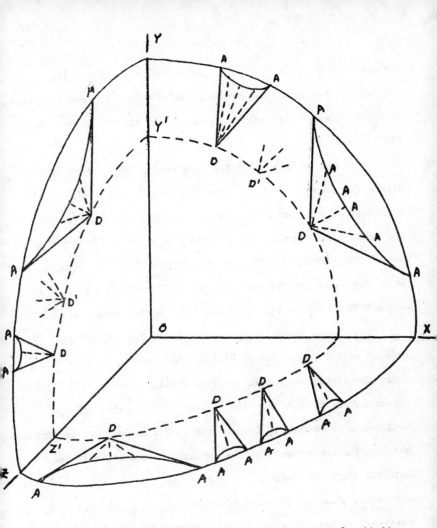

Figure 6-2e-1. Schematic diagram of typical cones of activity resulting from an energy source at points D along the subtectonospheric fracture system. O is the center of the Earth. OX, OY, and OZ are the 3 mutually-orthogonal fracture planes of the subtectonosphere, extended to the surface of the Earth. See text for details.

263

various vertex angles are shown in the schematic sketch (in addition to the idealized vertex angle of 90°). For most purposes, it may be assumed that vertex angles do not exceed approximately 120°.

It should be understood that the entire <u>volume</u> contained within the cones ADA, rather than just the conical shells or surfaces, is considered "active", with <u>possible</u> maxima along the edges and at certain other areas (to be defined later when paths of preferential heat flow are discussed). The cones centered at D' represent incomplete cases (e. g., batholithic intrusions, <u>subsurface</u> fractures without surficial manifestations, etc.).

Fig. 6-2e-2 shows the geometry of a typical, idealized cone of activity centered at an energy source, O, at some point along the subtectonospheric fracture system. The point O is approximately 5400 km from the center of the Earth, and the tectonosphere is roughly 1000 km deep. The angle, DOD, is approximately 90°, but varies with the composition and state of the material above the point O. A <u>theoretical</u> minimum value for DOD is 0°, which would exist if there were a <u>radial</u> path of preferential heat flow, or a fracture extending <u>vertically</u> upward from O (i. e., the case of the "degenerate cone"). A <u>theoretical</u> maximum value for DOD is 180°, which would exist if the tectonospheric material directly above the point O were <u>completely</u> <u>impenetrable</u> to the effects produced by the energy source at O. In such case, it can be expected that the energy

264

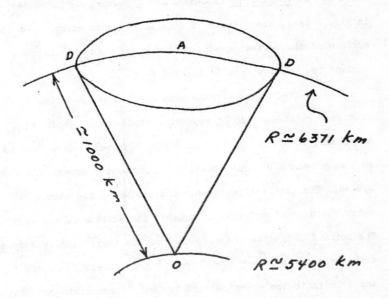

Figure 6-2e-2. Schematic diagram showing the geometry of a
typical, idealized cone of activity centered at an energy source
O located at some point along the subtectonospheric fracture
system. See text for details.

source would move along the subtectonospheric fracture system until it encountered a location where the overlying surface would not be completely impenetrable. However, as indicated in Fig. 6-2e-1, the effects of an energy source existing on the surface of the subtectonosphere need never reach the Earth's surface (e. g., the case D' in Fig. 6-2e-1).

Fig. 6-2e-3 shows a transverse cross-sectional view of the predicted relationships of several tectonospheric features of a typical, idealized cone of activity. The point O represents an energy source at some point of the subtectonospheric fracture system. The line FAF represents a diameter of the circle in which the cone of activity intercepts the surface of the Earth. The point A indicates the epicenter of the theoretically deepest deep-focus earthquakes (See Chapter 8 for details). Points B and C indicate the epicenters of typical intermediate and shallow-focus earthquakes. The model does not exclude A as a possible epicenter for intermediate and shallow-focus earthquakes, since the hypocenters may lie anywhere along the line AO. The converse, however, is not necessarily true since B and C would normally not be expected to have hypocenters deeper than those along the line DO and, therefore, should not normally be expected to produce deep-focus earthquakes.

Point D of Fig. 6-2e-3 represents the approximate region of most-intense expected present orogenic-magmatic activity (i. e., new islands, new mountains, new ridges, new volcanoes,

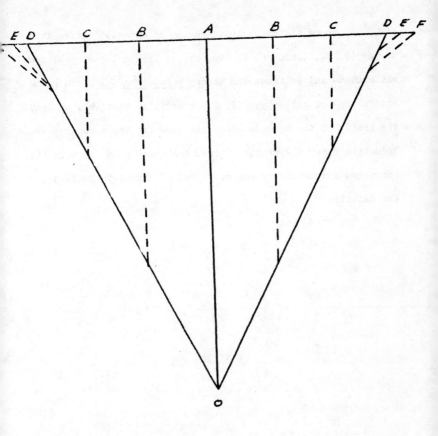

Figure 6-2e-3. Schematic diagram showing a transverse cross-sectional view of the predicted relationships of several tectonospheric features of a typical, idealized cone of activity. The point O represents some point of the tectonospheric fracture system. See text for details.

etc.). E shows the approximate region of <u>expected</u> active fracturing (i. e., trenches, faults, etc.); however, the model does
not exclude active fractuing at <u>any</u> <u>point</u> <u>along</u> <u>FAF</u>, since the
entire cone of activity, FOF, is potentially "active". F shows
the limits of the expected Andesite Line for those regions where
andesitic flows are predicted from the model (See Chapters 10,
Intrusive and Extrusive Activity; and 12, Mountain Building,
for details).

Global Tectonospheric Wedge-Belts of Activity.

When local cones of activity are formed for every point on the subtectonospheric fracture system, the surfical trace of their vertexes consists of 3 mutually-orthogonal great circles, aggregating 3 x 2 pi x 5400 ≐ 102,000 km in length. The resulting cones of activity form an infinity of cones whose "envelope" may be described as 3 global tectonospheric "wedge-belts of activity". They aggregate a length of 3 x 2 pi x 6371 ≐ 120,000 km where they intercept the surface of the Earth. This is shown schematically in Fig. 6-2f-1 and Fig. 6-2f-2.

Fig. 6-2f-3 shows the above in spherical perspective. Local cones of activity are shown at C, based on points along the subtectonospheric fracture system F. The tectonosphere and subtectonosphere are shown at T and S, respectively.

Since at any given time, "active" or effective energy sources can be expected along about 50% of the subtectonospheric fracture system (i. e., about 51,000 km), activity would be expected within about half of the global wedge-belt system (i. e., about 60,000 km) at any one time (See Chapter 7, The Orogenic-Cratonic Structure of the Continents; and Chapter 10, Intrusive and Extrusive Activity).

As indicated in subsequent chapters, all types of activity (tectonic, metamorphic, magmatic, etc.) would not necessarily be expected to occur simultaneously in all "active" sigments of the global tectonospheric wedge-belts of activity. For

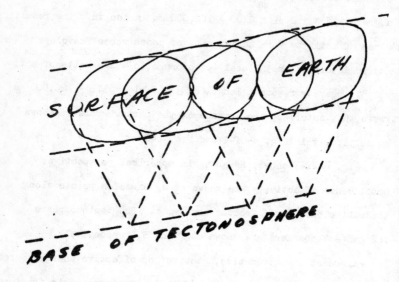

Figure 6-2f-1. Schematic diagram showing the envelope of an infinity of tectonospheric "cones of activity" forming a tectonospheric "wedge-belt of activity". See text for details.

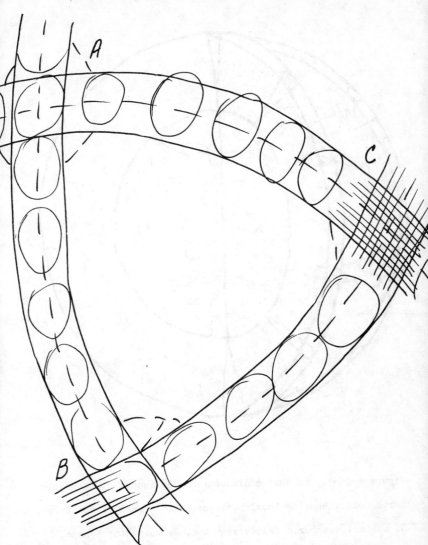

Figure 2-2f-2. Schematic diagram showing plan view of the Earth's surficial traces produced by 3 mutually-orthogonal "wedge-belts of activity", intersecting the Earth's surface at A, B. and C. See text for details. (Not to scale).

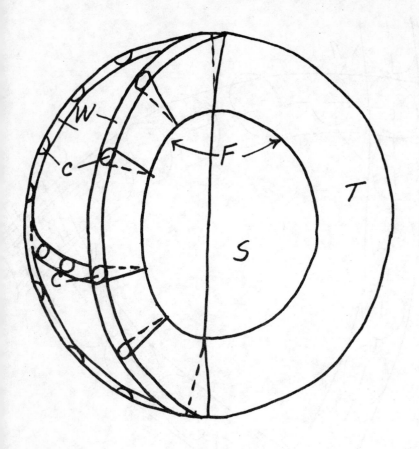

Figure 6-2f-3. Schematic diagram of idealized tectonospheric
Earth model, showing local tectonospheric "cones of activity",
C; subtectonospheric fracture system, F; subtectonosphere, S;
tectonosphere, T; and global tectonospheric "wedge-belts of
activity", W.

example, certain types of magmatic activity normally would not accompany certain types of tectonic activity. Nor would more than one type of magmatic activity be expected at a given location at a given time (See Chapter 10).

Paths of Preferential Flow of Heat within the Tectonosphere.

If an energy source at some point along the subtectonospheric fracture system consisted of heat and, if the tectonosphere above that point were homogeneous, then it might be a relatively simple matter to determine the most probable paths the heat might follow through the tectonosphere to the surface of the Earth. Also, these "preferential" paths of heat flow would be expected to lie somewhere within a typical cone of activity.

Since, however, the tectonosphere is not homogeneous, the "preferential" path of heat flow (from the surface of the sub-tectonosphere to the surface of the Earth) might be quite cir-cuitous. For example, Fig. 6-2g-1 shows, schematically, possible paths the heat might follow, while circumventing heterogeneities within the tectonosphere, before reaching the surface of the Earth. It might, for example, follow any one of a number of paths (shown schematically on the figure) before reaching level B, where it would form subsidiary cones of activity at B and B_1 before proceeding to level C; etc.

Eventually, the heat might reach level J, at any one of 4 locations, J_1 through J_4. Thus, from a source at O, the pre-ferential heat-flow path through the tectonosphere might lead to a surficial manifestation (e. g., a volcano) at any one of several fairly widely-separated locations on the surface, such as S_1 and S_4. On the other hand, it might lead to a surficial manifestation at S_2 or S_3 which are almost epicentral to the

274

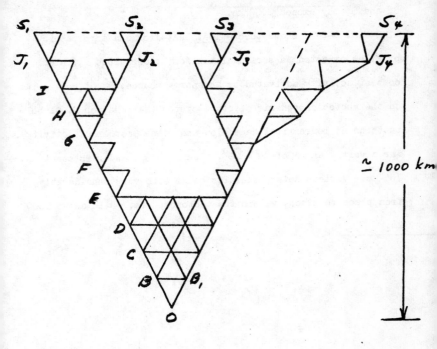

Figure 6-2g-1. Schematic diagram showing possible paths of preferential heat flow from a point, O, on the surface of the subtectonosphere to the surface of the Earth. See text for details. (Not to scale).

275

original source at O. (See Chapter 10, Intrusive and Extrusive Activity). For a vertex angle of 90°, the separation between S_1 and S_4 may exceed 2000 km.

Fig. 6-2g-2 shows <u>theoretical</u> heat-flow and topographic profiles expected as surficial manifestations above a typical cone of activity motivated by an energy source, O, at some point on the subtectonospheric fracture system (See Chapter 9, Global patterns of Terrestrial Heat Flow and Other Geothermal Activity). For a vertex angle of $2\theta = 90°$, then $\theta = 45°$, and r is about 785 km. Various heterogeneities cause θ to vary considerably, from place to place, within the tectonosphere. D is about 1000 km.

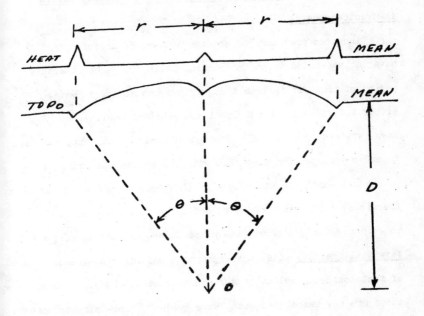

Figure 6-2g-2. Schematic diagram showing theoretical, idealized heat-flow and topographic profiles expected as surficial manifestations above typical cone of activity motivated by an energy source at some point, O, on the subtectonospheric fracture system. See text for details. (Not to scale).

277

Planes, Surfaces, and Shells of Preferential Rheidity within the Tectonosphere.

Since the previously discussed 1000-km of accretionary material was added to the primordial Earth over a considerable period of time, it follows that various accreted fragments, either individually or in global and regional "shells", would have retained some of their freedom of motion, both with respect to other accreted fragments and with respect to the basic 5400-km primordial Earth (tectonosphere of the present Earth). It follows, therefore, that such freedom of motion would most likely be in evidence along various planes, surfaces, and shells of pre-ferential rheidity which developed as a natural consequence of tectonospheric evloution during the past 4.6 b. y. Some of these are now associated with "wave guides", "low-velocity zones", and similar features and phenomena within the tectonosphere.

In connection with the above, it is well to recall that, at a rate of 1 m/yr, a depth of 1000 km is accreted in 1 m. y.; at a rate of 1 cm/yr, it is accreted in 100 m. y. This is impor-tant not only in connection with the development of rheid zones, surfaces, areas, and shells within the tectonosphere during the past 4.6 b. y. but also in connection with the development of the Earth's deep fracture system (See, e. g., Chapter 8, The Earth's Deep Seismicity; or Chapter 18, Plate Tectonics). In either case, assuming the accretionary process lasted 100 m. y., the energy required in any one year would be only that needed to fracture through (or maintain rheidity within) a vertical

278

thickness of about 1 cm of tectonospheric material. Assuming
the accretionary process lasted 1 b. y., the energy required
in any one year would be only that needed to fracture through
(or maintain preferential rheidity within) a vertical thickness
of about 1 mm of tectonospheric material.

From the standpoint of mechanical feasibility of _maintaining_
rheidity within the model, it is necessary to recall that, in
a homogeneous sphere of 6371 km radius, the basic inertia of
the inner 5400-km portion normally exceeds that of the outer
971-km shell by at least 60%. Thus, for most purposes, it may
be assumed that the _dis_equilibrating inertia of the inner 5400-km
portion of the model exceeds (and has _always_ exceeded) the _equili-
brating_ inertia of the outer 971-km portion by a factor of at
least 60%. For a non-homogeneous body such as the _actual_ Earth,
this inertia-difference factor would normally be even greater
than 60%.

During the _earliest_ stages of accretion of the tectonosphere
onto the subtectonosphere, the above indicated inertia-difference
factor would have been extremely large. As accretion proceeded,
the factor decreased from an infinite value and approached about
1.6 as an asymptote.

Other aspects of tectonospheric planes, surfaces, and shells
of preferential rheidity are discussed in Chapter 16 (Sea-Floor
Spreading), Chapter 17 (Continental Drift and Polar Wandering),
and Chapter 18 (Plate Tectonics).

Families and Stacks of Tectonospheric Plates.

Assuming that the subtectonosphere consists essentially
of 8 blocks separated by a fracture system of 3 mutually orthogonal planes, what can be said about the internal structure
of the tectonosphere? What is its structure today, and has
it always been roughly the same as it is today?

Theoretically, the tectonosphere could consist of 8 blocks,
each about 1000 km deep and with surficial dimensions of about
9,000 km, with blocks separated by a system of 3 mutually orthogonal tectonospheric wedge-belts. However, mechanico-geometrical
considerations associated with the evolutionary process of the
tectonospheric accretion dictate against the reality of such a
simple structure for the tectonosphere. On the other end of
the dimensional scale, the tectonosphere could, theoretically,
consist of an infinity of spherical shells (with radii ranging
from about 5400 km to 6371 km), each shell, in turn, being fractured into an infinity of tectonospheric plates.

The actual tectonosphere is defined somewhere between the
two extremes (i. e., between a single block and an infinity
of infinitely fractured tectonospheric plates). Although a
good physical analogue is not available, an approach to such
is afforded by 8 decks of cards having the following characteristics: (1) each card is shaped like the surficial octant of
a sphere: (2) the bottom card of each deck is 15% smaller than
the top card, with intermediate cards gradational between the

two; (3) each card is severed (or "fractured") into many pieces of various shapes and sizes; (4) each deck consists of 1000 cards (the number is arbitrarily selected for convenience, since the tectonosphere is 1000 km thick); (5) if the top card of each stack measures 10 cm on the side, the bottom card of each stack will measure 8.5 cm, and the 8 stacks will fit neatly onto a sphere of 54-cm radius; (6) the top card of each of the 8 stacks will combine to form the surface of a new sphere with a radius of about 64 cm; (7) the n-th card in each of the 8 stacks will combine to form the surface of a new sphere with a radius of about $(54 + 0.01 n)$ cm.

The "analogue" model described above has the following additonal features which are of interest to an analysis of tectonospheric evolution and behavior during the past 4.6 b. y.: (1) the top cards of each of the 8 stacks combine to form a spherical shell which is free to move along a spherical surface of "preferential rheidity" that separates it from card number 999; (2) any hemispheric part of shell number 1000 is free (within certain constraints) to move along lines of preferential rheidity formed by the 3 mutually orthogonal great circles separating the top card of each stack; (3) similar behavior may be attributed to any or all of the other 999 8-card spherical shells; (4) more complicated behavior within the 8 stacks develops when the 54-cm "base" sphere behaves in a manner analogous to that of the Earth's subtectonosphere as defined by the tectonospheric

Earth model.

Subsequent chapters will describe the details of the above and other features of the model. Suffice it here merely to point out several salient features of the tectonospheric-Earth-model behavior which may be derived from the above analogue model. For example, Fig. 6-2i-1 shows that, in a heterogeneous dynamic tectonosphere, the uppermost element of a stack of tectonopsheric plates need not necessarily be <u>vertically</u> above the lower elements of the same stack. Thus, a given stack may form a <u>rectilinear</u> pattern, SS', on the surface of the <u>subtectonosphere</u> but a <u>non-linear</u> one at the surface of the <u>Earth</u>. Also, the pattern is not necessarily linear within any intervening layer, or family, between the subtectonosphere and the surface of the Earth. Elements of intervening layers have been omitted from Fig. 6-2i-1. Intervening layers normally may be expected to occupy positions intermediate between the positions of those at the surface of the subtectonosphere and those of the surface of the Earth. However, as is pointed out in subsequent chapters, there are several notable exceptions.

Fig. 6-2i-2 shows the basic "causal" relationship between fractures within tectonospheric plates and the subtectonospheric fracture system. In most cases, the degree of fracturing within a given octant of the Earth model increases in each layer as one proceeds upward from the subtectonosphere to the surface of the Earth. Thus, a plate at the surface of the Earth normally

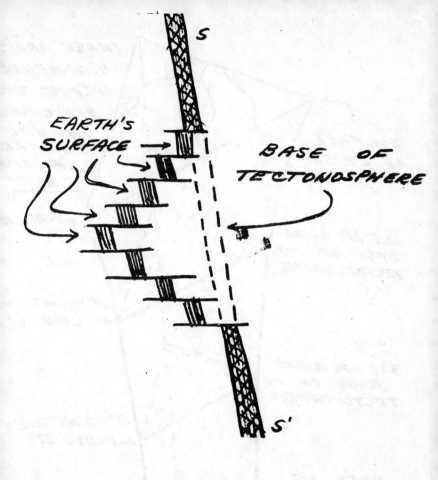

Figure 6-2i-1. Schematic diagram showing how, in a heterogeneous
dynamic tectonosphere, the uppermost elements of a stack of
tectonospheric plates need not necessarily be vertically above
the lower elements of the same stack. See text for details.

283

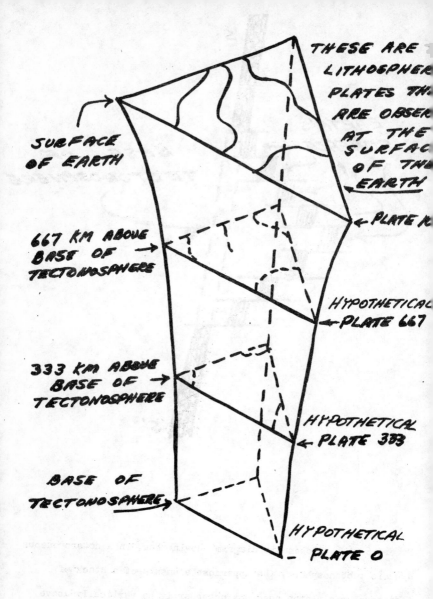

THESE ARE
LITHOSPHER
PLATES TH
ARE OBSER
AT THE
SURFAC
OF TH
EARTH

SURFACE
OF EARTH

← PLATE N

667 KM ABOVE
BASE OF →
TECTONOSPHERE

HYPOTHETICAL
← PLATE 667

333 KM ABOVE
BASE OF →
TECTONOSPHERE

HYPOTHETICAL
← PLATE 333

BASE OF
TECTONOSPHERE →

HYPOTHETICAL
PLATE O

Figure 6-2i-2. Basic relationship between fractures within tectonospheric plates and the subtectonospheric fracture system. See
text for details. (Schematic).

284

is more fractured than one at 667 km from the base of the tectono-sphere; one at 333 km from the base of the tectonosphere is more fractured than the base plate at 0 km; etc. (See Chapter 18, Plate Tectonics).

Low-Velocity Zones, Wave Guides, and Similar Features within the Tectonosphere.

Low-velocity zones, wave-guides, and similar tectonospheric features are formed by the combination of (1) stacks and families of tectonospheric plates, and (2) planes, surfaces, and shells of preferential rheidity. The resulting low-velocity zones, wave guides, and related features may form at any angle, ranging from horizontal to vertical, depending upon the specific location of the fractures between the stacks and families of tectonospheric plates involved, as well as upon which fractures are in a rheid state at the time under consideration (See Chapter 18, Plate Tectonics).

Lithospheric Plates.

Lithospheric plates comprise the uppermost members of tectonospheric plates. The evolution and present behavior of the
lithospheric plates are closely associated with those of the
underlying families and stacks of tectonospheric plates. These
plates developed their present state and behavior (fractures,
motions, etc.) as a result of the basic driving mechanism of
the model. For example, their fracture system depends primarily
upon the basic fracture system and motions of the underlying
stacks and families of tectonospheric plates, which in turn
depend upon the driving mechanism (Chapter 18, Plate Tectonics).

Similarly, their magmatic and metamorphic behavior, or
activity, is dependent upon the underlying layers of tectonospheric plates. This does not mean that the basic magmatic
(and other) activity dictated by the underlying layers might
not be enhanced by local sources of heat within the lithosphere
itself (e. g., by local pockets of radiogenic materials). (See
Chapter 9, Global Patterns of Terrestrial Heat Flow and Other
Geothermal Activity).

Asthenospheric Structure and Behavior.

The asthenosphere comprises one or more layers of tectono-spheric plates immediately below the lithospheric plates, together with the planes, surfaces, and shells of preferential rheidity existing within, above, and below these asthenospheric layers. The asthenosphere is not, strictly speaking, a separate entity, but rather a "combination" of tectonospheric plates with the rheidity provided by the flow of heat, and possibly of volatiles, along paths of preferential heat flow. In a sense, the charac-teristic behavior of the asthenosphere is a function of its environment. It has evolved into its present form and position because of what lies above and below it and what has been there in the past. Since both its substrata and epistrata have under-gone cyclically aperiodic changes during the past 4.6 b. y., the position, structure, and behavior of the asthenosphere have varied in a similar manner during that time. (Chapter 18, Plate Tectonics, contains other details).

Tectonospheric Structure beneath the Asthenosphere.

The tectonospheric structure beneath the asthenosphere is much the same as it is within and above the asthenosphere, except as it might be modified by (1) greater temperatures and pressures; and (2) nearness to the subtectonosphere, to its fracture system, and to the driving mechanism of the model (Chapter 18, Plate Tectonics).

As in the case of the asthenosphere, the tectonospheric structure beneath the asthenosphere has undergone cyclically aperiodic variations for similar reasons.

The Interface between the Tectonosphere and the Subtectonosphere.

As would be expected from basic considerations, the interface between the tectonosphere and the subtectonosphere does not exist as a detectable entity. After 4.6 b. y. of close association, the tectonosphere and the subtectonosphere are almost completely blended one into the other insofar as physically measurable effects are concerned.

The nearest physical analogue to the interface between the tectonosphere and the subtectonosphere would probably be a fluid-drive type of automotive transmission with (1) an almost-zero-viscosity fluid and (2) an extremely long time constant. As a consequence, this interface may not be detectable with present instruments since no discontinuity, as usually defined, exists there. After 4.6 b. y., it would be quite difficult to determine exactly where the subtectonosphere begins and the tectonosphere ends when evaluated geometrically. However, from mechanical considerations, the subtectonosphere contains the Earth's driving mechanism, while the tectonosphere behaves in response thereto. The interface serves as the "coupling" between the "driving" subtectonopshere and the "driven" tectonosphere. Therefore, the interface performs a highly essential function in the evolution and behavior of the tectonosphere.

Barrell, J., 1914a. The strength of the Earth's crust, part 5. *Jour. Geol.*, 22: 441-468.

Barrell, J., 1914b. The strength of the Earth's crust, part 6. *Jour. Geol.*, 22: 655-683.

Barton, D. C., Ritz, C. H., and Hickey, M., 1963. Gulf coast geosynclines. *Bull. Am. Assoc. Petrol. Geol.*, 17: 1446-1458.

Beloussov, V. V., 1969. Interrelations between the Earth's crust and upper mantle. In: P. J. Hart (Editor), *The Earth's Crust and Upper Mantle*. Am. Geophys. Union, Washington, pp. 698-712.

Bentley, C. R., 1964. The structure of Antarctica and its ice cover. In: H. Odishaw (Editor), *Research in Geophysics*, vol. 2. MIT Press, Cambridge, pp. 225-389.

Crittenden, M. D., 1963. Effective viscosity of the Earth derived from isostatic loading of Pleistocene Lake Bonneville. *J. Geophys. Res.*, 68: 5517-5530.

291

Drake, C. L., 1970. A long-range program of solid Earth studies. Trans. Am. Geophys. Union, 51: 152-159.

Ewing, J., 1969. Seismic model of the Atlantic Ocean. In: P. J. Hart (Editor), The Earth's Crust and Upper Mantle. Am. Geophys. Union, Washington, pp. 220-225.

Gutenberg, B., 1941. Changes in sea level, post glacial uplift, and mobility of the Earth's interior. Bull. Geol. Soc. Am., 52: 721-772.

Hamilton, R. A., Brooke, F. R., Peacock, S. D., Bowater, S., and Bull, C., 1956. British North Greenland Expedition, 1952-54: scientific results. Geographical J., 122: 203-241.

Harrison, J. C., 1955. An interpretation of gravity anomalies in the eastern Mediterranean. Phil. Trans. Roy. Soc. London, Ser. A, 248: 283-325.

Hart, P. J., (Editor), 1969. The Earth's Crust and Upper Mantle. Am. Geophys. Union, Washington, 735 pp.

Haskell, N. A., 1937. The viscosity of the asthenosphere.

Am. J. Sci., 33: 22-28.

Hospers, J., 1965. Gravity field and structure of the Niger Delta, Nigeria, West Africa. Bull. Geol. Soc. Am., 76: 407-422.

Khain, V. E. and Muratov, M. V., 1969. Crustal movements and tectonic structure of continents. In: P. J. Hart (Editor), The Earth's Crust and Upper Mantle. Am. Geophys. Union, Washington, pp. 523-538.

Knopoff, L., 1969. Continental drift and convection. In: P. J. Hart (Editor), The Earth's Crust and Upper Mantle. Am. Geophys. Union, Washington, pp. 683-689.

Lawson, A. C., 1942. Mississippi Delta: a study in isostasy. Bull. Geol. Soc. Am., 53: 1231-1254.

Lyustikh, E. N., 1969. Problem of convection in the Earth's mantle. In: P. J. Hart (Editor), The Earth's Crust and Upper Mantle. Am. Geophys. Union, Washington, pp. 689-692.

Mehnert, K. R., 1969. Petrology of the Precambrian basement

complex. In: P. J. Hart (Editor), The Earth's Crust and Upper Mantle. Am. Geophys. Union, Washington, pp. 513-518.

Molnar, P. and Oliver, J., 1969. Lateral variations of attenuation in the upper mantle and discontinuities in the lithosphere. J. Geophys. Res., 74: 2648-2682.

Popelar, J., 1968. Gravity field and isostasy in the area of the Czechoslovak West Carpathians. Sbornik Geologickych Ved, Uzita geofyzika, 7: 7-24.

Ronov, A. B. and Yaroshevsky, A. A., 1969. Chemical composition of the Earth's crust. In: P. J. Hart (Editor), The Earth's Crust and Upper Mantle. Am. Geophys. Union, Washington, pp. 37-57.

Runcorn, S. K., 1969. Convection in the mantle. In: P. J. Hart (Editor), The Earth's Crust and Upper Mantle. Am. Geophys. Union, Washington, pp. 692-698.

Stacey, F. D., 1969. Physics of the Earth. Wiley, New York, 324 pp.

Vogt, P. R., Schneider, E. D., and Johnson, G. L., 1969. The crust and upper mantle beneath the sea. In: P. J. Hart (Editor), The Earth's Crust and Upper Mantle. Am. Geophys. Union, Washington, pp. 556-617.

Walcott, R. I., 1968. The gravity field of Northern Saskatchewan and Northeastern Alberta with maps. Gravity Map Series Nos. 16 - 20, Dom. Obs. Can.

Walcott, R. I., 1970. Flexural rigidity, thickness, and viscosity of the lithosphere. J. Geophys. Res., 75: 3941-3954.

Chapter 7

THE OROGENIC—CRATONIC STRUCTURE OF THE CONTINENTS

When the structure and evolution of the continents are considered on a global scale over a span of 3.6 b. y., the characteristic orogenic—cratonic structure obvious in some continents is revealed to be present also in other continents. The continents appear to have grown outward by accretion onto a central shield or nucleus. Exactly what caused the continents to follow this pattern of quasi—organized accretionary growth is not known. Most modern hypotheses do not extend back as far as 3.6 b. y. during which observational evidence supports such growth—from—the—center evolution for the continents.

More specifically, an analysis of the continents over a long span of time reveals that the orogenic belts are long and rectilinear, i. e., the accretionary growth appears to have occurred along segments of great circles, which, unlike small circles, plot onto the surface of the Earth as rectilinear features (See any text in spherical geometry). The continents appear to be old, and the process of accretion from the center seems to have begun at least 3.6 b. y. ago. The oldest orogenic belts are nearest the centers of the continental nuclei; but a certain amount of overprinting by orogenic belts of different ages may be present farther from the centers. The orogenic belts generally appear to have been active on a worldwide basis in a cyclical but aperiodic manner during at least the past 3.6 b. y. The predominance of radioactive elements in continental nuclei suggests that chemical differentiation or an equivalent mechanical process has occurred in the mantle sometime during the past 3.6 b. y. If such a chemical or mechanical process in fact did occur, then the entire continental areas may be of ultimate mantle origin, and the primordial Earth might have been completely devoid of crust or continents as we know them today.

Characteristic orogenic—cratonic structure is fairly well preserved in most continents (See, for example, Douglas, 1969; Wynne–Edwards, 1969; Martin, 1969; Hurley, 1969; Drake, 1970). In general, what is seen is an old Precambrian shield successively surrounded by long more—or—less rectilinear orogenic or geosynclinal belts, often broken or otherwise modified by overprinting from subsequent orogenies, with at least three orogenies fairly well preserved in most continents, e. g., the Appalachian, Innuitian, and Cordilleran orogenies of the Canadian Shield.

Present hypotheses do not explain why the orogenic belts are rectilinear and of different ages within any one continent nor why they appear to extend worldwide with their activity having occurred in a cyclical but aperiodic manner containing concentrations at about 0.9 b. y. intervals for at least the

297

past 3.6 b. y. Also not understood is why the spatio-temporal relationships within all orogenic belts are remarkably similar, regardless of where found or to which period they belong. This chapter examines the evolution of orogens and cratons and explores tentative answers to these and related questions.

Continents, Platforms, Shields, and Cratons.

The major continents, North America, South America, Africa, Eurasia, Australia, and Antarctica, are large land masses rising more-or-less abruptly above the deep ocean floor and occupying about 1/3 of the Earth's surface.

Platforms, also called cratons, are large portions of the continental crust bounded by folded belts, or sometimes by oceans, and characterized by the following specific features: (1) tectonic stability over very long periods (at least several hundred million years); (2) general tectonic homogeneity (despite a complex inner structure and different times of formation of the separate parts); and (3) an absence of high gradients of movement.

Shields are elements of platforms (or of cratons) that display a tendency toward slow uplift (or subsidence) over periods of hundreds of millions of years. All shield areas have almost identical thicknesses: Baltic, 34 km; African, 35 km; Canadian, 36 km; Australian, 35 km; and East Antartic, 35 km.

Two Types of Platforms.

Platforms, for the purpose of this analysis, may be divided into two types, true and "quasi". True platforms comprise systems of extremely old stabilized segments of Archean platforms that later underwent tectonic quiescence. These Archean segments are separated by large deep faults and are welded with more recently consolidated areas along Early Proterozoic geosynclines. These welded zones, in some cases, were reactivated and reconsolidated during the Middle and Upper Proterozoic. In other cases, new geosynclines were formed in recently collapsed areas within the Archean or Lower Proterozoic basement (e. g., the Delhi Belt of India; the folded areas of Damara, Kantanga, and Kibara of Africa; and possibly the Red Sea area). These true platforms, or cratons, are always surrounded by much younger folded belts or by oceans.

The second type of platform, or quasicraton, comprises segments of a folded basement truncated by a pleneplain formed after the folding. The structure of the quasicratons is not interrupted by collapse zones but forms a single integrated system with the synchronous folded orogens. Unlike true cratons, quasicratons do not lose their mobility; rather, repeated geosynclinal phases alternate with the cratonic quiescence in a cyclical but aperiodic manner. All true cratons belong to the the pre-Phanerozoic regimes.

300

Continental Accretion.

The concept of the orogenic–cratonic evolution of the continents by accretion, as a result of geosynclinal activity along the peripheries of a central craton, was suggested by Dana (1873) and developed subsequently by many others including Bertrand (1887), Suess (1909), Kraus (1928), Born (1933), Stille (1940), Arkhangelsky (1941), Kober (1942), Bondarchuk (1944), Shatsky (1946), Pavlovsky (1953), Wilson (1957), Gilluly (1955), Vassilkivsky (1960), Magnitsky (1965), and Gnibdenko and Shaskin (1970).

The principle of continental accretion initially meant simply the transformation of a mobile geosynclinal region into a stable continental (i. e., platform) zone. With the development of geophysics, the concept was extended to include a "qualitative jump" whereby "oceanic" crust is transformed into continental crust (Bucher, 1950; Gilluly, 1955; Wilson, 1957; and Hess, 1955). Other analyses suggested that the process of continental accretion is complicated by regeneration of the geosynclinal regime (Born, 1933; Stille, 1940, 1944; Arkhangelsky, 1941). Stille (1940) concluded that only quasi-cratonic regions, but not true cratons, could be regenerated. Later the concept of geosynclinal regeneration was developed (Van Bemmelen, 1966; Beloussov, 1962) to a retransformation of continental crust into oceanic crust, thereby suggesting the hypothesis of reversibility of geological processes.

Orogens and Cratons.

The two fundamental elements of the continents, orogens and cratons, developed at a definite stage in the evolution of the Earth (See, e. g., Salop and Scheinmann, 1969). Because the oldest parts of the cratons are at least 3.6 b. y. old, the particular developmental stage for the orogenic-cratonic structure of the continents must have begun about 3.6 b. y. ago. It is well to take a brief look at the evolution of the orogens and cratons since that time.

The Archean.

During the Archean, a considerable portion of the Earth's surface was covered by a shallow sea, and the land was represented by low, frequently flooded islands or shoals of unstable outline. Sedimentation was accompanied by intense subaqueous effusions of predominantly basic lavas.

The Archean complexes of Eurasia are highly metamorphosed, differing in this respect from the synchronous complexes in North America and South Africa. However, the gneiss-granulite complexes of Labrador and the Grenville province, as well as Madagascar, Natal, Tanganyika, Uganda, Zambia, and a few other regions of Africa, can be correlated with the Archean of Eurasia.

The style of Archean tectonics suggests a high plasticity of the material involved, as well as a close relationship between

302

folding and other tectonomagmatic processes associated with the diapiric ascent of granitic masses. In most Archean complexes, plastic deformations do not appear to form any specific patterns, perhaps because the crust had not yet differentiated into areas with characteristic tectonic features. Consequently, Archean tectonic processes, unlike those of today, occurred everywhere with more-or-less equal intensity, causing the Archean to be referred to as the "permobile" stage of crustal development.

The Proterozoic.

The first identifiable orogens developed during the latter Archean or perhaps the early Proterozoic, at which time there also occurred extremely intense fragmentation and partial submergence of the Archean gneissic basement, previously denuded and even pleneplained in certain areas. At this time, the orogenic-geosynclinal and cratonic-massif systems developed in the following areas of Eurasia: Fennoscandia, Saksagan, Central-Asiatic, North-Kazakhstan, Yenissei, Taimyr, Baikal, Stanovoy, Mid-Asiatic, Arovelli, Satput, Shimoga-Dharwar, and a few lesser ones. Some of these early orogenic-cratonic systems completed their earliest recorded cycles by the middle of the Precambrian and became dormant; others contined into successive cycles or regenerations.

When closely analyzed, the oldest orogenic systems resemble the youngest orogens known today. Eugeosynclinal and miogeosynclinal details are distinguishable and can be divided into structural-formational zones on the basis of their volcano-sedimentary and plutonic series. As in the newer systems, the oldest systems display distinct areas of source and of accumulation of terrigenous material. In eastern Siberia, for example, in the Olekma River basin, the Lower Proterozoic deposits (older than 2.1 b. y.) are of the platform type.

The Lower Proterozoic rocks, like those of later geosynclinal

304

complexes, are characterized by linear folds grouped into elongated areas or into large arcs that are convex toward the entrant angles between the platform blocks.

Deep faults, thousands of km long, served as the boundaries of geosynclinal belts and structural zones. These intrageosynclinal faults were associated with massive subaqueous lava effusions (initial volcanism) and with accompanying ophiolitic intrusions. Faults located along the edges of platforms had rectilinear outlines and combined into systems of two or more intersecting directions. In this respect these faults resemble younger tectonofers (Salop and Scheinmann, 1969).

The faults not only determined the original configuration of the platforms and geosynclines, but, also, they controlled the location of the volcano-sedimentary formations and numerous plutonic bodies and they affected the general orientation of folded structures. The role of the faults as zones of high permeability and as heat "vents" of the crust has been discussed by Salop and Scheinmann (1969). Faults are usually associated with belts of higher metamorphism. The formation of belts and areas of granitization near faults and the subsequent ascent of rheomorphosed granitic masses along them led to an intense development of metamorphogenes and magmatic folding. Thus, faults controlled the location of folding near the faults and farther away.

The appearance of deep fractures at the beginning of the Proterozoic resulted from: (1) the greater thickness and rigidity of the crust; and (2) a drastic change in the thermal regime at the end of the most extensive Archean diastrophism.

The boundary separating the Archean and the Proterozoic identifies one of the most important events in the evolution of the orogenic-cratonic structure of the continents: it marks the transition of the Earth from a permobile system to a truly orogenic-cratonic structure. It is not known whether this transition occurred rapidly or over a comparatively long span of time; most estimates are approximately 200 m. y.

The oldest orogenic belts formed, not on a primary basaltic or mantle substratum but, on a fragmented and subsided granite-metamorphic Archean basement. Most likely, the earliest orogenic-cratonic structure was not a simple accretion around a greenstone core, but rather an intense mobilization of penetrating differential movements of the Archean granito-gneissic basement (Salop and Scheinmann, 1969).

The Later Precambrian.

During the later Precambrian the orogenic-cratonic evolution of the continents consisted basically of geosynclinal accretions onto the Archean platforms. This was, however, not a simple evolutionary process but one interrupted periodically by collapses and subsidences of separate parts of stabilized blocks, suggesting the influence of a causal mechanism deep within the tectonosphere.

The process of platform expansion began during the latter part of the Early Proterozoic when geosynclines began to attach themselves to the cratons. This is evident, for example, in the early Precambrian folded systems of Canada and Siberia. Substantial parts of the miogeosynclinal area of the Baikal system became accreted to the Angara and Adlan platforms. Also the Early Proterozoic Stanovoy geosynclinal system attached itself to the Aldan platform. Simultaneously, in some geosynclinal belts, Lower Proterozoic structures were rearranged by a collapse along new and some old faults, again suggesting a long-lived causal mechanism deep within the tectonosphere.

In all continents, the Middle Proterozoic consisted of a wide development of platforms and miogeosynclinal activity in which quartzites, quartzite-sandstones, quartz conglomerates, and chemogene dolomites predominated. The platforms were eroded to pleneplains with thick crusts of chemical weathering. In

307

some miogeosynclinal areas there appeared the first embryonic marginal depressions infilled with thick masses of red clastics and subaerial porphyries.

By the end of the Middle Proterozoic (approximately 1.65 b. y. ago) many continents were intuded by characteristic platform-type differentiated plutons of gabbro-norites, anorthosites, granophyric rapakivi granites, and alkaline rocks. These post-orogenic intrusions were separated from the synorogenic plutonism of the Middle Proterozoic (approximately 1.85 b. y. ago) by a period of tectonic quiescence in which there were local surface effusions of acid lavas.

During the second half of the Proterozoic (Upper Proterozoic and epi-Proterozoic) the tectonic behavior was very complex in all parts of the Earth. Large internal uplifts occurred in many geosynclinal belts, dissecting them into a zoned aspect. Many marginal depressions became isolated near the deep faults that separated platforms from miogeosynclines. Following the epi-Proterozoic diastrophism, detritic, molasse-like sediments accumulated in the marginal depressions (e. g., the Ashinsk series of the Urals). Within the active geosynclinal belts, the facies zonality intensified, and the amount of rudaceous rocks increased significantly. On the platforms, particularly in aulacogens, substantial masses of terrigenous and carbonate sediments accumulated. Basaltic volcanism was widespread on nearly all platforms.

On some platforms, collapse occurred, and new geosynclines formed on the base of the collapse structures. Many of these new geosynclines completed their development before the end of the Proterozoic (e. g., the Damara-Kibara-Katanga belt in Africa)..

Intense deastrophism near the end of the Late Proterozoic and epi-Proterozoic caused inversion within many geosynclinal belts. In most primary geosynclines (i. e., those formed during the early Proterozoic) endogenetic processes consisted of small intrusions of ultrabasic magma plus highly intense granitic plutonism. Granitic mountains formed at this time occupied areas of hundreds of thousands of km^2 per platform (e. g., the Baikal mountain region contains 120,000 km^2 of formations of this type).

Late Precambrian structures, like those of the Early Proterozoic and the Phanerozoic, do not display specific characteristics. Rather they contain all genetic types of folds, including gravity folds that suggest a greater amplitude of primary tectonic microstructures. Also, magmatogenic folding is widely developed near large granitic plutons. Late Precambrian metamorphism was linear, and granitization occurred locally in narrow zones of deep faults and at their intersections.

The Phanerozoic.

In almost all continents, an angular or stratigraphic unconformity separates the epi-Proterozoic from the Phanerozoic
(i. e., Eocambrian). Generally, the Eocambrian deposits show
a more-or-less gradual transition to the Cambrian. In some
continents, the latest Proterozoic strata occur primarily in
aulacogens whereas the earliest Phanerozoic strata form a vast
sheet at the bottom of the Paleozoic.

The Phanerozoic, from a tectonic standpoint, continued
the constantly increasing differentiation and amplitude of tectonism observed in the latter part of the Proterozoic. During
the Cambrian, the first frontal depressions, or foredeeps, appeared as a result of the growth and displacement of miogeosynclinal marginal depressions onto the cratons. The depth and
subsidence gradient of the foredeeps increased with time, thereby
creating a greater thickness of molasse (from 4 km in the epi-
Proterozoic to 20 km in the Cenozoic) plus more rudaceous rocks.
Average rates of sedimentation during the Early Proterozoic,
epi-Proterozoic, Cambrian, and Pliocene were 17, 37, 140, and
500 m per m. y. Evaporites and well-graded, mature sediments
accumulated only in the platforms during the Phanerozoic, thereby
creating a greater differentiation and variety of sediments.

Some polycyclic geosynclinal belts existed for as long as
2 b. y. (e. g., the Baikalian); younger monocyclic belts, for

only tens of m. y. Many details remain unclarified. The periodicity of tectonic processes over long periods of time, as well
as the apparent directivity and orderliness of the orogenic-
cratonic evolution of the continents, suggests a deep-seated
causal mechanism. However, many questions remain unanswered. Why,
for example, does the tectonic activity occur in cycles and why
are the cycles of unequal length (i. e., why is the behavior
cyclical but aperiodic)? What is the nature of the hypothetical
causal mechanism that controls these cycles? When and how did
such a causal mechanism originate?

Other details regarding the behavior of orogens and oratons
are contained in Salop and Scheinmann (1969), Beloussov (1970),
Gnibidenko and Shashkin (1970), Drake (1970), Benes (1968),
Misik (1968), and Glikson (1970), to name a few of the more
recent ones.

A Typical Continent: North America.

To say that some continents are typical and that others are not is perhaps an artificial categorization as well as an admission that some continents have been studied in greater detail than have others. In any case, the North American continent serves as a good base from which to analyze the orogenic-cratonic structure and evolution of a typical continent during the past 3.6 b. y.

Archean Protocontinents of North America.

The Canadian Shield contains a large number of volcanic and sedimentary belts enclosed in granitic rocks. In the southern part of the Shield these belts are arranged in linear patterns of alternating volcanic and sedimentary belts. In the northern part the volcanic and sedimentary belts appear more randomly arranged but still display sub-linear patterns.

The predominant trends and patterns discernible in the Archean belts of the Canadian Shield suggest primary features of an evolving crust rather than secondary deformational and metamorphic features. The evolution suggests protocontinental development. For example, the prevailing trends of Archean volcanic and sedimentary belts are easterly in the southern and central parts of the Shield (i. e., Superior and Charchill provinces) and northerly in the northwestern part (i. e., the

312

Slave province). Closer examination suggests that the Archean trends form the peripheries of three triangular or quasi-quadrate areas: (1) northwestern Ontario and the adjoining part of Manitoba; (2) the northwestern part of Hudson Bay; and (3) the central part of the Slave province. This distribution of Archean trends around separate nodal areas serves to delineate roughly the nuclei of three postulated ancient Archean protocontinents, named the Superior, Hudson, and Slave protocontinents.

In highly generalized form each protocontinent of the Canadian Shield comprises a craton, or nucleus, and orogens, or marginal orogenic belts. Other shields of the world exhibit a similar Archean structure. In the Canadian Shield, as in other shields, the exact number of Archean cratons is indeterminate. For example, a fourth Archean protocontinent (Ungava) is inferable in the northeastern part of the Superior province. Of the four postulated protocontinents of the Canadian Shield, the Superior is by far the largest and best developed.

In addition to the surficial accretions, the Archean protocontinents probably experienced vertical accretion of material to their undersides. This sub-crustal deposition, or sialic underplanting undoubtedly contributed to the growth and stabilization of Archean protocontinents. As the Archean protocontinents of North America grew horizontally and vertically, they merged one with the other to lose their identities and to become

313

what is now known as the Canadian Shield. The other continental shields were produced simultaneously by similar processes during this period (See, for example, King (1962), Holmes (1965), Dunbar (1966), and Goodwin (1968)).

The Superior Protocontinent.

The Superior craton, like most other cratons of the Earth, contains a number of alternating belts of volcanic and sedimentary material. In the southern part of this craton, three volcanic belts trend easterly: the Uchi, Keewatin, and Abitibi belts, separated by intervening sedimentary belts now composed of metasediments, migmatite, paragneiss, and granitic rocks.

In the northwestern part of the Superior craton, a similar though less well-defined series of easterly to southeasterly-trending volcanic belts is present: the Windigo, Amisk, and La Ronge belts. Thus, six orogens are distributed about this Archean nucleus. Other Archean nuclei, when examined closely, show a similar orogenic-cratonic structure.

Closer scrutiny of the three main easterly-trending volcanic belts in the southern part of the Superior craton (the Uchi, Keewatin, and Abitibi belts) reveal even greater detail regarding their physical structure and chemical composition: (1) the volcanic belts increase in width from the craton outward; (2) individual volcanic areas within the belts show a

314

systematic increase in size, the largest being in the eastern part of the outer belt (i. e., the Atibiti); and (3) the acidic volcanic content of the volcanic assemblages increases outward from the craton and reaches a maximum in the eastern part of the outer belt. The causal mechanism for these quasi-orderly variations within the orogenic-cratonic structure of the continents is examined in Chapter 10 (Intrusive and Extrusive Activity).

In the northwestern part of the Superior protocontinent, the three main volcanic belts (Windigo, Amisk, and La Ronge) show a similar but less-striking progressive change in relative size and acidic volcanic content to the northwest (i. e., outward from the center of the craton).

Summarizing, the Superior craton suggests a long and complex orogenic history involving many periods of granitic emplacements along linear or quasilinear peripheries of a central triangular or quadrate craton. The craton-to-periphery increase in size and complexity of volcanic components appears to be in the direction of younger rock assemblages.

The Hudson and Slave Protocontinents.

The Hudson and Slave cratons are not as well-defined as the typical Superior protocontinent primarily because of limited geological mapping, some metamorphic overprinting, and the presence of Proterozoic cover rocks. In the Hudson protocontinent, easterly trends predominate whereas in the Slave, the predominance is northerly. The quasilinear pattern combined with the presence of granitic masses suggests the presence of a cratonic mass in the northwestern part of the Hudson Bay area. Similar orogenic patterns in the Slave protocontinent suggest the presence of a cratonic mass also in that area.

Along and across the regional trends of the Hudson and Slave protocontinents, sedimentary belts alternate with relatively narrow and discontinuous volcanic belts.

Growth of the Canadian Shield during the Proterozoic.

The Proterozoic rocks, in contrast to those of the Archean, form long linear belts and broad basinal elements. The predominant northeasterly and northwesterly Proterozoic trends are rotated as much as 30° from the predominant northerly and easterly Archean trends. Most Proterozoic assemblages display flay-lying to gently-dipping strata, commonly overlying Archean basement rocks with a profound structural unconformity. This is particularly evident in the Superior province. In the Churchill province the Hudsonian orogeny during the Proterozoic obscured most of the Archean-Proterozoic age relationships, thereby placing in doubt the age of many rock units such as those in northern Saskatchewan and the south shore of Baffin Island and other Lower Proterozoic (Aphebian) rocks of the Canadian Shield.

The form and pattern of Proterozoic rocks is exemplified by the Labrador trough, a northwesterly-trending belt almost 1500 km long and containing a complex assemblage of sediments including iron formations, mafic volcanics, and associated intrusions. The shallow-water shelf facies on the west contains typical granular cherty iron oxide formation, orthoquartzite, and dolomite. The deep-water facies on the east contains fine-grained clastic rocks including pelites, mafic volcanics, and associated tuffs. Substantial proportions of mature, cross-bedded, and ripple-marked quartzite in the Lake Superior and

317

other regions of the Shield suggest stable platformal environments of deposition.

Volcanic rocks of the Proterozoic are represented mainly by thick, non-sequential, tholeiitic basalts, such as the Labrador trough volcanics, the Keweenawan flows of the Lake Superior region, and the Coppermine series of the Northwest Territories. Unlike Archean volcanic assemblages, these show no substantial evidence of compositional differentiation, thereby resembling the voluminous flood basalts of northwestern U. S. and the Deccan traps of India.

Early Proterozoic rocks of the Canadian Shield contain, among others, major deposits of U, Th, and Pb. Some Early Proterozoic igneous suites continued the late Archean alkalic intrusions with substantial K_2O. These include at least 60 alkalic and alkaline ring complexes tentatively indientified as plugs intruding the Archean Shield and representing at least 3 ages: 2.6, 1.6, and 1.0 b. y. ago (Wilson, 1966).

The widespread appearance of lamprophyre dikes and related activity reflect a substantial increase in K_2O content of the Precambrian crust. The appearance of K, together with large quantities of other heat-producing elements (i. e., U, Th, and Pb), during the late Archean and Early Proterozoic marks a degree of thermal migration and activation not detected in early periods of the Earth's crustal evolution.

Besides this thermal event, the Archean and Proterozoic of the Canadian Shield show sharp contrasts lithologically, orogenically, and tectonically. Particularly in the southern part of the Canadian Shield, the Archean-Proterozoic boundary is marked by a quiescent period (the Eparchean Interval) extending from the end of the Kenoran orogeny (2.5 b. y. ago) to about 2.35 b. y. ago. Very little is known about this 150 m. y. quiescent period except that typical Proterozoic crustal conditions characterize the period following the conclusion of the Eparchean Interval. What little information there is suggests that the Interval was marked by some first-order, mantle-crust differentiation associated with the widespread distribution of the major heat producing elements, U, Th, K, and Pb.

Orogenic-Cratonic Structure of North America: Greenland.

At the time of disintegration of Laurasia, block faulting, marine transgression, and basalt extrusion marked the geologic evolution of northern West Greenland. Downfaulting of the western seaboard probably began in the Cretaceous, and great thicknesses of limnic arkose accumulated rapidly in a newly-formed depression, while coarse conglomerate was deposited at the foot of the fault scarps.

In late Turonian time, a marine gransgression brought in a North American ammonite fauna; the sea later joined that covering western Europe (Chapter 15). Intercalations of tuff in Danian marine sediments are the first indication of volcanism which later led to extrusion of subaqueous basalts. There was intermittent faulting throughlut Cretaceous and Tertiary times, but the greatest movements occurred at the end of the Maestrichtian, in mid-Danian, and after extrusion of the basalts. The net result was a downthrow of several hundred meters on the seaward side of the faults, thereby creating the characteristic appearance of the Greenland coast in that area.

Unanswered Questions Regarding the Precambrian of the Canadian Shield.

When Hutton analyzed the evolution of the Earth on a global scale, he concluded that the surface was being destroyed in certain places and renewed in others in a system which he called "a circulation in the matter of the globe". The fragmentary remains of this circulation, that is the preserved supracrustal rocks, hold the answers to many question regarding the early history of the Earth. With one or two exceptions, the oldest rocks of the Canadian Shield are supracrustal rocks, i. e., statified volcanic, pyroclastic, and sedimentary rocks.

A study of the rocks and sediments of the Canadian Shield has answered many questions regarding the early history of that area. However, certain questions remain unanswered. What, for example, is the motive power for driving Hutton's "circulation in the matter of the globe"? What is the source for the early Proterozoic sediments that underlie at least part of Hudson Bay (Belcher Island)? What accounts for the differences between rocks of different ages in Hutton's recirculation system and between rocks of one age in different areas of the Canadian Shield? What is the true nature of the greenstone belts and their relationship to the "original" sedimentary basins?

Whether considering Hutton's hypothesis or later hypotheses such as plate tectonics, these and related questions must be answered.

Modern Hypotheses for the Orogenic-Cratonic Structure of the Continents.

Classic hypotheses regarding the orogenic-cratonic structure of the continents have been adequately analyzed in the literature (See, e. g., Runcorn, 1962; Holmes, 1965; and Hart, 1969). Rather than dwell upon the details of the traditional hypotheses, salient features of the more recent hypotheses are discussed in the remainder of this chapter.

Wilson's Froth-and-Slag Hypothesis.

The continents, according to Wilson's (1967) hypothesis, are like the islands of froth or slag and have collected along the mountains that form where convection currents descend. The ocean floors correspond to the clear soup or iron and reveal, by chains of islands and lateral ridges, the approximate direction of flow. In most cases the flow is away from the rising current beneath the mid-ocean ridges and toward the down-currents beneath the mountain systems.

To provide continental uplift Wilson's concept uses a combination of isostatic forces and mantle currents at depths of about 50 km. Continental splitting and spreading occur when the ascending currents form under a continental mass. In contrast with Wilson's model, Menard (1964) suggested that a convection current that was ascending under the East Pacific Rise at that time also continued beneath California and Nevada to cause uplift in the Colorado Plateau and the Basin and Range province. This uplift, in Wilson's model, ended at the Rocky Mountain Trench which, until Miocene time, would have been a large right-lateral transform fault. During the Miocene, this activity ceased farther inland but withdrew to the San Andreas fault which formed at that time and has been active ever since.

Wilson summarizes his concept: "Thus we begin to glimpse a much more complex geological history than we had previously envisioned and one in which continents have slowly been carried about breaking apart at one time and coalescing at another and all the time growing by uplift from below and by accretion at the continental margins which changed with the moving mosaic of the continents".

Wilson's hypothesis lacks the geometrical and mechanical details to explain why continents break apart and certain places and coalesce at others.

Yates' Hypothesis of Opposing Accretionary Processes.

Yates (1968) has suggested that accretionary growth along continental margins is the algebraic sum of two opposing accretionary processes: one increases the area of the continental block by outbuilding over oceanic crust; the other increases the thickness of the marginal area of the continent by underthrusting the continental waste and marine volcanic rocks that have accumulated along the margins.

The process that increases the thickness of the continental margins may operate similar to that discussed by Dietz (1966). If this thickening proceeds at a rate greater than the horizontal accretion, the continental margin shrinks in area but grows in thickness along its margin. If marginal accretion is greater, the continent grows in area. The operation of such counter processes can be used to account for the difference in distribution patterns, for example, of the elements north and south of the Trans—Idaho Discontinuity. These differences in pattern result from important differences in the tectonic history of the continent north of the discontinuity from that of the continent south of the discontinuity.

Although interesting, Yates' hypothesis fails to provide any details useful on a global scale.

Vroman's Hypothesis of Subcrustal Contraction.

In Vroman's (1968) hypothesis the main tectonic features of the Earth are explained by contraction below the crust. With special emphasis on rift-valleys, Vroman proposes that an expansion of the Earth was followed by a cracking of the crust with sufficient widening of the cracks to form the oceans. Simultaneous mantle-surface contraction left the radius of the Earth intact. Because the surface area remained constant as the oceans widened, the continents were reduced thereby causing compression and folding in continental areas.

Vroman concluded also that mantle convection is inadequate to cause all the Earth's tectonic behavior. He felt that rifting of a subcrustal layer is required to cause rifting of the crustal layer.

Vroman recommended abandonment of the rigid principle of mantle convection wherein it is postulated that (1) an ocean-ridge corresponds to an upward convection current and (2) a geosyncline corresponds to a downward current. Although opposed to the convection hypothesis as a driving mechanism for his model, he did not propose a suitable alternate driving mechanism.

Almost in exact antithesis to Vroman's subcrustal contraction hypothesis is Jordan's expansion hypothesis, described in the following section. These two hypotheses together indicate that data support both a contracting and an expanding Earth model.

325

Jordan's Expansion Model.

Jordan (1969) concluded from global analyses that (1) if there is no expansion of the Earth, then there must exist some compensating compressional feature within the tectonosphere; or (2) if there is insufficient compensation, then expansion must be inferred from sea-floor spreading. The initial Earth, according to Jordan, had a radius of only about 65% that of the present, and the entire Earth was covered by a sialic layer of spherical symmetry and uniform thickness.

In this model, elastic expansion occurred until approximately 350 m. y. ago (the Carboniferous). Post-Carboniferous expansion is attributed to segregation of dissolved material from the core into the mantle beneath the Byerly sphere.

Jordan' model, although global, lacked a suitable driving mechanism.

<u>Barnett's</u> <u>Hypothesis</u> <u>of</u> <u>Tensional</u> <u>Features</u> <u>Peripheral</u> <u>to</u> <u>Shield</u>
<u>Areas.</u>

Barnett (1969) suggested that in the early phase of Earth
history the distribution of the oceanic ridges was somewhat
reversed from what it is today, i. e., all tension rifts that
evolved into the present ridges were mid-oceanic and, because
the oceans were originally narrow, all lay "along circles around
continental shields". Similarly, today the newer ridges form
in midoceanic areas and, because the oceans are now wide, they
lie far from the continental margins. Because of continental
drift "occasioned by these new ridges", many of the older ridges
are now near or even beneath the continental margins between
which the formerly lay. No details are provided regarding the
geometry and mechanics of a suitable causal mechanism to drive
Barnett's model.

Similar models place geosynclines completely around the
Pacific and Arctic oceans (See, e. g., Churkin, 1969).

327

Beloussov's Hypothesis of Oceanization.

Beloussov's (1969) hypothesis of oceanization postulates a basification process whereby oceans form when the continents split into blocks, sink, and "dissolve" in the mantle. Basic to this hypothesis are periods of long, moderate radioactive heating that produces partial melting in the upper mantle and subsequent differentiation into "continental" type material. Specifically, Beloussov postulates that at the end of the Paleozoic or beginning of the Mesozoic, an episode of increased radioactive heating produced complete melting of ultrabasic material in some regions of the mantle. This heating, of unspecified origin, was accompanied by violent volcanic processes during which basic and ultrabasic intrusions penetrated the crust, broke into fragments, and initiated metamorphism accompanied by the expulsion of water and consequent increase in density. As a result of the density increase, blocks of the crust sank into the upper mantle, gradually melting and dissolving. As the blocks sank, they were replaced near the surface by water and basaltic extrusions.

Beloussov (1970), in providing further arguments, has suggested his oceanization process as a counter proposal to continental drift and sea-floor spreading.

An analysis of the process of oceanization leads one to ask whether there is an antithetical process of "continentalization".

328

If so, how are these two opposing processes related spatially and temporally? At what depth within the tectonosphere is this relationship effected? And what specific surficial manifestations do we have of the hypothetical continentalization?

Boldizsar's Hypothesis of Successive Episodes of Radioactive Heat Production.

Boldizsar's (1969) hypothesis suggests that 3 b. y. ago there was no continental crust, the entire Earth being covered by primitive ocean about 2.6 km deep. According to his model, the evolution of the continental crust began with the development of successive radioactive heat sources in the mantle. Boldizsar's model, however, does not specify how or why the "successive radioactive heat sources" developed nor why they formed at certain places but not at others.

The growth and emergence of continental crust was balanced by subsidence of the primitive ocean floor, i. e., the 2.6 km depth separated continental and transitional crust from oceanic crust. Erosion of silicic continental crust then carried the radioactive heat sources laterally into the oceans. Subsequent orogenic cycles caused by this heat enlarged the continents.

All of the orogenic heat of the Archean cycles, according to Boldizsar's model, has been dissipated. Some orogenic heat remains in the regions of the Paleozoic cyles. Considerable amounts remain in the Meso-Cenozoic (Alpine) regions. Since the present Quaternary orogeny may last another 100 m. y., thickening of oceanic and transitional crust of the Mediterranean and Black seas by sedimentation plus mantle differentiation may completely eliminate the gaps between Europe, Africa, and Asia.

Jacoby's Hypothesis of Crustal Regeneration on a Global Scale.

Jacoby (1970) postulates an unstable upper mantle to produce (1) magmatic diapirism under the mid-ocean ridges and (2) lithospheric subduction under the island arcs. Forces from these two processes move the lithospheric plate by overcoming the viscous drag due to the asthenosphere.

Jacoby uses order-of-magnitude estimates of forces available and of energy required to establish the plausibility of his concept on a global scale. He thereby determines the drift rate from the resistance the plate encounters at its edges, particularly beneath the island arcs where it plunges into the mesosphere. He proposes that his model be tested by studying the deep structure of the ridges, the stress field in the plates, and the gravity field over ridges and trenches.

Jacoby concedes that his model is over-simplified and that there are large uncertainties in his assumptions. His model, he admits, does not take into consideration the Earth's spherical

shape nor the complicated geometry of the individual plates.
In defense of his model, he shows that it is consistent with
present ideas of sea-floor spreading and with the assumed structure of the upper mantle.

Christensen's Hypothesis of Partial Melting.

Partial melting within the upper mantle is often cited
as one of the possible causes for the characteristics of the
asthenosphere (See, e. g., Sammis et al., 1970). If the entire
upper mantle were in such a state of incipient melting, widespread volcanism would be more prevalent and would not be limited
to certain linear belts. It follows, therefore, that some stabilization mechanism deep within the mantle must be controlling
the time and place where such incipient melting occurs.

According to Christensen's (1970) model, partial melting
of mantle peridotite under midoceanic ridges marks the initial
stages in the development of the oceanic crust. Tholeiitic
magma, which escapes to the ocean floor, forms layer 2. Layer
3, composed of amphibolite and hornblende gabbro, originates
beneath the ocean floor by crystallization of tholeiitic magma
under hydrous conditions and by subsequent metamorphism in the
vicinity of the ridge crests. Once formed, the ocean floor
is transported laterally by the horizontally spreading upper
mantle. Disposal of the ocean crust is accomplished by the

331

downward movement in the vicinity of "descending limbs of convection cells". Partial melting of the oceanic crust in the regions of downward convection produces the calc-alkaline suite of rocks and eclogite.

Christensen's model does not provide the geometrico-mechanical components of the causal mechanism for the "descending limbs of convection cells".

Earlier hypotheses based on incipient melting were discussed by Berry and Knopoff (1967), Toksöz and associates (Toksöz et al., 1967), Oxburgh and Turcotte (1968), and Ringwood (1969).

Morgan's Hypothesis of Narrow Plumes of Ascending Deep Material.

Morgan (1970) proposed a scheme of deep-mantle convection in which narrow plumes of deep material rise and spread out radially in the asthenosphere. The spreading currents produce stresses on the bottoms of the lithospheric plates, thus providing the driving mechanism for continental drift. All island chains and aseismic ridges are formed by plate motion over such "hot spots", according to Morgan.

Morgan presents three examples to support his conclusions: the Hawaiian-Emperor, Tuamotu-Line, and Austral-Gilbert-Marshall island chains. All three show a marked parallelism and can be generated by the same motion of the Pacific Plate over three fixed "hot spots".

332

Elsasser's Hypothesis of Non-Uniform Crustal Growth.

Elsasser's (1970) model postulates that the asthenosphere serves as a storage bin to support non-uniform crustal growth. Using Menard's 6 km^3/yr aggregated growth rate of ridge basalts as a lower limit for crustal growth, Elsasser found that present crustal material has cycled through the asthenosphere 6 times during the Earth's existence.

The cyclical return of the crustal material to the asthenosphere in Elsasser's model corresponds to a gradual rise of the center of gravity within the mantle of crustal-type chemical components. Thus, this model explains the ready availability of crustal material for the epeirogenetic risings observed in more recent geological times.

Other Hypotheses Regarding Continental Structure and Evolution.

Other hypotheses bearing on the orogenic-cratonic structure of the continents are those by Volvovskii and associates (Volvovskii et al., 1966), Wilson (1968), Egyed (1969), Dewey and Bird (1970), Julian (1970), MacGillavry (1970), Johnson and Smith (1970), and Gnibidenko and Shashkin (1970).

Besides these individual hypotheses, consolidated efforts have been made toward a better understanding of continental structure and evolution. Some of these have been extended projects, such as the Upper Mantle Project (See, e. g., Hart, 1969; Drake, 1970); the Deep Sea Drilling Project (See, e. g., Hammond,

1970; Anon, 1971); and the Inter-Union Commission on Geodynamics (Drake, 1970). Others were symposiums examining specific processes related to continental structure and evolution (See, e. g., Sykes, Kay, and Anderson, 1970).

Concepts of the new global tectonics have not been developed in sufficient geometrical, mechanical, and chemical detail to postulate either the origin of the continents 3.6 b. y. ago or their present orogenic-cratonic structure. However, salient applicable hypotheses of this new unifying concept are discussed later under the generally included major heading of the new global tectonics: (1) sea-floor spreading (Chapter 16); (2) continental drift (Chapter 17); and (3) plate tectonics (Chapter 18).

Causal Mechanisms for Continental Evolution and Structure.

An analysis of the above hypotheses indicates that they provide little in the way of meaningful causal mechanisms to explain the orogenic-cratonic structure and evolution of the continents during the past 3.6 b. y. Synthesis of the geology of ancient orogenic belts, such as the Appalachian-Caledonian system, suggests that a deep-seated global causal mechanism was operating long before the most recent "drift episodes". Consequently, terms such as "pre-drift configuration" of the continents are undefined both spatially and temporally.

334

Similarly, since the oldest ocean has been in existence less than half of the Phanerozoic, sea-floor spreading and related concepts cannot be considered as suitable causal mechanisms for the characteristic structure and evolution of the continents during the Archean and Proterozoic.

A closer scrutiny of modern hypotheses for continental structure elicits several questions. For example, since all continents are more-or-less similar, it is reasonable to assume that a single global causal mechanism, operating during at least the past 3.6 b. y., was responisble for, or at least contributory to, the orogenic-cratonic structure of the continents. Also, since all oceans appear to be similar, it is reasonable to assume that a single global mechanism, operating during at least the past 0.3 b. y., was responsible for, or at least contirbutory to, the creation of the present oceans.

Therefore, since the period of operation for the continental causal mechanism (3.6 b. y.) is an order of magnitude greater than that for the oceans (0.3 b. y.), the question arises as to whether the oceans have been created and destroyed approximately 10 times during the existence of the continental cratons. If so, what exactly is the spatio-temporal relationship between the continental and oceanic causal mechanisms?

With specific reference to hypotheses regarding the new global tectonics, two other major questions remain unanswered:

335

(1) why did the plates assume their observed sizes and shapes, and (2) why does the oceanic-crust-forming magma ascend along linear or quasi-linear belts.

Answers to these and related questions will be explored in the following section, where the orogenic-cratonic structure of the continents according to the tectonospheric Earth model is examined.

The Structure of the Continents as Interpreted by the Dual Primeval Planet Hypothesis.

In analyzing the orogenic-cratonic evolution of the continents as interpreted by the dual primeval planet hypothesis, the cratons are identified with the 8 geoblocks of the tectonospheric Earth model; the orogens, with the three mutually orthogonal wedge-belts of activity. Fig. 7-7-0-1 shows the surficial traces of the 8 geoblocks and of the three wedge-belts of activity associated with them. The eight points of the model are marked: AL, Aleutians; BV, Bouvet; GA, Galapagos; GI, Gibraltar; BN, Bengal; and KR, Kermadecs. In this projection, the Antarctic geoblock (BV-KR-BN) is hidden from view.

The surficial trace of each geoblock is an equilateral (right) triangle with sides of 10,000 km. Each wedge-belt of activity is roughly centered by a great circle about 40,000 km long on the surface of the Earth. The apexes of each wedge-belt of activity lie roughly along a great circle approximately 34,000 km long on the surface of the sub-tectonosphere, 1000 km beneath the surface of the Earth.

To facilitate association of the features of the model with the actual features of the Earth, Table 7-7-0-1 lists the 8 geoblocks of the model together with the corresponding shields of the actual Earth. According to the model, each of these shields has evolved in roughly the same manner during the past

337

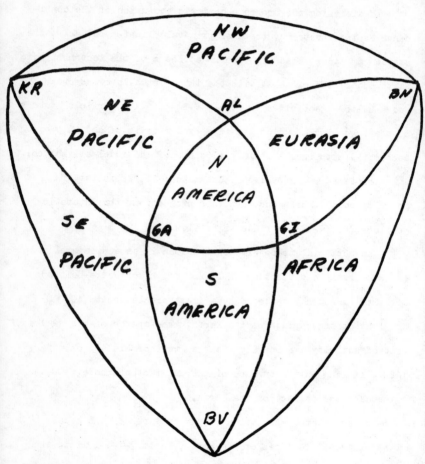

Figure 7-7-0-1. Schematic sketch of the Earth showing the surficial traces of the 8 geoblocks of the model and the wedge-belts of activity associated with them. See text for details.

3.6 b. y. and in accordance with the integrated global mechanism described in the next section.

TABLE 7-7-0-1. The 8 geoblocks of the tectonospheric Earth model and the Precambrian shields identified therewith. The three hypothetical submerged "shields" shown in parentheses are discussed in Chapter 15.

TEM Geoblocks	Corresponding Shields
Al-Ga-Gi	North America
Al-Gi-Bn	Eurasia
Al-Bn-Kr	(NW Pacific)
Al-Kr-Ga	(NE Pacific)
Bv-Ga-Gi	South America
Bv-Gi-Bn	Africa
Bv-Bn-Kr	* Antarctica
Bv-Kr-Ga	(SE Pacific)

* Hidden from view, on back of Fig. 7-7-0-1.

<u>The</u> <u>Evolution</u> <u>of</u> the <u>Orogens</u> <u>as</u> <u>Interpreted</u> <u>by</u> the <u>Tectonospheric</u>
<u>Earth</u> Model.

The most essential element in the evolution of the orogenic-
cratonic structure of the continents was the development of
the orogens. These originated, according to the tectonospheric
Earth model, about 4.6 b. y. ago when the Earth was formed.

The orogens, from a geometrico-mechanical standpoint, may
be analyzed in cross-section by studying the behavior of relative tension and compression within two areas of an idealized
wedge-belt of activity. The basic elements of this behavior
are shown in Fig. 7-7-0-2. The geanticlines A and the geosynclines S, according to the model, are formed by the dynamic
reaction arising within the tectonosphere from the conditions
of relative compression K and of relative tension T existing
along the left and right faces of the idealized wedge-belt of
activity LOR.

Basic to the formation and development of the geanticlines
and geosynclines is the behavior of subwedges A and S as the
relative pressure changes in the left and right faces of the
wedge-belt of activity LOR. Under conditions of relative compression along the face LO, subwedge A is "wedged" upward to
produce a geanticline. Simultaneous with the formation of the
geanticlines, the geosyncline forms. Relative tension existing
along the face RO permits subwedge S to drop down into the "sink"

340

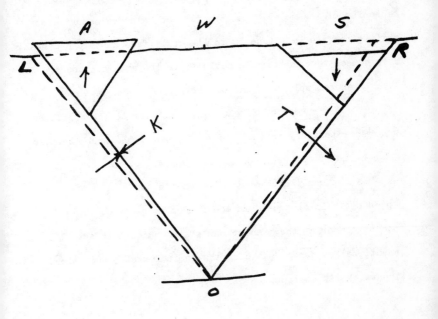

Figure 7-7-0-2. The evolution of geanticlines, A, and geosynclines, S, by conditions of relative compression, K, and of
relative tension, T, along the left and right faces of an idealized
wedge-belt of activity. See text for details.

and thereby to provide the basis for the formation of the geo-syncline.

With the passing of time and in response to the basic driving mechanism of the model, the condition of relative compression shifts from the left face LO to the right face RO of the system. This causes tectonomagmatic activity in the area of subwedge S. Later, general uplift at the center W produces mountains.

This 3-step cycle of behavior, according to the model, has occurred repeatedly in the active regions of the Earth's three wedge-belts of activity during the past 4.6 b. y.

How the continents evolved under the conditions of this 3-step cycle is discussed in the following section. This is done by comparing the predictions of the model with actual ob-servational evidence regarding the evolution of one of the con-tinents, North America.

<u>The Evolution of the North American Continent as Interpreted</u>
<u>by the Tectonospheric Earth Model</u>.

The evolution of a continent, according to the tectonospheric
Earth model, involves one geoblock and one ninety-degree segment of each of the three wedge-belts of activity. In the case
of North America, the geoblock is that identified by the points
Aleutians, Galapagos, and Gibraltar on Fig. 7-7-0-1. The three
associated $90°$ wedge-belt segments are those lying peripheral
to the North American geoblock: (1) Aleutians-Galapagos, (2)
Galapagos-Gibraltar, and (3) Gibraltar-Aleutians. See Fig.
7-7-2-1.

The Canadian Shield, according to the model, has remained
within the confines of the quasi-rigid equilateral (right) triangle defined by the 3 ninety-degree wedge-belt segments during
most of the past 3.6 b. y. The three wedge-belt segments, during
this time, have provided the tectonomagmatic activity that has
successively activated the various Precambrian and Phanerozoic
orogenic belts that have occurred along the peripheries of the
central craton during at least the past 3.6 b. y.

It is difficult to determine exactly when the first nuclei
of the central craton formed or where they came from. According
to the model, this occurred sometime during the first 1 or 2
b. y. of the Earth's existence, i. e., during the period 4.6
to 2.6 b. y. ago. There are two possible sources for the material

343

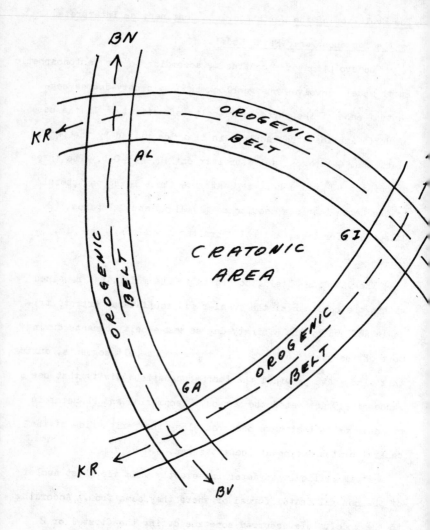

Figure 7-7-2-1. Schematic diagram of Aleutians-Galapagos-Gibralta
geoblock and associated segments of wedge-belts of activity that f
the basis for the evolution of the North American continent. See
for details.

344

that formed the initial cratonic nuclei: (1) granitic-type material accreted onto Primordial Earth from the fragments of Earth Prime; and (2) material produced by the tectonomagmatic activity of the 3 associated wedge-belt segments.

In either case, the oldest platforms, as interpreted by the model, lie above the approximate centers of the geoblocks. Observational evidence shows that, beginning with the Middle Proterozoic, some small quasi-stable platforms had developed. These grew and stabilized as the wedge-belts of activity added new material to both the bottoms and sides of the embryonic platforms. In time, the central cratons tended to act as giant "limit stops" that restricted the wedge-belt activity to the horizontal peripheries of the cratons.

As the central cratons grew in size from this continual accretion to their peripheries, some became so large that they were designated super-continents, which split apart when they became juxtaposed above a particularly active wedge-belt. One such example is that of South America and Africa, shown in Fig. 3-8-1. Another example is the breakup of Pangaea. In other cases, two platforms caught within a system of wedge-belts were driven toward each other and welded together (e. g., the East European and Siberian platforms).

The various spatio-temporal distributions of the North American orogenic belts with respect to the central craton may

345

be plotted from a study of the ages of the rocks involved. See Fig. 7-7-2-2 for such distributions during the past few b. y. Immediately peripheral to the central craton in this schematic sketch are the oldest identifiable orogenic belts. Farther out and roughly parallel to them are the more recent orogenic belts: Innuitian, Cordilleran, and Appalachian.

A closer scrutiny of the geology of North America permits the correlation of specific regional predictions of the model with actual observational data. One such correlation involves the older Appalachians of New England as they existed prior to the Taconic orogeny. A restored section of the Cambrian and Ordovician rocks of this region is shown in Fig. 7-7-2-3. Also shown is the predicted geometry of the wedge-belt of activity that activated the Taconic orogeny, including the relative positions of the geosynclines and the geanticline within this wedge-belt at that time. The cross-section shown is part of the Ordovician position of the Galapagos-Gibraltar wedge-belt segment.

Another correlation between specific predictions of the model and actual observational data is provided by the major tectonic elements of the ancestral Rocky Mountains, i. e., the Uncompaghre and Front Range uplifts separated by the Colorado Trough during the late Paleozoic. See Fig. 7-7-2-4 for a restored section of this region, together with the basic elements

346

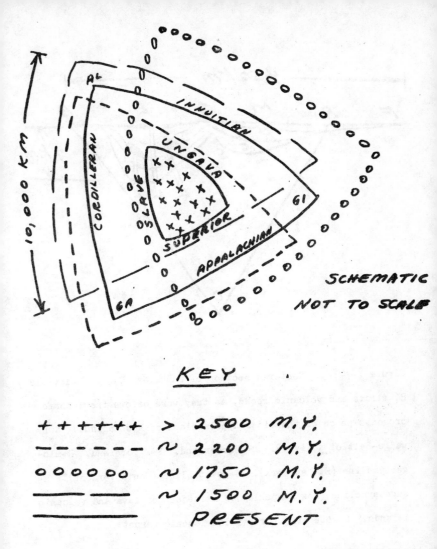

KEY

+ + + + + + > 2500 M.Y.
- - - - - - ~ 2200 M.Y.
oooooo ~ 1750 M.Y.
———— —— ~ 1500 M.Y.
———————— PRESENT

Figure 7-7-2-2. Various spatio-temporal positions of the North American orogenic belts with respect to the central craton during the past 2.5 b.

347

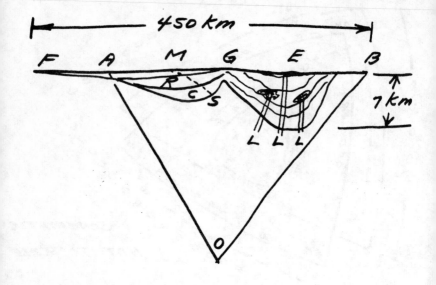

Figure 7-7-2-3. Restored section of Cambrian, C, and Ordovician, R, strata and volcanic rocks, as they were before the Taconic orogenic phase of the older Appalachians of New England. AOB= wedge-belt of activity. B= borderland. F= foreland. M=mio-geosyncline (E. New York). G=geanticline (South Vermont). E= eugeosyncline (New Hampshire). L=feeders of lava and volcanic islands. S=sole of U. Ordovician Taconic thrust.

348

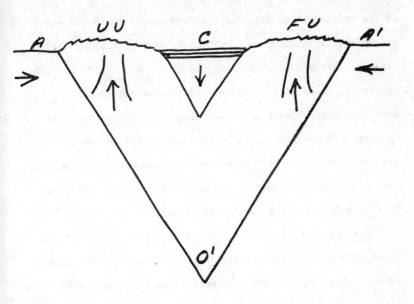

Figure 7-7-2-4. Major tectonic elements in the ancestral Rocky Mountains (late Paleozoic), looking northwesterly along the right edge of the Galapagos–Aleutians wedge-belt of activity. AO'A' = subsidiary wedge-belt of activity. UU = Uncompaghre uplift. FU = Front Range uplift. C = Colorado trough. Section AA' extends from about 38°N, 108°W to about 40°N, 105°W.

of the wedge-belt of activity involved according to the model.
The section shown is that of the right edge of the Galapagos-
Aleutians wedge-belt of activity as it existed during the late
Paleozoic. Many of the present northwest-trending uplifts later
developed along the axis of the Colorado Trough as the area C
reversed its downward trend to an upward trend in response to
the behavior of the Galapagos-Aleutians wedge-belt of activity
during the late Paleozoic. The Laramide intrusions followed
this same linear trend during the Cenozoic.

A third correlation between specific predictions of the
model and actual observational data is provided by the post-
Paleozoic behavior of the Galapagos-Gibraltar wedge-belt. This
behavior manifested itself in the post-Paleozoic activity of
the Gulf Coast geosyncline. The Paleozoic-to-Pleistocene strata
of this system are shown in Fig. 7-7-2-5, which also shows the
geosynclinorium, AO'A'; the geosyncline, G; and the anticlino-
rium, U. The miogeosynclinorium is off the figure to the right.
Intrusive salt is indicated by S.

A more generalized representation of Mesozoic North America,
according to the model, includes both geosynclines and geanticlines as shown in Fig. 7-7-2-6. In this representation, the
sections RR' and PP' are quite similar to the sections predicted
by the idealized wedge-belt shown in Fig. 7-7-0-2, with cross-
sectional width of 1000 pi/2, or about 1570 km. According to

350

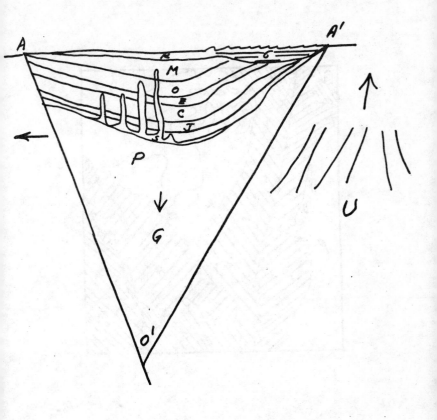

Figure 7-7-2-5. Post Paleozoic orogenic behavior of the Gulf Coast geosyncline, looking northeasterly along the left edge of the Galapagos-Gibraltar wedge-belt of activity. Approximate scale: 1" ≐ 225 km horizontally; 10 km vertically. See text for details.

351

Figure 7-7-2-6. North American geosynclines, G, and geanticlines, A, during the late Mesozoic. See text for details.

the model, the Rocky Mountain (GA-AL) and Appalachian (GA-GI) wedge-belts of activity intersected each other in northern Mexico during the late Mesozoic. This point of intersection has migrated in a southeasterly direction since the Mesozoic and is now in the vicinity of the Galapagos point of the model.

The Pleistocene orogenic-cratonic structure of the Basin and Range region of the U. S. A. provides a more-recent example of correlation between the predictions of the model and observational data for that area. This structure is shown schematically in Fig. 7-7-2-7, which is a transverse cross-section made along the right edge of the Galapagos-Aleutians wedge-belt of activity. Also shown are (1) the subsidiary wedge-belt of activity, (2) the uplifts upon which the Pleistocene glaciations formed, and (3) the ancestral lakes Lahontan and Bonneville.

Other details regarding the orogenic-cratonic structure and evolution of North America during the past 3.6 b. y. may be analyzed in a similar manner by comparing applicable spatio-temporal predictions of the model with corresponding observational data for the areas being analyzed. The same type of analyses, also, may be used for studying the orogenic-cratonic structure and evolution of the other continents during the past 3.6 b. y.

353

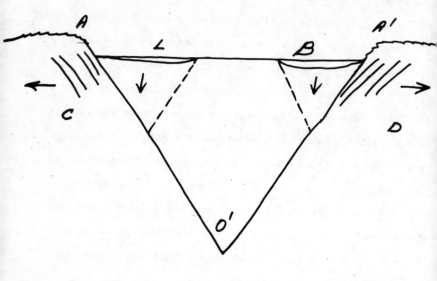

Figure 7-7-2-7. Pleistocene orogenic-cratonic structure of the Basin and Range region of southwestern U. S. A., looking northwesterly along the right edge of the Galapagos-Aleutians wedge-belt of activity. AOA' = subsidiary wedge-belt of activity. A and A' = uplifts upon which the Pleistocene glaciation formed. L = Lake Lahontan. B = Lake Bonneville.

Anon., 1971. News notes. Geotimes, 16(1): 27 and 30.

Arkhangelsky, A. D., 1941. Geological Structure and Geological History of the U. S. S. R. Gostoptekhizdat, Moscow, 376 pp.

Barnett, C. H., 1969. Oceanic rises in relation to the expanding Earth. Nature, 221 (5185): 1043-1044.

Beloussov, V. V., 1962. Basic Problems in Geotectonics. McGraw-Hill, New York, 816 pp.

Beloussov, V. V., 1969. Interrelations between the Earth's crust and upper mantle. In: P. J. Hart (Editor), The Earth's Crust and Upper Mantle. Am. Geophys. Union, Washington, pp. 698-712.

Beloussov, V. V., 1970. Against the hypothesis of ocean-floor spreading. Tectonophysics, 9: 489-511.

Benes, K. (Editor), 1968. Upper Mantle (Geological Processes). Academia, Prague, 260 pp.

Berry, M. J. and Knopoff, L., 1967. Structure of upper mantle beneath western Mediterranean basin. J. Geophys. Res., 72: 3613-3626.

Bertrand, M., 1887. La chaine des Alpes et la formation du continent europeen. Bull. Soc. Geol. France, 15 (3): 423-477.

Boldizsár, T., 1969. Terrestrial heat flow and Alpine orogenesis. Bull. Volcanol., 33 (1): 293-297.

Bondarchuk, V. G., 1944. Geomorphology in geosynclines. *Izv. Acad. Sci. U. S. S. R., Geol. Ser.*, 1: 107-112.

Born, A., 1933. Uber Werden und Zerfall von Kontinental-schollen. *Forsch. Geol. Paleontol.*, 10 (32): 348.

Bucher, W. H., 1950. Megatectonics and geophysics. *Trans. Am. Geophys. Union*, 31: 495-507.

Christensen, N. I., 1970. Composition and evolution of oceanic crust. *Marine Geology*, 8 (2): 139-154.

Churkin, M., Jr., 1969. Paleozoic tectonic history of the Arctic basin north of Alaska. *Science*, 165 (3893): 549-555.

Dana, J. D., 1873. On some results of the Earth's contraction from cooling, including a discussion of the origin of mountains and the nature of the Earth's interior. *Am. J. Sci.*, Ser. 3, 5: 423-443; 6: 6-14, 104-115, 161-172.

Dewey, J. F. and Bird, J. M., 1970. Mountain belts and the new global tectonics. *J. Geophys. Res.*, 75: 2625-2647.

Dietz, R. S., 1966. Passive continents, spreading seafloors, and collapsing continental rises. *Am. J. Science*, 264 (3): 177-193.

Dietz, R. S. and Holden, J. S., 1970. East Indian basin (Wharton basin) as pre-Mesozoic ocean crust. *Geol. Soc. Am. Abs. w. Programs*, 2: 537.

Douglas, R. J. W., 1969. Orogeny, basement, and geological map of Canada. *Geol. Soc. Can. Spec. Paper* 5, pp. 1-6.

356

Drake, C. L., 1970. A long-range program of solid Earth studies. Trans. Am. Geophys. Union, 51: 152-159.

Dunbar, C. O., 1966. The Earth. World, New York, 252 pp.

Egyed, L., 1969. Physik der festen Erde. Akademiai Kiado, Budapest, 368 pp.

Elsasser, W. M., 1970. Non-uniformity of crustal growth. Trans. Am. Geophys. Union., 51: 823.

Gilluly, J., 1955. Geologic contrasts between continents and crustal structure. In: A. Poldervaart (Editor), Crust of the Earth. Geol. Soc. Am., pp. 7-18.

Glikson, A. Y., 1970. Geosynclinal evolution and geochemical affinities of early Precambrian systems. Tectonophysics, 9: 397-433.

Gnibidenko, H. S. and Shashkin, K. S., 1970. Basic principles of the geosynclinal theory. Tectonophysics, 9: 5-13.

Goodwin, A. M., 1968. Archean protocontinental growth and early crustal history of the Canadian shield. In: K. Benes (Editor), Upper Mantle (Geological Processes). Academia, Prague, pp. 69-89.

Hammond, A. L., 1970. Deep Sea Drilling: A giant step in geological research. Science, 170: 520-521.

Hart, P. J. (Editor), 1969. The Earth's Crust and Upper Mantle. Am. Geophys. Union, Washington, 735 pp.

Hess, H. H., 1955. Serpentines, orogeny, and epeirogeny.

357

In: A. Poldervaart (Editor), The Crust of the Earth. Geol. Soc. Am. Spec. Paper 62, New York, pp. 391-407.

Hurley, P. M., 1969. Some observations on the geological history of Laurasia. Geol. Soc. Am. Abs. w. Programs, 7: 112.

Jacoby, W. R., 1970. Instability in the upper mantle and global plate movements. J. Geophys. Res., 75: 5671-5680.

Johnson, H. and Smith, B. L. (Editors), 1970. The Megatectonics of Continents and Oceans. Rutgers, New Brunswick (N. J.) 284 pp.

Jordan, P., 1969. On the possibility of avoiding Ramsey's hypothesis in formulating a theory of Earth expansion. In: S. K. Runcorn (Editor), The Application of Modern Physics to the Earth and Planetary Interiors. Wiley-Interscience, London and New York, pp. 55-63.

Julian, B. R., 1970. Regional variations in upper mantle structure in North America. Trans. Am. Geophys. Union, 51: 359.

King, L. C., 1962. The Morphology of the Earth. Hafner, New York, 699 pp.

Kober, L., 1942. Tectonische Geologie. Borntraeger, Berlin, 492 pp.

Kraus, E., 1928. Das Wachstum der Kontinente nach der Zyklustheorie. Geol. Rundschau, 19: 353-481.

MacGillavry, H. J., 1970. Turbidite detritus and geosyncline history. _Tectonophysics_, _9_: 365-393.

Martin, H., 1969. Problems of age relations and structure in some metamorphic belts of southern Africa. _Geol. Assoc._ _Can._ _Spec._ _Paper 5_, pp 17-26.

McBirney, A. R., 1970. Cenozoic igneous events of the Circum-Pacific. _Geol._ _Soc._ _Am._ _Abs._ _w._ _Programs_, _2_: 748-751.

Menard, H. W., 1964. _Marine Geology of the Pacific_. Mc Graw-Hill, New York, 271 pp.

Misik, M. (Editor), 1968. _Orogenic Belts._ Academia, Prague, 327 pp.

Morgan, W. J., 1970. Plate motions and deep mantle convection. _Trans._ _Am._ _Geophys._ _Union_, _51_: 822.

Ocola, L. C. and Meyer, R. P., 1970. Regional upper crustal structure of mid-continent of U. S. A. _Geol._ _Soc._ _Am._ _Abs._ _w._ _Programs_, _2_: 638.

Oxburgh, E. R. and Turcotte, D. L., 1968. Mid-ocean ridges and geotherm distribution during mantle convection. _J._ _Geophys._ _Res._, _73_: 2643-2661.

Pavlovsky, E. V., 1953. On some general regularities of the Earth's crust development. _Izv._ _Acad._ _Sci._ _U._ _S._ _S._ _R._, _Geol._ _Ser._ _5_: 82-89.

Ringwood, A. E., 1969. Composition and evolution of the upper mantle. In: P. J. Hart (Editor), _The Earth's Crust and Upper Mantle_. Am. Geophys. Union, Washington, pp. 1-17.

Runcorn, S. K. (Editor), 1962. Continental Drift. Academic, New York, 338 pp.

Salop, L. I. and Scheinmann, Y. M., 1969. Tectonic history and structures of platforms and shields. Tectonophysics, 7: 565–597.

Sammis, C., Jordan, T., and Anderson, D. L., 1970. Inhomogeneity in the upper mantle. Trans. Am. Geophys. Union, 51: 828.

Shatsky, N. S., 1946. The Wegener hypothesis of geosynclines. Izv. Acad. Sci. U. S. S. R., Geol. Ser. 4: 7–21.

Stille, H., 1940. Einführung in den Bau Nordamerikas. Borntraeger Verlagsbuchhandlung, Berlin, 717 pp.

Stille, H., 1944. Geotektonische Gliederung der Erdsgeschichte. Abh. Preuss. Akad. Wiss., Math.-Nat. Klasse, 3, 80 pp.

Suess, E., 1909. The Face of the Earth. Clarendon, Oxford, 2 vols., 400 and 673 pp.

Sykes, L. R., Kay, R., and Anderson, D. L., 1970. Mechanical properties and processes in the mantle. Trans. Am. Geophys. Union, 51: 874–879.

Toksöz, M. N., Chinnery, M. A., and Anderson, D. L., 1967. Inhomogeneities in the Earth's mantle. Geophys. J., 13: 31–59.

Van Bemmelen, R. W., 1966. On mega-undulations: A new model for the mantle's evolution. Tectonophysics, 3(2): 83–127.

Vassilkovsky, N. P., 1960. Study of geosynclines in the light of modern geology. Sibirsk. Nauchn. Issled. Inst. Geol.,

Geofiz., i Mineral. Syrya, 13: 5-56.

Volvovskii, I. S., Garetskii, R. G., Shlezinger, A. E., and Shraibman, V. I., 1966. Trudy, 165, 288 pp.

Vroman, A. J., 1968. The main tectonic features of the Earth explained by contraction below the crust. In: K. Benes (Editor), Upper Mantle (Geological Processes). Academia, Prague, pp. 19-29.

Wilson, H. D. B., 1966. Alkalic and alkaline ring complexes in the Archean. Can. Geophys. Bull. 19: 182.

Wilson, J. T., 1957. Origin of the Earth's crust. Nature, 179 (4553): 228-229.

Wilson, J. T., 1967. Theories of building of continents. In: T. F. Gaskell (Editor), The Earth's Mantle. Academic, London and New York, pp. 445-473.

Wilson, J. T., 1968. Static or mobile Earth: The current scientific revolution. Proc. Am. Phil. Soc., 112: 309-320.

Yates, R. G., 1968. The Trans-Idaho Discontinuity. In: K. Benes (Editor), Upper Mantle (Geological Processes). Academia, Prague, pp. 117-123.

Chapter 8

THE EARTH'S DEEP SEISMICITY

A study of the evolution of the Earth's tectonosphere should include a brief analysis of the Earth's seismicity, because there appears to be a relationship between the Earth's seismicity and the dynamics of the tectonosphere. Since both appear to be deep-seated, a close scrutiny of the Earth's deep seismicity seems particularly important.

Although the exact relationship between seismic activity and tectonospheric dynamics is not understood, the observed correlation between the two is such that any hypothesis for the evolution of the tectonosphere must account for those areas of seismic behavior that correlate with the dynamics of the tectonosphere. For example, strong tectonic activity is concentrated into a relatively few narrow mobile belts, aggregating about 60,000 km (a distance equivalent to about 1 1/2 global circumferences). Almost all present seismic and volcanic activity occurs within those segments of the mobile belts that comprise the oceanic and continental rift systems, the island-arc complexes, and the young fold belts.

These observations suggest that the narrow active belts are due to interactions between a relatively small number of very large blocks. Not understood is the cause of the concentration of material into such large blocks and the resulting localization of tectonic, magmatic, seismic, and metamorphic activity into long narrow belts that lie between the hypothesized blocks. Although the cause for such concentration is not understood, it suggests that seismic belts are part of a single global system.

On the assumption, then, that seismicity is a measure of present-day tectonospheric behavior, the following sections examine some recent observations bearing on global and regional patterns of seismicity in the hope that such might give an indication of the causal relationship between seismic activity and tectonospheric behavior.

Fig. 8-1-1 shows a worldwide distribution of earthquakes. A study of the data upon which Fig. 8-1-1 is based (Barazangi and Dorman, 1969; ESSA, 1970) shows that the deeper earthquakes are restricted to narrow more-or-less central portions of those belts that contain earthquakes of all depths. Where deep earthquakes do not occur, seismic activity is concentrated primarily along the edges of lithospheric plates.

Global characteristics of the Earth's seismicity may be considered in terms of tectonospheric structure, correlations of seismic activity with tectonomagmatic activity, and seismic behavior at various depths.

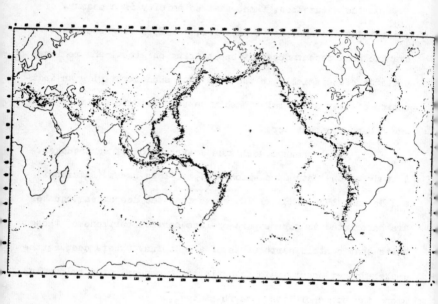

Figure 8-1-1. Worldwide distribution of earthquakes. (Based on Barazangi and Dorman, 1969; ESSA, 1970).

364

Seismic Behavior as a Function of Tectonospheric Structure.

Since seismic behavior appears to correlate with tectono-spheric dynamics, this suggests that the former is also related to tectonospheric structure. Consequently, it is well to review briefly evidence indicating that seismic behavior is a function of this structure.

Observations suggest that, in both oceanic and continental areas, both the crust and the mantle are thinly stratified (Kosminskaya and Zverev, 1969). Other studies indicate that, in addition to such vertical variations in structure, the tectonosphere contains horizontal variations in both the crust and upper mantle (See, e. g., Ringwood, 1962a, 1962b; Cook, 1962; Clark and Ringwood, 1964; Hart, 1969).

Although the oceanic crust is thinner than the continental crust and differs in other respects, such differences do not extend into the upper mantle. Consequently, the upper mantle is approximately the same under both oceanic and continental areas (Dorman, 1969). Wang (1970) concluded that lower seismic velocities beneath tectonically active areas are caused by differences in degree of melting, rather than from lateral differences in temperature, density, or composition; but his analysis did not reveal the cause for the differences in degree of melting under different parts of the globe.

From these and other analyses (See, e. g., Bolt and Nuttli, 1966; Cleary and Hales, 1966; Hales and Doyle, 1967; Toksöz and Arkani-Hamed, 1967; Herrin and Taggart, 1968; Alterman, 1969; Raitt, 1969; Julian, 1970; Vanek, 1969; Drake and Nafe, 1969; Miyamura, 1969; Ritsema, 1969), it is clear that seismic behavior is definitely a function of tectonospheric structure.

Correlation of Seismic Behavior with Tectonomagmatic Activity.

It has long been observed that seismic behavior correlates generally with certain tectonomagmatic activity when viewed on a global scale (See, e. g., Gutenberg and Richter, 1954; Holmes, 1965; Hart, 1969; Drake, 1970). In certain areas of the Earth, more specific correlations have been observed between (1) heat flow and seismic delay time (Horai, 1969); (2) alignments of seismic epicenters, gravity trends, and volcanoes (Danes, 1970); and (3) deep earthquakes, shallow earthquakes, and volcanism occurring in common structures (Latter, 1969).

In spite of these correlations, there are distinct cases of non-correlation. Some segments of mid-ocean ridges are tectonically active but seismically inactive. Some volcanic segments are aseismic. Some highly seismic segments are not volcanic. Other highly seismic segments showing no tectonomagmatic activity (at least not on the Earth's surface) can be explained by seismic stress release and fracture at depth, without the occurrence of surficial fracture. But other non-correlations of seismic activity with tectonomagmatic activity are not so easily explained.

From the above, it may be seen that any hypothesis regarding the Earth's seismic behavior must explain cases of both correlation and non-correlation of seismic behavior with tectonomagmatic activity.

Seismic Behavior at Various Depths within the Tectonosphere.

In analyzing the Earth's seismic behavior, it is necessary to consider a complete spectrum of earthquake depths, from the surface of the Earth to over 700 km into the tectonosphere. The problem is complicated by differences in seismic behavior at various depths. For example, the upper mantle appears to attenuate S waves more than does the lower mantle; the difference amounts to an order of magnitude (Knopoff, 1969). The upper mantle also can be separated into several layers, the lower layers having lower values of Q. Average values of Q from Knopoff's model are 120, from 0 to 325 km; 75, from 325 to 650 km; etc.

A study of the Earth's seismic behavior reveals that some shallow earthquakes are produced by the same mechanism that produces deep earthquakes. Those shallow earthquakes that do not have a common motivation with the deep earthquakes are probably produced by surficial activity such as (1) interactions between lithospheric plates, (2) gravitational sliding within the lithosphere, (3) volcanic activity, and (4) isostatic adjustments and certain other second-order activity near the surface of the Earth.

Not completely understood is the motivation for those shallow earthquakes that seem to share a causal mechanism with the deep earthquakes. Much work remains to be done in this area. The task is made more difficult by the paucity of deep earthquakes and by the inapplicability of most laboratory "laws" at depths of 700 km. Certain observations, if considered on a global basis, may, however, provide at least tentative answers. These are considered in the subsequent section on hypotheses. Suffice it here to say merely that deep-focus earthquakes are fewer in total number than shallower ones; they are more limited in geographical distribution; and they have more complex energy distributions.

Regional Characteristics of the Earth's Seismicity.

Certain regions of the Earth show distinct seismic patterns. The two most conspicuous of these are the Circum-Pacific belt and the Trans-Eurasian (or Tethyan) belt, lying orthogonal to to each other. Segments of these belts are approximately great-circular arcs: 180° for the Trans-Eurasian belt; 270° for the Circum-Pacific, as shown in Fig. 8-1d-1.

The Trans-Eurasian seismic belt, as usually defined, extends approximately along most of the great-circular segment EAB, aggregating about 180°. The Circum-Pacific seismic belt, as usually defined, extends approximately along the great-circular segment A'ACDF', aggregating roughly 270°. The two belts are usually described as intersecting in the vicinity of A. The Circum-Pacific seismic belt accounts for about 75% of the earth-quakes that have occurred within recorded history; the Trans-Eurasian (Tethyan) belt accounts for all except about 2% of the remainder.

Fig. 8-1d-2 shows the same information as Fig. 8-1d-1 but uses a Mercator projection instead of the Mollweide.

Other general characteristics of the Circum-Pacific and Trans-Eurasian seismic belts have been described by Bullen, 1963, 1967; Gaskell, 1967; Gutenberg, 1959, 1960; Gutenberg and Richter, 1954; Holmes, 1965, Richter, 1958; Steinhart and Smith, 1966; and other standard texts.

Specific portions of these two seismic belts have been

370

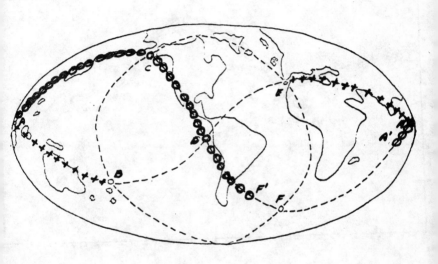

Figure 8-1d-1. The Circum-Pacific and Trans-Eurasian (Tethyan) seismic belts superimposed on a Mollweide projection of the Earth.

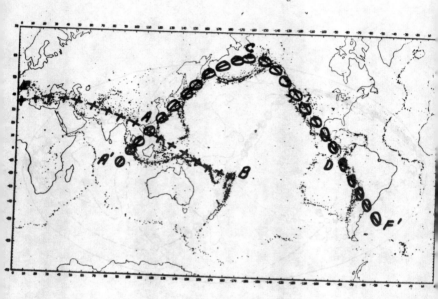

Figure 8-1d-2. The Circum-Pacific and Trans-Eurasian (Tethyan)
seismic belts superimposed on a Mercator projection of the Earth.
(Based on Barazangi and Dorman, 1969; ESSA, 1970).

studied to determine their nature and, possibly, the cause of their unique characteristics (See, e. g., Sutton et al., 1967; Anderson, 1968; Isacks et al., 1968; Brune, 1968; Saavarensky and Golubeva, 1969; Solomon and Toksöz, 1970; Molnar and Oliver, 1969; McGinley and Anderson, 1969; Sykes et al., 1969; Dorman, 1969; Wellman, 1969; Northrop et al., 1970; Danes, 1970).

A close scrutiny of these two belts, which account for almost all of the Earth's present seismicity, should provide many details regarding tectonospheric dynamics. Any analysis of these two belts should note the relatively short-lived existence of the geographical features associated with them. Thus, since neither the Pacific Ocean nor the Tethyan zone have been in existence over a long period of time, one would suspect that the two seismic belts presently associated with these geographical features are also relatively short-lived.

Confirmation of the suspected short-lived existence of these and other regional seismic patterns may be found in the relationships existing between the Earth's seismic behavior and other geophenomena. It has long been suspected, for example, that the Earth's seismic behavior correlates with certain tectonomagmatic activity and that both are related to the deep structure and behavior of the Earth on a global scale (See, e. g., Hart, 1969; Drake, 1970). Most evidence indicates that tectonomagmatic activity has occurred in a cyclically aperiodic manner in various parts of the Earth during at least a major portion of the past 3.6 b. y. 373

If the Earth's seismic behavior has always been associated with its tectonomagmatic activity, both could easily be attributable to a single, deep-seated global mechanism that has existed during a considerable portion of the Earth's life. The section on hypotheses discusses this in greater detail.

Low-Velocity Zones, High-Velocity Areas, and Other Seismic Anomalies.

In considering the Earth's seismic behavior and its relationship to the evolution of the Earth's tectonosphere, it is well to include a brief analysis of the salient seismic anomalies within the tectonosphere. Most of these anomalies are related to seismic velocity gradients and are usually referred to as "low velocity zones", "wave guides", "high velocity areas", etc. Generally, low-velocity zones correlate with areas that exhibit tectonomagmatic activity; high-velocity areas, with stable areas. Thus, low-velocity anomalies are usually associated with high-temperature, high-attenuation environments, whereas high-velocity anomalies are usually associated with low-temperature, low-attenuation environments within the tectonosphere. A study of seismic velocity anomalies, therefore, provides a means of evaluating some of the "fine structure" characteristics within the tectonosphere.

The existence of low-velocity zones within the tectonosphere has been confirmed for both P and S waves in most tectonic configurations (See, e. g., Beloussov, 1969; Anderson, 1967; Tryggvason, 1964; Fedotov et al., 1964; Pakiser, 1963). Furthermore, the most probable explanation for the low-velocity zone is incipient melting of tectonospheric material (perhaps eclogite or peridote) in the presence of water (See, e. g., Lambert and Wyllie, 1970; Kushiro et al., 1968; Ringwood, 1969; Hill and Boettcher, 1970).

The exact configuration of the low-velocity zone varies with the tectonic environment and probably coincides with a layer of reduced viscosity (Archambeau et al., 1969; Turcotte and Oxburgh, 1969). Some evidence indicates that the low-velocity zone may consist of several layers, one above the other (Beloussov, 1969).

Simple explanations for the cause of the melting are not available (Ringwood, 1969); but it appears that melting occurs where the geotherm rises above the dry solidus (ibid.; Lambert and Wyllie, 1970). Unexplained is the limited vertical extent of the zone from 100 to 250 km under continents; from 50 to 400 km under oceans. Also, the low-velocity zone may not be produced by the effects of high temperature alone (Ringwood, 1969).

In some areas, the low-veolcity zone lies below a discontinuity. For example, between Tampa, Florida, and San Louis Potosi, Mexico, an apparent velocity of 8.7 km/sec was found beneath the deep ocean, suggesting a discontinuity or a rapid velocity change at a depth of about 60 km. The low-velocity layer, if it exists along this profile, must lie beneath the discontinuity, since the crustal velocity in Mexico is 6.0 km/sec and is 7.8 km/sec beneath the M discontinuity. The depth to the M discontinuity there is 29 km beneath the coastal plain and 35 km beneath the plateau (Hales et al., 1970).

In other regions, such as the Canadian shield, there is

no low-velocity zone for P waves but only an interval of depth where such velocities have little or no gradient (Magnitsky and Zharkov, 1969). Other areas show variations of the intensity or gradient in velocity decrease.

Seismic attenuation appears to correlate with low-velocity regions. Values are about 3 times higher in the semi-stable regions of the U. S. than in the Basin and Range province. Also, regions of low velocity are often areas of high heat flow, particularly on mid-ocean ridges and in analogous continental regions such as the Basin and Range province, where heat-flow values are well above average.

There appears to be a correlation between regional variations in crustal thickness and the Earth's stress field (See, e. g., McBirney, 1969). Basically, the crust appears thinner than average beneath regions in tension (e. g., the Basin and Range province, "active" regions in both continental and oceanic areas, and mid-ocean ridges). On the other hand, it appears thicker than average under regions in compression (e. g., continental shields, ocean basins, the low-density "roots" of fold mountains, and the crust beneath islands in island-arc complexes). Under the Kurile volcanic island arc, a projection of the low-velocity zone reaches the base of the crust; lava flows through this projection to the volcanoes (Beloussov, 1969).

Liebermann and Schreiber (1969) found evidence that a low-velocity layer for S waves can exist without one for P waves and that shear velocity may decrease with depth in the transition region of the mantle.

The tectonosphere displays a myriad of heterogeneities, and wave behavior within heterogeneous material may differ considerably from that within a homogeneous, idealized, non-statified material without pressure and temperature gradients. For example, one phenomenon that does not appear in homogeneous media is the interaction of P and S waves, such as is observed within heterogeneous material even in the absence of boundaries of discontinuities. Among other things, there is a conversion of P to S waves during propagation as well as during internal reflection. Also, higher degrees of heterogeneity cause a higher degree of conversion from P to S waves (Acharya, 1970). Generally, low-velocity zones and other seismic anomalies appear to be related to physical parameters of the travel path, rather than solely to parameters of the environments at the source or receiving stations (Weetman et al., 1970).

Other characteristics of seismic-velocity anomalies have been analyzed by Evernden (1970), Fitch (1970), Gumper and Pomeroy (1970), Beaudet (1970), Solomon and Toksöz (1970), Hasegawa (1970), Stauder and Nuttli (1970), Julian (1970), Hales (1969), and others.

When all the characteristics of the seismic-velocity anomalies of the tectonosphere are considered, it is clear that any hypothesis for the Earth's seismic behavior must embody global seismic-velocity patterns as well as the cause of their correlation with tectono-magmatic activity on a global scale.

Seismic Behavior within the Subtectonosphere.

A satisfactory model for subtectonospheric regions does not exist (Cleary, 1969). Certain information, however, is available regarding the core-mantle interface. The S waves that graze the interface appear to travel about 1000 km within the lowest 100 km of the mantle. Other information suggests that the radius of the core may actually be larger by almost that amount (Sachs, 1966, 1967).

Toksöz et al. (1970) found evidence indicating changes in velocity gradients in the lower half of the liquid core. Since a homogeneous fluid under adiabatic conditions cannot explain this behavior, they suggested that the velocity profile might be explained by solid inclusions of a few percent within the liquid core. This concept fails to explain how stability is maintained between the solid inclusions and the liquid core. Since this is critically dependent upon a well-behaved balance between viscosity, phase parameters, composition, and other factors affecting the state variables of the core, their hypothesis can be accepted only with reservations.

Although there are some variations within the lower mantle and the core, they are not significant compared with variations within the tectonosphere (Engdahl and Felix, 1970). Consequently, the seismic behavior within the subtectonosphere is not considered important to an analysis of the behavior and evolution of the tectonosphere.

Hypotheses for Explaining the Earth's Seismicity.

Many hypotheses have been developed for the cause of earthquakes, but none gives a completely satisfactory explanation. This arises primarily from the observed complexity of the Earth's seismic behavior.

The Earth's seismic activity appears to be related to tectonospheric structure and behavior. Any hypothesis for the cause of the Earth's seismicity must, therefore, consider tectonospheric structure and dynamics: their relationship to, and their influence upon, the seismic behavior of the Earth. Causal relationships between the Earth's internal behavior and the surficial manifestations thereof must, therefore, be studied on both global and regional scales.

If there are regional patterns of seismicity, there must be regionally controlled causes for such. If there are global patterns of seismicity, there must be globally controlled causes. If there are patterns of seismicity at certain depths but not at other depths, then there must be either: (1) different causal mechanisms operating at different depths, or (2) local influences at certain depths but not at others. From these and related considerations, it appears that the mechanics and geometry of the understructure exert a significant influence upon the causal mechanism controlling the creation of earthquakes at all depths.

In some earthquakes, more energy is released in the upper mantle than in the crust, within a given period of time (Karnik,

1969). In other areas, earthquakes that are laterally separated, but of similar depths, appear to be temporally and genetically related. Strong intermediate shocks in Rumania, for example, follow the same sequence as intermediate shocks in the Tyrrhenian Sea.

An explanation for the above observations must, of necessity, be an integral part of any hypothesis for explaining the Earth's seismic behavior.

Causal Mechanisms for the Earth's Seismic Behavior.

The dynamic nature of the tectonosphere is expressed today in terms of "new global tectonics" or "plate tectonics" (Chapter 18). Basic to the hypothesis (See, e. g., Isacks et al., 1968) is the concept of a rigid upper layer (the lithosphere), composed of the crust and uppermost part of the mantle. The lithosphere has considerable strength, is roughly 100 km thick, and rests, or floats, on a second layer (the asthenosphere), with practically no strength and extending from the base of the lithosphere to several hundred km below.

Under the plate-tectonics concept, the floating lithosphere is divided into vast plates bounded by ocean ridges, certain faults, and quasi-arcuate structures. The plates spread apart at the ridges, grind against each other at the faults, and are underthrust at "island arcs" and similar structures. In some areas, the ocean ridges are broken into sections of fault lines, striking normal to the ridges. Small earthquakes result from movements along these faults. Large earthquakes occur where the leading edge of one plate underthrusts another plate. The underthrusting, by some undefined means, creates large, linear ocean trenches, as at Tonga, the Kuriles, Japan, the Marianas, the Peru-Chile coast, Puerto Rico, etc. Earthquakes in these trench areas are produced by the descent of the plates to depths exceeding 700 km, roughly along the Benioff planes.

Motive power for the descent, it is hypothesized, is provided

382

by systems involving: (1) dragging or carrying, (2) pulling,
(3) sliding, (4) pushing, or (5) a combination of two or more
of these modes. In any case, the seismic behavior in the vicinity
of the descending plate would be roughly the same.

The plate-tectonics concept is generally satisfactory for
oceanic areas, but it leaves some unanswered questions regarding
the Earth's seismic behavior in continental areas. In fact,
it is unclear whether any of the common parameters of plate
tectonics are operative in intra-continental blocks (See, e.
g., Fitch, 1970). The Himalayan-Burmese mountain belts exem-
plify the inherent complexity of seismic behavior in those areas
where one continent may be underthrusting another.

Although the plate-tectonics concept is the most frequently
cited causal mechanism for the Earth's seismic behavior, it is
well to review salient features of other hypotheses. Among the
early hypotheses is Reid's (1911) "elastic rebound" hypothesis.
Reid summarized the salient features of his concept as follows:

a. The fracture of the rock, which causes a tectonic earthquake, is the result of elastic strains, greater than the strength
of the rock can withstand, produced by the relative displacements of neighboring portions of the Earth's crust.

b. The relative displacements are not produced suddenly
at the time of the fracture, but attain their maximum amounts
gradually during a more or less long period of time.

c. The only mass movements that occur at the time of the

earthquake are the sudden elastic rebounds of the sides of the fracture towards positions of no elastic strain; and these movements extend to distances of only a few miles from the fracture.

d. The earthquake vebrations originate in the surface of the fracture; the surface from which they start has at first a very small area, which may quickly become very large, but at a rate not greater than the velocity of compressional elastic waves in the rock.

e. The energy liberated at the time of the earthquake was, immediately before the rupture, in the form of energy of elastic strain of the rock.

Reid's hypothesis explained many seismic observations, but it did not specify the exact process by which elastic strain develops within the tectonosphere. The development of strain at any depth, however, suggests that a stress (or force) is operating at that depth. The occurrence of earthquakes to depths exceeding 700 km makes the problem even more complex.

In modern applications of Reid's concept, stress develops across a fracture or fault in a temporally linear manner. When the stress exceeds a critical value, corresponding with the inherent strain limit for a given depth and structure, an earthquake is produced by the resulting rupture. Observational evidence, however, refutes this concept, since there is no regularity in the interval between major earthquakes at a given point on a fault.

It is probably inadequate (except for shallow earthquakes) to assume that "fault movement" consists of simple sliding across fault planes within the face of the fault. This problem can be circumvented, in most cases, by Orowan's (1960) mechanism of "unstable creep", whereby a zone of weakness is made selectively weaker by creep and deformation concentrated in zones of weakness to the end that the creep accelerates to a catastrophic rate (Griggs and Baker, 1968; Byerlee and Brace, 1968).

Related to stress buildup is the unexplained observation that tilts develop in the vicinity of an impending shock, beginning several hours before the earthquake. Local magnetic changes precede creep increments indicating that deep-seated stress changes occur as much as several weeks before surface movements (Breiner and Kovack, 1967).

Most "source" mechanisms proposed for earthquakes are necessarily complex bacause the stress field must be in static equilibrium both before and after the stress release. This requirement eliminates all simple couples and many "double" couples with the result that each earthquake requires a carefully tailored geometrico-mechanical explanation.

Some evidence refutes the validity of even the most complex "fault plane" solutions as being unrealistic in terms of the seismic behavior actually observed when an earthquake occurs. Evison (1967) has discussed a focal mechanism embodying volume change accompanying a sudden phase transition as an alternate approach.

385

Woollard (1969) concluded that the present pattern of seismicity is related to stress _relief_ rather than to the mechanism responsible for the global fracture pattern. Under his concept, the difference between the stress-relief pattern (i. e., the present pattern of seismicity) and the global fracture pattern represents the temporal lag between the two.

In a related study, Griggs and Baker (1968) concluded that it is unlikely that deep-focus earthquakes can be caused by any of the known types of rupture, in spite of evidence indicating that seismic waves have the charactueristics of a shear-fracture source. They suggested a mechanism embodying "shear-melting instability" due to thermally activated flow in materials subject to shear strain at high pressure and temperature.

Summarizing, none of the current hypotheses regarding causal mechanisms for the Earth's seismic behavior accounts for earthquakes at all depths in all oceanic and continental areas.

Causes for Variations in Seismic Velocities within the Tectonosphere.

Molnar and Oliver (1969) have demonstrated the existence and determined the pattern of lateral variations of seismic wave attenuation within the uppermost mantle on a worldwide basis. From their observations, they concluded (1) that S_n, which does not penetrate the low-velocity channel, propagates very efficiently across the stable regions of the Earth, the continental shields, and the deep ocean basins, but very inefficiently when the paths cross the crests of the mid-ocean ridge system or the concave sides of most island arcs; (2) that attenuation is more pronounced in the uppermost mantle near the ridge crests and the island arcs than in the more stable regions; (3) that, if low attenuation, or high Q, correlates with high strength, the uppermost mantle is considerably weaker under the ridge crests and the concave sides of the island arcs than it is elsewhere; (4) that, consequently, part of the strong outer shell, or lithosphere, is discontinuous with gaps at the ridges and island arcs; and (5) that the gaps at the transform faults are very narrow.

Knopoff (1969) found that the upper mantle appears to attenuate S waves considerably more than does the lower mantle, the values of Q being about 160 and 1450, respectively, or about an order of magnitude difference. No hypothesis has been proposed to account for such large differences in attenuation factors. Nor have any data been obtained regarding variations in Q due to

387

lateral heterogeneities; present values of Q are computed on the assumption of radial symmetry (an assumption contrary to fact). Further data are needed to permit greater resolution regarding the distribution of Q with depth.

Jacob (1970) found that P velocities in the "descending lithospheric slab" are 5% to 10% higher than in the "regular" mantle. He also found a region with 5% lower velocities associated with the volcanic belt with an indication that this feature acts as a lateral low-velocity channel.

Although these and other analyses indicate that the exact causes for variations in seismic velocities within the tectonosphere are not known, it appears that such variations are related to tectonomagmatic activity.

Earthquake Prediction, Control, and Prevention.

Some geologists feel that it may be possible to predict earthquakes using the relatively simple method of monitoring the air at the Earth's surface for traces of gases of deep-seated origin. It is generally conceded, however, that specific prediction of time, place, and magnitude of earthquakes is impossible with present techniques. The inability is primarily a function of improper understanding of the mechanics and geometry involved (Stacey, 1969).

We know when and where earthquakes occur, but we do not know how or why. Swarms of very small shocks occur in the epicentral regions of earthquakes several months prior to the main shock. Ground tilt correlates strongly with the growth and decay of seismic activity, and anomalous tilt is observed immediately prior to the occurrence of some earthquakes. Magnetic fluctuations are often observed prior to earthquakes. By closely correlating all these anomalous observations, long-range predictions are possible within the foreseeable future. Closer monitoring of the same parameters should also permit short-range predictions. These depend upon the development of a technique for measuring, or otherwise determining, the stress (rather than the strain that is produced by the stress). Direct measurement of stress is impractical; indirect determinations are suggested by the effects of stress upon the electromagnetic properties of rocks and upon seismic velocities.

389

Piezomagnetic effects appear particularly promising for determining stress from strain (See, e. g., Breiner and Kovach, 1967). Earlier studies using the stress dependence of seismic velocity were made by Tocher (1957). Another approach for predicting earthquakes is based on the possible time-space correlations between earthquakes. Shocks are not strictly independent events. The probability of the occurrence of an earthquake is greater, in a given area, immediately following a shock, than after a long quiescent period. This procedure is useful only in areas where shocks have occurred recently. Some major shocks are preceded by foreshocks, suggesting the possibility of earthquake prediction on the basis of statistical analysis of microshock sequences (Press and Brace, 1966; Oliver, 1966; Grover, 1967).

Much work remains to be done prior to the development of a practical procedure for earthquake prediction. Similar conclusions may be drawn for earthquake control and prevention. There is conflicting evidence that manmade shocks (e. g., nuclear explosions) may trigger earthquakes (See, e. g., Healy and Marshall, 1970). Earlier studies indicated that the filling of reservoirs, rivers behind new dams, and wells may trigger earthquakes.

Earthquakes may be triggered by relatively small unbalanced forces. This suggests the possibility of earthquake control, except that a problem arises in determining exactly when and where to place the planned "stress release".

The Earth's Seismic Behavior According to a Dual Primeval Planet Hypothesis.

Earlier sections of this chapter explained how earthquakes arise through sudden releases of energy within some confined regions of the Earth. The energy involved may be any one or more of the following: gravitational potential, kinetic, chemical, and elastic strain. The release of the energy may be regarded as the immediate cause of an earthquake.

Unspecified in current hypotheses are answers regarding (1) the nature of the processes creating the accumulation of the energy that, upon release, triggers the earthquake, and (2) the ultimate source, or origin, of the energy-accumulation processes that precede the earthquakes. Tentative answers to these two questions are provided by the tectonospheric Earth model derived from the dual primeval planet hypothesis.

Earthquake Casual Mechanisms Derived from the Tectonospheric Earth Model.

A successful analysis of the Earth's seismicity depends upon an understanding of the interrelationships of the geometrico-mechanical parameters associated with earthquakes. This suggests a categorization of earthquakes based on these parameters.

The tectonospheric Earth model separates earthquakes into 18 categories based on fundamental geometrico-mechanical parameters of (1) depth, (2) location, and (3) pressure-status of the hypocentral volume at the time of the earthquake. Table 8-4c-1 summarizes these 18 categories of earthquakes. The depth parameter is expressed in terms of the type of tectonospheric plate in which the earthquake occurs: L, for lithospheric plates; A, for asthenospheric plates; S, for subasthenospheric plates. The location parameter is expressed in terms of the region of the tectonospheric Earth model in which the earthquake occurs: C, within a current wedge-belt of activity; F, within a fossil wedge-belt of activity; S, within a central craton. The pressure-status parameter is specified in terms of the lateral force experienced by the hypocentral volume at the time of the earthquake: K, for relative compression; T, for relative tension.

Thus, an earthquake occurring at a lithospheric (L) depth within a fossil (F) wedge-belt in relative compression (K) at the hypocentral depth is of category LFK. Similarly, an earthquake occurring at an asthenospheric (A) depth within a current

TABLE 8-4c-1. Summary of the 18 basic categories of earthquakes derivable from a first-order tectonospheric Earth model based on the dual primeval planet hypothesis. See text for details.

	WITHIN CURRENT WEDGE-BELT OF ACTIVITY (C)		WITHIN FOSSIL WEDGE-BELT OF ACTIVITY (F)		WITHIN CENTRAL CRATON OR SHIELD (S)	
	Compressive (K)	Tensile (T)	Compressive (K)	Tensile (T)	Compressive (K)	Tensile (T)
Litho-spheric (L)	LCK	LCT	LFK	LFT	LSK	LST
Astheno-spheric (A)	ACK	ACT	AFK	AFT	ASK	AST
Sub-astheno-spheric (S)	SCK	SCT	SFK	SFT	SSK	SST

(C) wedge-belt in relative tension (T) at the hypocentral depth is of category ACT. The remaining 16 categories of earthquakes follow in a similar manner.

To facilitate analysis, the mechanism causing a given earthquake may be assigned the same category designation as that of the earthquake. Thus, an earthquake occurring at a subasthenospheric (S) depth within a current (C) wedge-belt in relative tension (T) at the hypocentral depth is motivated by a causal mechanism of category SCT. For certain purposes, earthquakes may be grouped into C, F, and S types, with 6 categories within each group; e. g., C-type comprises categories LCK, ACK, SCK, LCT, ACT, and SCT.

Fig. 8-4c-1, which has been adapted from Chapter 7, shows the relationship between the 18 categories of earthquake mechanisms. Type-C earthquakes (those with C as the second letter of the category designation) occur within the current wedge-belts indicated by the solid lines in the figure. Type-F earthquakes occur within the fossil wedge-belts indicated by broken lines. Type-S earthquakes occur within the center of the figure.

To visualize the global distribution of the 18 categories of causal mechanisms, it is well to recall that the idealized wedge-belt of activity has a surficial width of 1570 km (for a vertex angle of 90°, equivalent to a Benioff plane of 45°). Since current wedge-belts of activity intercept approximately 7% of the Earth's surface, epicenters for C-type earthquakes

394

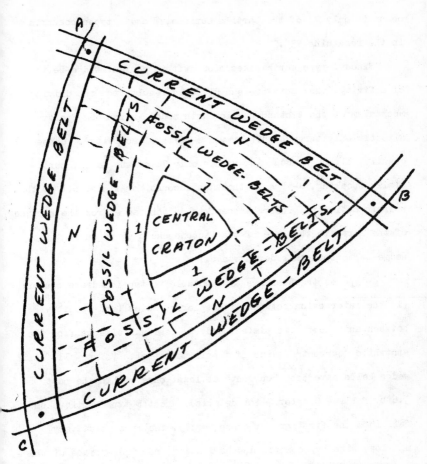

Figure 8-4c-1. Idealized orogenic-cratonic structure of the
continents according to a first-order tectonospheric Earth model
based on the dual primeval planet hypothesis. See text for
details. (Adapted from Chapter 7).

occur in only 7% of the Earth's surface, F and S types occurring in the remaining 93%.

Since C-type earthquakes occur within current wedge-belts of activity, they more-or-less _directly_ result from the driving mechanism of the tectonospheric Earth model (which can operate directly only through current wedge-belts of activity). Consequently, although occurring within only 7% of the Earth, surficially speaking, C-type earthquakes account for almost all of the seismic activity. Their hypocentric depths range from the surface of the Earth to the base of the tectonopshere since current wedge-belts span this depth range.

F-type earthquakes are attributable to two immediate causes: (1) the interaction between lithospheric plates (and possibly between asthenospheric plates) and (2) residual effects from the erstwhile "current" status of fossil wedge-belts (since all fossil wedge-belts have been "current" at least once within the past 3.6 b. y.; see Chapter 7 for details). Most F-type earthquakes fall into the first group; consequently, they are relatively shallow, with hypocentric depths ranging from the surface of the Earth to the base of the local asthenosphere.

S-type earthquakes are attributable to the inherent mobility of the central cratons. Briefly, these earthquakes are caused when, in response to the driving mechanism of the tectonospheric Earth model, the central cratons move along shells, planes, surfaces, and zones of preferential rheidity as they continually reposition themselves to minimum—energy configurations on the globe. (See Chapter 17 for details). Consequently, S-type earthquakes are few in number and relatively shallow in depth. Being farthest from current wedge-belts of activity, they are even less numerous than F-type earthquakes.

<u>Correlations between Predictions of the Model and Actual Observation</u>
<u>of the Earth's Seismic Behavior.</u>

Comparison of the predictions of the tectonospheric Earth model with actual observations of the Earth's seismic behavior permits an evaluation of the model as a possible working hypothesis. Although it is not possible to consider all predictions of the model, a few may be selected as being representative in both global and regional areas (See, e. g., Tatsch, 1966, 1969, 1970).

Fig. 8-4d-1 outlines the global areas in which the 3 basic types of earthquakes are predicted according to the first-order tectonospheric Earth model. C represents current wedge-belts of activity, in which C-type earthquakes are predicted. F represents areas of fossil wedge-belts of activity, in which F-type earthquakes are predicted. S represents central cratons, in which S-type earthquakes are predicted. For ease of orientation, continental outlines are included.

Correlations between predictions of the model and actual observations are particularly noteworthy in the prominent Circum-Pacific and Trans-Eurasian (Tethyan) seismic belts, previously indicated in Fig. 8-1d-1 and Fig. 8-1d-2.

A significant regional correlation occurs near the KR (Kermadecs) point of the the subtectonospheric fracture system of the model. This correlation is shown in Fig. 8-4d-2. The solid lines KR-BN and KR-BV represent surficial traces of segments of the subtectonospheric fracture system. Their correlation

398

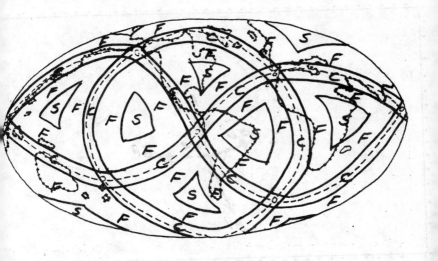

Figure 8-4d-1. Mollweide projection of the Earth superimposed upon the idealized wedge-belt system of a first-order tectonospheric Earth model to show the predicted interrelationships of C, F, and S type earthquakes.

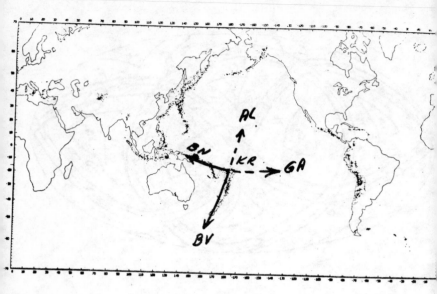

Figure 8-4d-2. Correlation between predictions of tectonospheric Earth model and actual observations of Earth's seismicity: orthogonal fracture system epicentral to KR (Kermadecs) point of the model. See text for details. (Based on Barazangi and Dorman, 1969; ESSA, 1970).

with the distribution of deep earthquakes in that region suggests the existence of a deep-tectonospheric orthogonal-fracture system within a few hundred km of the KR (Kermadecs) subtectonospheric point of the first-order model (Tatsch, 1963, 1964). The arcs KR-AL and KR-GA represent surficially dormant segments of current wedge-belts of activity. The characteristics of these dormant segments are discussed when intrusive and extrusive activity is considered in Chapter 10.

Another specific correlation between prediction and observation appears in two prominent belts of deep-focus earthquakes in South America, shown schematically in Fig. 8-4d-3 and based on observational data (ESSA, 1970). The correlation involves the parallel belts, which are 700 km apart and lie within latitudes $22^{o}S$ to $29^{o}S$ (belt S on the figure) and $7^{o}S$ to $12^{o}S$ (belt N). O is on the surface of the subtectonosphere. ABCD is on the surface of the Earth. The line GA-BV is the centerline of the GA-BV wedge-belt of activity.

A comparison of Fig. 8-4d-3 with a world seismicity map (ESSA, 1970) suggests that South America is rotated about 35° with respect to the GA-BV tectonospheric arc and that the rotation occurred at great depth, possibly exceeding 600 km. Chapter 17 discusses the rotation of South America.

Fig. 8-4d-4 is a map of global seismic observations annotated with predictions of the tectonospheric Earth model: C for C-type predictions; F for F-type. Nearly all deep earthquakes

401

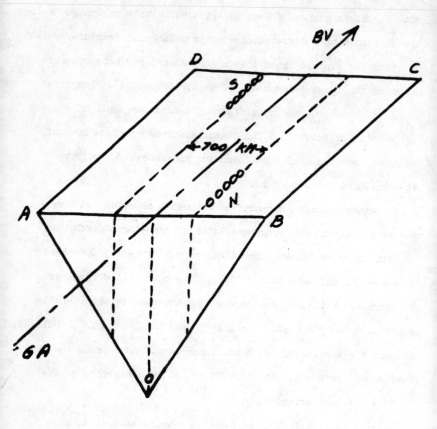

Figure 8-4d-3. Schematic sketch showing how 2 parallel belts
of deep-focus earthquakes in South America are interpreted by
the dual primeval planet hypothesis.

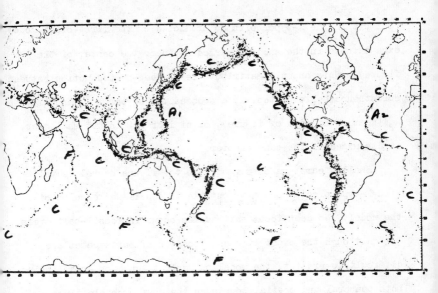

Figure 8-4d-4. Observed earthquake belts annotated with earthquake causal mechanism designations derived from the tectonospheric Earth model. C represents earthquake causal mechanism of C-type; F, of F-type. A_1 and A_2 are anomalous belts discussed in Chapter 17. See text for details. (Based on Barazongi and Dorman, 1969; ESSA, 1970).

fall within the C-type belts as predicted by the first-order model except those in the anomaly marked A_1 (near the center of the figure). This anomaly, attributable to deep-seated motion between subasthenospheric plates, is discussed in Chapter 18. Anomaly A_2 (near right edge of figure), of similar origin but shallower, is discussed in the same chapter.

Shallow earthquakes marked F in the figure correlate well with F-type predictions attributable to fossil wedge-belts of the model. No deep-focus earthquakes occur in fossil-belt areas.

Although the exact time and location of earthquakes are difficult to predict, the model suggests a basis for establishing both temporal and spatial bounds on the most probable times and locations of future earthquakes.

REFERENCES

Acharya, H. K., 1970. P and S interactions in inhomogeneous media. _Trans. Am. Geophys. Union_, 51: 361.

Alterman, Z., 1969. Higher-mode surface waves. In: P. J. Hart (Editor), _The Earth's Crust and Upper Mantle_. Am. Geophys. Union, Washington, pp. 265–272.

Anderson, D. L., 1967. Latest information from seismic observations. In: T. F. Gaskell (Editor), _The Earth's Mantle_. Academic, London, pp. 355–420.

Anderson, D. L., 1968. Chemical inhomogeneity of the mantle. _Earth and Planetary Sci. Ltrs._, 5: 89–94.

Archambeau, C. B., Flinn, E. A., and Lambert, D. G., 1969. Fine structure of the upper mantle. _J. Geophys. Res._, 74: 5825–5865.

Balakina, L. M., Misharina, L. A., Shirkova, E. I., and Vvedenskaya, A. V., 1969. The stress state in earthquake foci and the elastic stress field of the Earth. In: _A Symposium_ on Processes in the Focal Region, IUGG 14th Gen. Assy., pp. 194-198.

Barazangi, M. and Dorman, J., 1969. World seismicity map of ESSA Coast and Geodetic Survey epicenter data for 1961-1967. Bull. Seismol. Soc. Am., 59: 369-380.

Beaudet, P. R., 1970. Elastic wave propagation in heterogeneous media. Bull. Seismol. Soc. Am., 60: 769-784.

Beloussov, V. V., 1969. Interrelations between the Earth's crust and upper mantle. In: P. J. Hart (Editor), The Earth's Crust and Upper Mantle. Am. Geophys. Union, Washington, pp. 698-712.

Bolt, B. and Nuttli, O., 1966. P wave residuals as a function of azimuth, 1, observations. J. Geophys. Res., 71: 5977-5985.

Breiner, S. and Kovach, R. L., 1967. Local geomagnetic events associated with displacements on the San Andreas fault. Science, 158: 116.

Brune, J. N., 1968. Seismic moment, seismicity, and rate

of slip along major fault zones. *J. Geophys. Res.*, 73: 777-784.

Bullen, K. E., 1963. *An Introduction to the Theory of Seismology.* Cambridge Univ. Press, Cambridge, 380 pp.

Bullen, K. E., 1967. Basic evidence for Earth divisions. In: T. F. Gaskell (Editor), *The Earth's Mantle.* Academic, London, pp. 11-39.

Byerlee, J. D. and Brace, W. F., 1968. Stick slip, stable sliding, and earthquakes: effect of rock type, pressure, strain rate, and stiffness. *J. Geophys. Res.*, 73: 6031-6037.

Clark, S. P. and Ringwood, A. E., 1964. Density distribution and constitution of the mantle. *Rev. Geophys.*, 2: 35-88.

Cleary, J., 1969. The S velocity at the core-mantle boundary from observations of diffracted S. *Bull. Seismol. Soc. Am.*, 59: 1399-1405.

Cleary, J. R. and Hales, A. L., 1966. Azimuthal variation of U. S. station residuals. *Nature*, 210: 619-620.

Cook, K. L., 1962. The problem of the mantle-crust mix:

lateral inhomogeneity in the uppermost part of the Earth's mantle. *Advan. Geophys.*, 9: 295-360.

Danes, Z. F., 1970. Earthquake occurrence in the state of Washington. *Science*, 167: 396.

DeLoczy, L., 1969. Stratigraphic and paleogeographic problems of the Parana Gondwana basin. In: *Gondwana Stratigraphy*, IUGS Symposium, Buenos Aires, October 1967, UNESCO, Paris, pp. 967-969.

Dorman, J., 1969. Seismic surface-wave data on the upper mantle. In: P. J. Hart (Editor), *The Earth's Crust and Upper Mantle*. Am. Geophys. Union, Washington, pp. 257-265.

Drake, C. L., 1970. A long-range program of solid Earth studies. *Trans. Am. Geophys. Union*, 51: 152-159.

Drake, C. L. and Nafe, J. E., 1968. The transition from ocean to continent from seismic refraction data. In: L. Knopoff, C. L. Drake, and P. J. Hart (Editors), *The Crust and Upper Mantle of the Pacific Area*. Am. Geophys. Union, Washington, pp. 174-186.

Engdahl, E. R. and Felix, C. P., 1970. Array analysis of core phases. Trans. Am. Geophys. Union, 51: 360.

ESSA, 1970. World Seismicity. Mercator chart NEIC-3005. Coast and Geodetic Survey, Washington, D. C.

Evernden, J. F., 1970. Study of regional seismicity and associated problems. Bull. Seismol. Soc. Am., 60: 393-446.

Evison, F. F., 1967. On the occurrence of volume change at the earthquake source. Bull. Seismol. Soc. Am., 57: 9-25.

Fedotov, S. A., Matveyeva, N. N., Tarakanov, R. Z., and Yanovskaya, T. B., 1964. Velocities of longitudinal waves in the upper mantle in the area of the Japanese and Kuril Islands. Izv. Akad. Nauk SSSR, Ser. Geofiz., 1964, 1185-1191.

Fitch, T. J., 1970. Earthquake mechanisms and island arc tectonics in the Indonesian-Philippine region. Bull. Seismol. Soc. Am., 60: 565-591.

Gaskell, T. F., (Editor), 1967. The Earth's Mantle. Academic, London, 509 pp.

Gentili, C. A., 1969. Outcrop distribution of the Serra Geral formation in the provinces of Entre Rios and Corrientes Argentine Republic. <u>Gondwana</u> <u>Stratigraphy</u>, IUGS Symposium, Buenos Aires, October 1967, UNESCO, Paris, pp. 985-986.

Griggs, D. T. and Baker, D. W., 1968. The origin of deep-focus earthquakes. In: H. Mark and S. Fernbech (Editors), <u>Properti</u> <u>of</u> <u>Matter</u> <u>under</u> <u>Unusual</u> <u>Conditions.</u> Wiley-Interscience, New York, pp. 23-37.

Grover, J. C., 1967. Forecasting earthquakes: correlation between deep foci and shallow events in Melanesia. <u>Nature</u>, 213: 686.

Gumper, F. and Pomeroy, P. W., 1970. Seismic wave velocities and Earth structure on the African continent. <u>Bull.</u> <u>Seismol.</u> <u>Soc.</u> <u>Am.</u>, 60: 651-668

Gutenberg, B., 1959. <u>Physics</u> <u>of</u> <u>the</u> <u>Earth's</u> <u>Interior.</u> Academic, New York, 240 pp.

Gutenberg, B., 1960. Low-velocity layers in the Earth,

410

ocean and atmosphere. Science, 131: 959-965.

Gutenberg, B. and Richter, C. F., 1954. Seismicity of the Earth and Associated Phenomena, 2d. ed. Princeton Univ. Press, Princeton, 273 pp.

Hales, A. L., 1969. A seismic discontinuity in the lithosphere. Earth and Planetary Sci. Ltrs., 7: 44-46.

Hales, A. L. and Doyle, H. A., 1967. P and S travel time anomalies and their interpretation. Geophys. J. Roy. Astron. Soc., 13: 403-413.

Hales, A. L., Helsley, C. E., and Nation, J. B., 1970. Travel times for an oceanic path. Trans. Am. Geophys. Union, 51: 359.

Hart, P. J. (Editor), 1969. The Earth's Crust and Upper Mantle. Amer. Geophys. Union, Washington, 735 pp.

Hasegawa, H. S., 1970. Short-period P-coda characteristics in the eastern Canadian shield. Bull. Seismol. Soc. Am., 60: 839-858.

411

Healy, J. H. and Marshall, P. A., 1970. Nuclear explosions and distant earthquakes: a search for correlation. Science, 169: 176–177.

Herrin, E. and Taggart, J., 1968. Source bias in epicenter determination. Bull. Seismol. Soc. Am., 58: 1791–1796.

Hill, R. E. T. and Boettcher, A. L., 1970. Melting relationships in basalt-H_2O and basalt-H_2O-CO_2 to 30 kilobars. Trans. Am. Geophys. Union, 51: 438.

Holmes, A., 1965. Principles of Physical Geology. Ronald, New York, 1288 pp.

Horai, K., 1969. Cross-covariance analysis of heat flow and seismic delay times for the Earth. Earth and Planetary Sci. Ltrs., 7: 213–220.

Isacks, B., Oliver, J., and Sykes, L. R., 1968. Seismology and the new global tectonics. J. Geophys. Res., 73: 5855–5899.

Jacob, K. H., 1970. P-residuals and global tectonic structures investigated by three-dimensional seismic ray tracings

with emphasis on Longshot data. Trans. Am. Geophys. Union,
51: 359.

Julian, B. R., 1970. Regional variations in upper mantle
structure in North America. Trans. Am. Geophys. Union, 51: 359.

Karnik, V., 1969. Seismicity of the European areas. In:
P. J. Hart (Editor), The Earth's Crust and Upper Mantle. Am.
Geophys. Union, Washington, pp. 139-144.

Knopoff, L., 1969. Attenuation of seismic waves in the
mantle. In: P. J. Hart (Editor), The Earth's Crust and Upper
Mantle. Am. Geophys. Union, Washington, pp. 273-276.

Knopoff, L., Drake, C. L., and Hart, P. J. (Editors), 1968.
The Crust and Upper Mantle of the Pacific Area. Am. Geophys.
Union, Washington, 522 pp.

Kosminskaya, I. P. and Zveref, S. M., 1968. Deep seismic
sounding in the transition zones from continents to oceans. In:
L. Knopoff et al. (Editors), The Crust and Upper Mantle of the
Pacific Areas. Am. Geophys. Union, Washington, pp. 122-130.

Kushiro, I., Syono, Y., and Akimoto, S., 1968. Melting of a peridotite nodule at high pressure and high water pressures. J. Geophys. Res., 73: 6023-6029.

Lambert, I. B. and Wyllie, P. J., 1970. Low-velocity zone of the Earth's mantle: incipient melting caused by water. Science, 196: 764-766.

Latter, J. H., 1969. Time and space correlation in seismicity and volcanism. Geol. Soc. London Proc., 1658: 262-264.

Lehmann, I., 1967. Low-velocity layers. In: T. F. Gaskell (Editor), The Earth's Mantle, Academic, London, pp. 41-61.

Liebermann, R. C. and Schreiber, E., 1969. Critical thermal gradients in the mantle. Earth and Planetary Sci. Ltrs., 7: 77-81.

Magnitsky, V. A. and Zharkov, V. N., 1969. Low-velocity layers in the upper mantle. In: P. J. Hart (Editor), The Earth's Crust and Upper Mantle. Am. Geophys. Union, Washington, pp. 664-675.

McBirney, A. R., 1969. Andesitic and rhyolitic volcanism of orogenic belts. In: P. J. Hart (Editor), The Earth's Crust and Upper Mantle. Am. Geophys. Union, Washington, pp. 501-507.

McGinley, J. R., Jr. and Anderson, D. L., 1969. Relative amplitudes of P and S waves as a mantle reconnaissance tool. Bull. Seismol. Soc. Am., 59: 1189-1200.

Mitronovas, W., Isacks, B., and Seeber, L., 1969. Earthquake distribution and seismic wave propagation in the upper 250 km of the Tonga island arc. Bull. Seismol. Soc. Am., 59: 1115-1135.

Miyamura, S., 1969. Seismicity of the Earth. In: P. J. Hart (Editor), The Earth's Crust and Upper Mantle. Am. Geophys. Union, Washington, pp. 115-124.

Molnar, P. and Oliver, J., 1969. Lateral variations of attenuation in the upper mantle and discontinuities in the lithosphere. J. Geophys. Res., 74: 2648-2682.

Northrop, J., Morrison, M. F., and Duennebier, F. K., 1970.

Seismic slip rate versus sea-floor spreading rate on the Eastern Pacific Rise and Pacific Antarctic Ridge. J. Geophys. Res., 75: 3285-3290.

Oliver, J., 1966. Prospects for earthquake prediction. Sci. J., 2: 44-54.

Orowan, E., 1960. Mechanism of seismic faulting. In: Griggs and Handin (Editors), Rock Deformation, Geol. Soc. Am. Mem. 79: 323.

Pakiser, L. C., 1963. Structure of the crust and upper mantle in the western United States. J. Geophys. Res., 68: 5747-5756.

Press, F. and Brace, W. F., 1966. Earthquake prediction. Science, 152: 1575.

Raitt, R. W., 1969. Anisotropy of the upper mantle. In: P. J. Hart (Editor), The Earth's Crust and Upper Mantle. Am. Geophys. Union, Washington, pp. 250-256.

Reid, H. F., 1911. The elastic rebound theory of earthquakes.

Bull. Dept. Geol., Univ. Calif., 6: 413-420.

Richter, C. F., 1958. Elementary Seismology. Freeman, San Francisco, 768 pp.

Ringwood, A. E., 1962a. A model for the upper mantle, 1. J. Geophys. Res., 67: 857-867.

Ringwood, A. E., 1962b. A model for the upper mantle, 2. J. Geophys. Res., 67: 4473-4477.

Ringwood, A. E., 1970. Composition and evolution of the upper mantle. In: P. J. Hart (Editor), The Earth's Crust and Upper Mantle. Am. Geophys. Union, Washington, pp. 1-17.

Ritsema, A. R., 1969. Seismology and upper mantle investigations. In: P. J. Hart (Editor), The Earth's Crust and Upper Mantle. Am. Geophys. Union, Washington, pp. 110-115.

Sachs, S., 1966. Diffracted wave studies of the Earth's core, 1: amplitudes, core size, and rigidity. J. Geophys. Res., 71: 1173-1181.

Sachs, S., 1967. Diffracted P-wave studies of the Earth's

417

core, 2: lower mantle velocity, core size, lower mantle structure.

J. Geophys. Res., 72: 2589-2594.

Savarensky, E. F. and Golubeva, N. V., 1969. Seismicity

of continental Asia and the region of the Sea of Okhotsk, 1953-

1965. In: P. J. Hart (Editor), The Earth's Crust and Upper

Mantle. Am. Geophys. Union, Washington, pp. 134-139.

Sleep, N. H. and Biehler, S., 1970. Topography and tec-

tonics at the intersections of fracture zones with central rifts.

J. Geophys. Res., 75: 2748-2752.

Solomon, S. C. and Toksöz, M. N., 1970. Lateral variation

of attenuation of P and S waves beneath the United States. Bull.

Seismol. Soc. Am., 60: 819-838.

Stauder, W. and Nuttli, O. W., 1970. Seismic studies:

south central Illinois earthquake of November 9, 1968. Bull.

Seismol. Soc. Am., 60: 973-981.

Steinhart, J. S. and Smith, T. J., (Editors), 1966. The

Earth beneath the Continents. Am. Geophys. Union, Washington,

663 pp.

Sutton, G. H., Mitronovas, W., and Pomeroy, P. W., 1967.
Short-period seismic energy radiation patterns from underground
nuclear explosions and small-magnitude earthquakes. _Bull. Seismol._
Soc. Am., 57: 249-267.

Sykes, L. R., Isacks, B. L., and Oliver, J., 1969. Spatial
distribution of deep and shallow earthquakes of small magnitude
in the Fiji-Tonga region. _Bull. Seismol. Soc. Am._, 59: 1093-1113.

Tatsch, J. H., 1963. Certain seismological implications
of applying a dual primeval planet model to the Earth. _Trans._
Am. Geophys. Union, 44: 887.

Tatsch, J. H., 1964. Distribution of active volcanoes:
summary of preliminary results of three-dimensional least-squares
analysis. _Geol. Soc. Am. Bull._, 75: 751-752.

Tatsch, J. H., 1966. Certain correlations between seismological
observations in the African continent and deductions made from
applying a dual primeval planet model to that region of the Earth.
Trans. Am. Geophys. Union, 47: 490.

Tatsch, J. H., 1969. Sea-floor spreading, continental drift, and plate tectonics unified into a single global concept by the application of a dual primeval planet hypothesis to the Earth. *Trans. Am. Geophys. Union*, 50: 672.

Tatsch, J. H., 1970. Global seismicity patterns as interpreted in accordance with a dual primeval planet hypothesis. *Geol. Soc. Am. Abstracts with Programs*, 2: 153.

Tocher, D., 1957. Anisotropy in rocks under simple compression. *Trans. Am. Geophys. Union*, 38: 39.

Toksöz, M. N. and Arkani-Hamed, J., 1967. Seismic delay times: correlation with other data. *Science*, 158: 783-784.

Toksöz, M. N., Wiggins, R. A., and Husebye, E. S., 1970. Structure and properties of the Earth's core. *Trans. Am. Geophys. Union*, 51: 359.

Tryggvason, E., 1964. Arrival times of P waves and upper mantle structure. *Bull. Seismol. Soc. Am.*, 54: 727-736.

Tsuboi, C., Wadati, K., and Hagiwara, T., 1962. Prediction of earthquakes: progress to date and plans for further development. Earthquake Research Institute, Tokyo.

Turcotte, D. L. and Oxburgh, 1969. Convection in a mantle with variable physical parameters. $\underline{J.\ Geophys.\ Res.}$, 74: 1458-1474.

Vanek, J., 1969. Upper mantle structure and velocity distributions in Eurasia. In: P. J. Hart (Editor), $\underline{The\ Earth's}$ $\underline{Crust\ and\ Upper\ Mantle.}$ Am. Geophys. Union, Washington, pp. 246-250.

Volarovich, M. P., Balashov, D. B., Tomashevskaya, I. S., and Pavlogradskii, V. A., 1963. A study of the effect of uniaxial compression upon the velocity of elastic waves in rock samples under conditions of high hydrostatic pressure. $\underline{Bull.}$ $\underline{(Izv.)\ Acad.\ Sci.}$, USSR, $\underline{Geophys.\ Ser.}$, 728.

Wang, C. Y., 1970. The upper mantle beneath tectonic regions. $\underline{Trans.\ Am.\ Geophys.\ Union,}$ 51: 358.

Weetman, B. G., Davis, L. L., Foote, R. Q., and Hayes, W. W., 1970. Seismic anomaly at NRDS, Nevada. _Trans. Am. Geophys. Union_, 51: 359.

Wellman, H. W., 1969. Wrench (transcurrent) fault systems. In: P. J. Hart (Editor), _The Earth's Crust and Upper Mantle_. Am. Geophys. Union, Washington, pp. 544–549.

Woollard, G. P., 1969. Regional variations in gravity. In: P. J. Hart (Editor), _The Earth's Crust and Upper Mantle_. Am. Geophys. Union, Washington, pp. 320–341.

A detailed description of the Earth's thermal behavior
is, of necessity, a rough guess because the nature of the distribution of heat sources within the Earth is not known. Certain
inferences may be drawn, however, from an analysis of the Earth's
surficial features that reflect its thermal behavior. When this
is done, it is found that the Earth's surface may be divided
into active and inactive regions. The Earth's geothermal patterns may be analyzed in terms of the structural and orogenic
environments in which these active and inactive regions occur:

a. Continental Margins. Generally, there are no
significant changes in heat-flow rates at continental margins
that are inactive (Langseth, 1969), but there is a large discontinuity of heat-flow at continental margins that are active,
i. e., those bounded by an active trench.

b. Axial Zones of Mid-Ocean Ridges. The axial zones
of mid-ocean ridges have heat-flow rates of 2.5 to 3.0 HFU. The
width of the axial zone correlates with the spreading rate approximately as $W = 290\ S - 100$, where W is the width of the
axial zone in km; S, the spreading rate in cm/yr. Thus the
thermal flux varies inversely as W and S.

c. The Flanks of Mid-Ocean Ridges. Most active mid-ocean ridges contain extensive areas of low heat flow. In contrast to the above average values along the axial ridges, the heat-flow rates in the flanks of the ridges are only about one-half that of the world average.

     d. <u>Ocean Basins</u>. The ocean basins of the world display a remarkable uniformity of heat flow. Regardless of the size of the ocean, the average thermal flux is approximately 1.3 HFU.

     e. <u>Continental</u> <u>Shields</u>. The heat flow within continental shields, like that in ocean basins, is remarkably uniform throughout the world.

A preliminary analysis of the geothermal behavior within the active and inactive regions of the Earth permits a generalization: the "newest" regions of the Earth contain the largest geothermal anomalies. Because this rule has many exceptions, it is well to examine in greater detail the geothermal behavior related to certain particularly anomalous geophenomena and features, i. e., sea-floor spreading, low-velocity zones, downgoing lithospheric plates, and island-arc areas.

Global Geothermal Patterns Associated with Sea-Floor Spreading.

Any analysis of the geothermal behavior of the Earth's
tectonosphere must consider patterns associated with sea-floor
spreading and other concepts of the new global tectonics. Briefly,
the concept of sea-floor spreading (Hess, 1962), for which a rather
firm basis has been established with marine geomagnetic data
(Vine, 1966), has recently been extended to include the continents in the tectonic-plate hypothesis (Isacks et al., 1968).
Differential movements of large crustal plates occur at mid-
ocean ridges, at deep-sea trenches, and along fracture zones
of the sea floor. Crustal material, according to the tectonic-
plate hypothesis, is created (ascends) at the ridge crest and
is destroyed (sinks) at the trenches.

Other details regarding the new global tectonics are discussed in later chapters. Suffice it here merely to examine
the geothermal evidence for and against the concept of sea-
floor spreading.

Evidence to test this concept is difficult to collect because of the relatively low thermal conductivity and large thermal
inertia of rocks near the Earth's surface. Also, complications
of the environment near the sea floor probably contribute to
a large variability in oceanic heat-flow values (See, e. g.,
Von Herzen and Uyeda, 1963). Nevertheless, the heat-flow data
in some regions have been shown to fit reasonably well some

425

quantitative theories associated with the formation of new crust at ridge crests (Langseth et al., 1966; McKenzie, 1967).

Crustal underthrusting at the eastern margin of the Antilles (Chase and Bunce, 1969) that the relatively narrow (1700 km) region between the eastern Caribbean and the mid-Atlantic ridge is a region where crust is both being created and destroyed (Von Herzen et al., 1970). Based on the structure and dynamics of the Earth's crust along two transects from the Antilles to the mid-Atlantic ridge, including detailed studies of the ridge crest and of the Caribbean island arc, the crustal plate in this region appears to be bounded on the east by the mid-Atlantic ridge crest which is displaced by a series of equatorial east-west trending fracture zones. One of the larger and more northerly active fracture zones is the Vema fracture (Heezen et al., 1964).

The following are the salient conclusions made by Von Herzen and his associates (Von Herzen et al., 1970) from a study of 41 new heat-flow measurements between the Caribbean and the mid-Atlantic ridge: (1) a large but systematically-distributed variability is closely related to active tectonic elements of that region; (2) the data indicate a more complicated but systematic pattern of high and low heat flow with distance from the ridge crest for the region between active fracture zones south of 12° N; (3) the floor of the Vema fracture shows high and uniform heat flow; and (4) a suggested pattern of moderately high (2.0 HFU) heat flow on the Lesser Antilles platform, and moderately low (0.7 to 1.0 HFU) heat flow on the lower eastern flank of the Antillean platform, may be a result of tectonic processes or of the local sea-floor environment.

An analysis of the above shows that actual geothermal observations do not support all the predictions of sea-floor-spreading concepts in all regions of the Earth. Other such areas, including the Aleutians and Kermadecs-Tonga regions, are discussed in the chapter on sea-floor spreading.

<u>Geothermal</u> <u>Patterns</u> <u>Associated</u> <u>with</u> <u>Low-Velocity</u> <u>Zones.</u>

Geothermal patterns associated with low-velocity zones must be understood in order to understand the thermal behavior of the Earth's tectonosphere. Briefly, these highly-rheidic low-velocity zones appear to owe their salient geothermal characteristics to conditions of low internal friction (i. e., viscosity) resulting from high temperatures existing in some regions of the tectonosphere. All planes, areas, shells, and other surfaces of low-velocity appear to be also regions of preferential heat-flow.

The causal mechanism for regions of preferential heat flow appears to determine the locations of the Earth's low-velocity zone. Most analyses indicate that the low-velocity zone is too pronounced to be an effect solely of high-temperature gradients (Anderson and Sammis, 1969). Although partial melting is consistent with high temperature as the causal mechanism, other considerations require that there also be an abundance of water in the low-velocity zone in order to lower the solidus below the melting point of anhydrous silicates. From this, it appears that the causal mechanism for the low-velocity high-temperature anhydrous regions of the Earth must be able to supply both heat and water. Unanswered are questions regarding the nature of a causal mechanism that can supply both heat and water deep within the tectonosphere.

The problems of the low-velocity zone are complicated by the observation that it is possible to have low-velocity areas for shear waves without requiring them for compressional waves (Liebermann and Schreiber, 1969). Because the shear velocity has been observed to decrease with depth, this suggests that more water is available deep within the tectonosphere than near the surface.

It appears, therefore, that the causal mechanism for the low-velocity zone is capable of simultaneously providing both heat and water and that the source lies below the low-velocity zone. The nature of such a causal mechanism has not been described in the literature.

Geothermal Patterns in Hypothetical Downgoing Lithospheric Plates

Geothermal patterns associated with the hypothesis of down-going lithospheric plates are not completely understood. Initial observational evidence supported the hypothesis, but more-recent evidence leaves some unanswered questions regarding the general validity of the hypothesis as presently formulated. By way of background, anomalous heat-flow values are observed on the "continental" side of some island arcs near areas of Cenozoic activity (Uyeda and Horai, 1964). A low heat-flow zone is often observed over areas of shallow seismicity, whereas high-flow zones occur over areas of intermediate and deep seismicity (Uyeda and Vacquier, 1968). The volcanic front (i. e., the oceanward limit of volcanism) lies consistently above earthquakes at depths of about 100 km (Sugimura, 1968). These and similar observations define the geothermal pattern in the vicinity of hypothetical downgoing lithospheric plates.

From this thermal pattern, it has been postulated that under island arcs there is a downgoing mantle convection current (Oliver and Isacks, 1967; Utsu, 1967: Utsu and Okada, 1968). This model explains low heat-flow values over trenches (Langseth et al., 1966) but fails to account for the high heat-flow values inside the arcs. Neither of the two possible sources (i. e., volcanic intrusion and shearing motion) is adequate to account for the geothermal pattern actually observed in the vicinity of the

430

ypothetical downgoing lithospheric plates.

A hypothesis of dissipative heating beneath an island arc provides a quantity of heat that is consistent with observational data. Such a hypothesis does not, however, provide a satisfactory mechanism to explain how the dissipated heat is carried rapidly to the surface where the anomalous pattern is observed (Hasebe et al., 1970).

Other problems associated with the thermal patterns of hypothetical downgoing lithospheric slabs have been discussed (See, e. g., Utsu, 1967; Isacks et al., 1968; McKenzie and Sclater, 1968; Isacks and Molnar, 1969; Sykes et al., 1969; Griggs, 1970; Hasebe et al., 1970; Jacob, 1970; Minear and Toksöz, 1970a, 1970b; Hanks and Whitcomb, 1971; Luyendyk, 1971; McKenzie, 1971; and Toksöz et al., 1971). None of these authors has provided a completely satisfactory explanation for exactly how the hypothesized downgoing lithospheric slab produces the geothermal pattern actually observed.

<u>Correlations</u> <u>between</u> <u>Patterns</u> <u>of</u> <u>Geothermal</u> <u>Activity</u> <u>and</u> <u>Other</u>
<u>Geophenomena</u> <u>and</u> <u>Features</u>.

In spite of problems encountered interpreting the heat-flow
patterns associated with low-velocity zones, island arcs, and
other anomalous areas, strong correlations do exist between
geothermal patterns and other global geophenomena and features.
There is, for example, a general correlation of heat flow with
geology, the higher heat flows being associated with the more
recent orogenies.

The earliest correlations suspected between two geophenomena involved heat-flow and gravity patterns. Lee and MacDonald
(1963) concluded that high heat-flow values probably correlate
with low values of gravity. They found that continental and
oceanic averages for both heat flow and gravity are roughly
the same when adjustments are made for topography. Specific
geoidal anomalies, however, do not correlate with present locations of continents and oceans but, rather, with deep-seated
fundamental features of the tectonosphere (Chapter 13).

Tatsch (1963, 1964, 1967, 1969, 1970) found that thermal
patterns can be shown to correlate with global patterns of other
geophenomena and features provided that the low "signal" of
the observations (i. e., the paucity of observational data
regarding heat flow) can be offset by decreasing the "noise"
of the system. One approach for reducing the "noise" involves

a transformation of coordinates from ordinary geographical to the geotectonospheric coordinates described in Chapter 3.

Elsasser (1967, 1970) found that the equality of the average outward flow of heat from oceans and continents is the result of a dynamic balance in a convecting (or otherwise well-behaved) upper mantle. Although radioactivity is approximately the same beneath oceans and continents, some significant differences exist within the upper mantle to maintain this dynamic balance (Polyak and Smirnov, 1966). How this is done is not known.

Horai (1969) found a global correlation between heat flow Q and seismic delay time dt:

$Q = (0.69 \, dt + 1.67)$ HFU.

For a seismic delay time of 5 seconds, this rule predicts a heat flow of

$Q = 0.69 \times 5 + 1.67 = 5.12$ HFU.

For a "negative" delay time (i. e., early arrival) of 1 sec, the rule predicts a heat flow of

$Q = 0.69 \times (-1.0) + 1.67 = 0.98$ HFU.

Lubimova and Feldman (1970) found a correlation between geoelectrical parameters and geothermal activity in many areas. Other researchers found other correlations. However, of all the correlations reported, the most significant is that between heat flow and the age of the underlying basement rock. Lubimova (1969) found that heat-flow values are very uniform in all

433

shields (0.92 ± 0.38 HFU), as well as in all Meso-Cenozoic orogenic areas (1.92 ± 0.49 HFU), the general correlation being that heat flow decreases with increase in age of basement. Viewed in this light, the lack of exact spatial correlation between gravity and heat flow is explained by a temporal lag between the Earth's internal thermal behavior and the causal mechanism producing changes in gravity. This temporal lag could easily amount to millions of years between measured heat flow and measured gravity at the surface of the Earth. This lag, converted to spatial dimensions, would be of the order of hundreds of km, at 1 cm/yr. Later analyses regarding geothermal correlation with age of basement served to confirm Lubimova's findings (See, e. g., Verma et al., 1970; Fotiadi et al., 1970).

Lubimova found also that temporal variations in heat-flow rates are significantly different for regions of tectonogenesis of dissimilar ages provided the deepest levels of the upper mantle are assumed to be responsible for the formation of the tectonic zones and provided the tectonogenesis of dissimilar age is characterized by various intensities of sialic inflow from the mantle. Neither of these assumptions can be confirmed directly, but indirect evidence suggests that they are realistic assumptions.

Based on the above and other analyses, a generalization may be made: regions of modern orogenesis correlate with high heat-flow rates. This suggests that the same causal mechanism that controls the heat-flow rates on a global scale also controls the Earth's orogenic activity. What is this causal mechanism? When did it start functioning? What started it and what keeps it going?

Possible Sources of Energy Involved in the Earth's Observed Heat-Flow Patterns.

It has long been recognized that the Earth's heat-flow pattern is a surficial manifestation of the Earth's internal behavior. But what is the source of the energy that drives the Earth's internal behavior, and what is the nature of this source that causes it to portray a unique pattern on the surface of the Earth?

Many possible sources have been suggested, but none is completely satisfactory (Sykes et al., 1970; Elsasser, 1971; Ichiye, 1971; Magnitsky, 1971; Toksöz et al., 1971; Torrance and Turcotte, 1971). Among those proposed are: gravitational sinking of dense material in island arcs; shear forces on the bottoms of lithospheric plates being convected in the astheno-sphere; the decay of radioactive nuclides; chemical differences inhereted from the formation of protoplanets; the vertical trans-port of magma in a moving fracture beneath ocean ridges; and frictional heating along Benioff planes.

The great diversity of these possible sources, as well as their individual inadequacies, indicates just how very little is known about the actual source of energy involved in the Earth's characteristic heat-flow patterns. Most hypotheses suggest that, whatever the ultimate source, the energy is closely related to thermal-conductivity patterns within the Earth's tectonosphere.

Thermal Conductivity within the Earth's Tectonosphere.

Closely related to the problem of determining the Earth's energy sources is thermal conductivity within the tectonosphere. The two most important mechanisms for thermal-energy transfer within the Earth are (1) radiative transfer and (2) mass transport (Clark, 1969).

Insofar as it concerns the Earth's tectonosphere, radiative transfer involves the displacement of phonons of heat energy from one location to another. Mass transport involves the displacement of heated material from one location to another.

Significant amounts of thermal energy may be transferred from one region of the Earth to another by mass transport of heated material such as that involved in intrusive and extrusive activity. Generally, the higher the velocity of mass transport, the greater the quantity of thermal energy transferred.

In most cases involving mass transport of heated material within the tectonosphere, components of both vertical and horizontal mass displacements are involved. Within the low-velocity zone, the mass transport of heated material is largely horizontal with a small vertical component. The same is true of lenses and similar intrusive bodies near the Earth's surface during the final stages of their emplacement when, blocked from further upward movement, they spread outward horizontally. In stocks, volcanoes, diapirs, and similar bodies, the mass transport of

437

heated material is largely vertical with a small horizontal component. Because of the inherent structural and compositional heterogeneity of the tectonosphere, a truly vertical or truly horizontal transport situation is not likely.

The critical velocity for mass transport involving thermal-energy transfer is about 10^{-9} cm/sec, or about 0.03 cm/yr. At velocities greater than this, mass transport is the dominant mechanism for transferring thermal energy within the Earth (Clark, 1969). In fact, mass transport, in the form of intrusive and extrusive activity, represents the Earth's thermal "safety valve" as earthquakes represent the geometrico-mechanical safety valve.

When the thermal energy within a given region of the tectonosphere exceeds the ability of that region to contain the energy, there are two possible consequences: (1) the tectonosphere weakens to permit efficient flow of the excess thermal energy; or (2) that region of the tectonosphere partially melts and moves upward as intrusive or extrusive activity or moves sideward as part of a low-velocity zone. Through various combinations of these two mechanisms, excess thermal energy is removed from the Earth's tectonosphere. In a sense, the affected regions have merely been obeying the second law of thermodynamics during the past 4.6 b. y.

Several questions concerning the Earth's thermal conductivity remain unanswered. Why, for example, does more heat accumulate in some parts of the tectonosphere than in others? It has long been known that the Basin and Range area of the western U. S. is hotter, on average, than the rest of the U. S. Why has thermal energy accumulated under the Basin and Range area more than it has under the Atlantic seaboard? Does this mean that the subsurface geometry, mechanics, and chemistry of the Basin and Range area are significantly different from those of the Atlantic seaboard? If so, what causes these inherent differences between the Atlantic and Pacific seaboard areas of the U. S.?

Before considering answers to these and related questions, it is well to look briefly at existing models regarding the petrogenesis of the Earth's lithosphere, i. e., the origin of the top layer of the tectonosphere, with particular attention to the function of thermal energy in this process. Until the origin, evolution, migration, and present behavior of thermal energy within the process of lithospheric petrogenesis is understood, there is little hope of understanding the behavior of mass transport and radiative transfer at depths below the lithosphere.

439

Heat-Flow Patterns as Constraints on the Petrology of Lithogenesis

Heat-flow patterns appear to be related to the petrology involved in the genesis of the lithosphere. The nature of this relationship if not known. But one of the primary keys to understanding this relationship requires an understanding of how thermal energy participates in the process by which mantle material is converted to crustal material. The paucity of data prevents a complete analysis of this process. Partial analyses may be made, however, on the basis of constraints placed by thermal data upon certain idealized models regarding the evolution of the Earth's tectonosphere. One area facing such thermal-data constraints involves petrologic models for the upper part of the tectonosphere.

Specifically, the observed decrease in heat flow with distance from the crests of mid-ocean ridges provides a physical parameter with which to test existing petrologic models in the area of lithospheric-plate origin. Anderson (1971) used the observed low heat-flow anomalies on the flanks of some spreading ridges to hypothesize an endothermic chemical reaction associated with the initial formation of lithospheric plates.

Because heats of formation for the deserpentinization of chrysolite and for some amphibolite-pyroxene reactions do satisfy Anderson's hypothesis in some cases only, further studies are needed to develop a model that removes the inconsistencies between the observational data and predictions of proposed models. The next section takes a closer look at global geothermal patterns as a measure of heat-transfer variations within the Earth's tectonosphere as a means of doing this.

Global Geothermal Patterns as a Measure of Heat-Transfer Variations within the Earth's Tectonosphere.

One way of interpreting global geothermal patterns is to assume that they represent variations of heat-transfer efficiency within the Earth's tectonosphere. This somewhat oversimplified approach has not always been obvious nor is it completely acceptable today.

About 15 years ago, initial analyses of heat-flow data supported the concept of a worldwide convective pattern (Bullard et al., 1956; Von Herzen, 1959). With the increase of areas analyzed and of the quantity of data collected, several inconsistencies within this simple convective model became apparent.

First, the characteristics of the relatively broad band of high heat flow values on the East Pacific Rise were found to contrast sharply with those of the narrow and sometimes nonexistent heat-flow bands of the Mid-Atlantic and Mid-Indian ocean ridges (Von Herzen and Langseth, 1965; Langseth et al., 1966). Second, long narrow heat-flow patterns, coincident with ridge crests, were found to be incongruous with the simple

441

convective pattern previously hypothesized (Von Herzen and Lee, 1969). Third, profiles across oceanic ridges indicated an extremely variable pattern near the crests, including the relatively smooth topography of the East Pacific Rise (Von Herzen and Uyeda, 1963). Fourth, some high heat-flow regions did not appear to be associated with oceanic ridges; and, at the boundaries of some of these anomalous heat-flow regions, the transition to a region of different heat flow was found to occur within relatively short distances (Von Herzen, 1967). Alternate hypotheses were formulated on the basis of these complexities.

One of the first of the hypotheses that attempted to embody some of the anomalies of the geothermal patterns was that which associated a relatively shallow convective pattern with magmatic intrusions near the surface (Von Herzen and Uyeda, 1963; Von Herzen and Vacquier, 1966).

An alternate hypothesis suggested that the convective pattern and the upwelling of material beneath some ridges have been rejuvenated fairly recently after a period of quiescence (Langseth et al., 1966). Several questions remain unanswered. Why does this phenomenon occur only beneath some ridges? What initiates the periodic quiescence? What terminates the quiescence? What activates the rejuvenation?

Beloussov (1968) rejected the concept of a descending branch within the hypothesized convective system. He concluded that

442

observed high heat-flow values can be associated only with the rising of heated material from the mantle. Specifically, his analyses showed that the rising material occurs in the form of basalt asthenoliths fused out of the peridotite of the upper mantle within the seismic waveguide.

Lee (1968) suggested "selective fusion" as an alternative to convection for the Earth's observed geothermal pattern. He supported his hypothesis with the observation that pronounced fractionation of radioactive elements usually occurs in selective-fusion processes. At about that time, Horai and Uyeda (1969) concluded, from an analysis of geothermal patterns due to all causes, that nearly all anomalously high heat-flow values are related to either intrusive or extrusive magmatic activity. It can be shown, in this connection, that any heat produced below approximately 750 km could not have reached the surface only by heat conduction during the existence of the Earth (Tikhonov et al., 1969). Heat reaching the Earth's surface from depths greater than 750 km must, therefore, have followed paths of preferential heat flow such as those discussed in Chapter 6.

Simmons and Roy (1969) described the North American geothermal pattern as being composed basically of two provinces: (1) and eastern heat-flow province in which a few anomalies, clearly associated with crustal variations of radioactivity, are

443

superimposed on a rather uniform datum of about 1.2 HFU; and (2) a western province in which generally high heat-flow values greater than 2 HFU are partially attributable to upper mantle phenomena. Simmons and Roy did not specify the details of the particular "upper mantle phenomena" beneath the western part of the North American continent that account for the unique heat-flow province described.

In Eurasia, similar variations, according to Lubimova and Polyak (1969), were noted in areas of Paleozoic folding, except that the excess above average was not so great: 1.6 HFU. Part of this excess is attributable to (1) perturbations introduced by the effects of salt-dome tectonics in northern Germany, (2) more-active later tectonics in southern Germany and western Czechoslovakia, and (3) Cenozoic magmatism in all these areas. Average heat-flow values in the Alpine foldings are slightly higher (1.7 HFU) than those of the Paleozoic foldings and show three trends: (1) the foredeeps have a below-average value of 1.2 HFU; (2) the folded structures of the Alps, Carpathians, and Caucasus have a slightly-above-average value of 1.8 HFU; and (3) the Cenozoic volcanism in the Caucasus, in East Kamchatka, in the Hungarian depression, and in the southern Herz have a far-above-average value of 2.2 HFU.

The heat-flow values in the continental areas of the U. S., Canada, Australia, and India show a definite correlation with

444

age, the highest heat-flow values being associated with the youngest basements. Hamsa and Verma (1969) predicted that such correlation, when confirmed in detail, will permit a better understanding of the relationships between geology and the age of past tectonic events that motivated the geological features now observed. They did not specify a causal mechanism for the observed correlation.

A similar analysis (Makarenko et al., 1969) showed that heat-flow values reflect the time of crustal consolidation, i. e., the higher the heat-flow values, the more recent the geosynclinal activity. In their model, active magmatism is possible only in regions of high heat flow. Low heat-flow values reflect one or more of the following: (1) crustal and mantle quiescence; (2) negative topography; (3) thick sedimentary basins; (4) high isostatic anomalies; and (5) low gradients of vertical heat flow. An extension of their analysis revealed a close correlation between heat flow and depth of the high-conductivity layer.

In analyses related to paths of preferential heat flow within the mantle, Gupta and associates (Gupta et al., 1970) found that Pliocene-Miocene igneous intrusions into the crust account for the present high heat-flow values in certain areas, suggesting that the igneous intrusion provides a "preferential" path for the flow of heat from the interior of the tectonosphere

445

to the surface of the Earth. A similar situation was shown
to exist in Korea (Mizutani et al., 1970).

Not all paths of preferential heat flow follow intrusive
and extrusive activity. Faults in the Precambrian basement
also provide paths of preferential heat flow (Rao et al., 1970).

An analysis of these and other geothermal data indicates
that most geothermal patterns reflect a variation in heat-
transfer efficiency. That is, heat-flow anomalies generally
are low in regions of inefficient heat transfer; high, in regions
of efficient heat transfer.

Global Geothermal Patterns According to the Tectonospheric Earth Model.

Global heat-flow and other geothermal patterns, according to the tectonospheric Earth model, are identifiable with the wedge-belts of activity and with the paths of preferential heat flow associated with them (Chapter 6).

Using standard geographical coordinates, heat-flow and other geothermal data may be expressed as a function of their geographical distribution on the surface of the Earth:

$h = h(\Theta_k, \phi_k)$,

where Θ_k is colatitude and ϕ_k is longitude in standard geographical coordinates.

This equation may be expanded to any degree of resolution or order N:

$$h = \sum_{n=0}^{N} \sum_{m=0}^{m} \left[A_n^m \cos m\,\phi_k + B_n^m \sin m\,\phi_k \right] P_n^m (\cos \Theta_k)$$

for values of k from 1 to N (See, for example, Kreyszig, 1962; Merritt, 1962; Abramowitz and Stegun, 1964; Lee and Uyeda, 1965; or any reference discussing associated Legendre functions of the second kind).

According to the dual primeval planet hypothesis, geothermal patterns are "harmonic" with respect to the geotectonospheric coordinate system, rather than with respect to the geographical coordinate system based on the present rotational axis of the

447

Earth. That is, the dual primeval planet hypothesis and the tectonospheric Earth model postulate that the Earth's surficial geothermal pattern manifests its geotectonic structure and behavior during the past 4.6 b. y., rather than manifesting its undefined association with the Earth's present axis of rotation.

Preliminary analyses have verified this hypothesized "harmonic association" by means of transforming the coordinates from the standard geographical system to the geotectonospheric system described in Chapter 3 (Tatsch, 1963, 1964, 1967, 1969, 1970). This transformation, briefly stated, involves substituting geotectonospheric coordinates for the values of Θ_k and ϕ_k in the above equation.

Analyses made with these transformed data reveal the predicted correlation between geothermal data and other geophenomena and features when considered on a global scale; e. g., intrusive and extrusive activity (Chapter 10); the Earth's gravity field (Chapter 13); and the Earth's geomagnetic field (Chapter 14).

REFERENCES

Abramowitz, M. and Stegun, I. A. (Editors), 1964. Handbook of Mathematical Functions. National Bureau of Standards, Washington, 1046 pp.

Anderson, D. L. and Sammis, C., 1969. The low velocity zone. Geofis. Internac., 9: 3-19.

Anderson, R. N., 1971. Heat flow and the petrology of mid-ocean ridges. Geol. Soc. Am. Abs. w. Programs, 3(2): 72.

Beloussov, V. V., 1968. Some general aspects of development of the tectonosphere. In: K. Benes (Editor), Upper Mantle (Geological Processes), Report of 23rd Session, IGC. Academia, Prague, pp. 9-17.

Boldizsar, T., 1969. Terrestrial heat flow and Alpine orogenesis. Bull. Volcanol., 33(1): 293-297.

Bullard, E. C., Maxwell, A. E., and Revelle, R., 1956. Heat flow through the deep sea floor. Advan. Geophys., 3: 153-181.

Chase, R. L. and Bunce, E. T., 1969. Underthrusting of the eastern margin of the Antilles by the floor of the western North Atlantic Ocean, and origin of the Barbados Ridge. J. Geophys. Res., 74: 1413-1420.

Clark, S. P., 1969. Heat conductivity in the mantle. In: P. J. Hart (Editor), The Earth's Crust and Upper Mantle. Am. Geophys. Union, Washington, pp. 622-626.

Elsasser, W. M., 1967. Interpretation of heat flow equality. J. Geophys. Res., 72: 4768-4770.

449

Elsasser, W. M., 1970. Non-uniformity of crustal growth. *Trans*. *Am*. *Geophys*. *Union*, 51: 823.

Elsasser, W. M., 1971. Sea-floor spreading as thermal convection. *J*. *Geophys*. *Res*., 76: 1101-1112.

Fotiadi, E. E., Moiseenko, U. I., Sokolova, L. S., and Duchkov, A. D., 1970. Geothermal investigations in some regions of western Siberia. *Tectonophysics*, 10: 95-101.

Griggs, D. T., 1970. The sinking lithosphere and the focal mechanism of deep earthquakes. In: *Symposium* *on* *the* *Nature* *of* *the* *Solid* *Earth*, in Honor of Professor F. Birch. McGraw-Hill, N. Y., in press.

Gupta, M. L., Verma, R. K., Hamsa, V. M., Rao, G. V., and Rao, R. U. M., 1970. Terrestrial heat flow and tectonics of the Cambay Basin, Gujarat State. *Tectonophysics*, 10: 147-163.

Hamsa, V. M. and Verma, R. K., 1969. The relationship of heat flow with age of basement rocks. *Bull*. *Volcanol*., 33: 123-152.

Hanks, T. C. and Whitcomb, J. H., 1971. Comments on paper by J. W. Minear and M. N. Toksöz, "Thermal regime of a downgoing slab and new global tectonics". *J*. *Geophys*. *Res*., 76: 613-616.

Hasebe, R., Fujii, N., and Uyeda, S., 1970. Thermal processes under island arcs. *Tectonophysics*, 10: 335-355.

Heezen, B. C., Gerard, R. D., and Tharp, M., 1964. The Vema fracture zone in the equatorial Atlantic. *J*. *Geophys*. *Res*., 69: 733-739.

Hess, H. H., 1962. History of the ocean basins. In: A. E. J. Engle et al. (Editors), Petrological Studies. Geol. Soc. Am., New York, p. 599.

Horai, K., 1969. Cross-covariance analysis of heat flow and seismic delay times for the Earth. Earth and Planetary Sci. Letters., 7: 213-220.

Horai, K. and Uyeda, S., 1969. Terrestrial heat flow in volcanic areas. In: P. J. Hart (Editor), The Earth's Crust and Upper Mantle. Am. Geophys. Union, Washington, pp. 95-109.

Ichiye, T., 1971. Continental breakup by nonstationary mantle convection generated with differential heating of the crust. J. Geophys. Res., 76: 1139-1153.

Isacks, B., Oliver, J., and Sykes, L. R., 1968. Seismology and the new global tectonics. J. Geophys. Res., 73: 5855-5899.

Isacks, B. and Molnar, P., 1969. Mantle earthquake mechanisms and the sinking of the lithosphere. Nature, 223: 1121-1124.

Jacob, K. H., 1970. P-residuals and global tectonic structures investigated by three-dimensional seismic ray tracings with emphasis on Longshot data. Trans. Am. Geophys. Union, 51: 359.

Kreyszig, E., 1962. Advanced Engineering Mathematics. Wiley, New York, 856 pp.

Lambert, I. B. and Wyllie, P. J., 1970. Low-velocity zone of the Earth's mantle: incipient melting caused by water. Science, 196: 764-766.

Langseth, M. G., 1969. The heat flow through the surface of the oceanic lithosphere. Tectonophysics, 7: 545.

Langseth, M. G., LePichon, X., and Ewing, M., 1966. Crustal structure of the mid-ocean ridges; 5: Heat flow through the Atlantic Ocean floor and convection currents. J. Geophys. Res., 71: 5321-5355.

Lee, W. H. K., 1968. Effects of selective fusion on the thermal history of the Earth's mantle. Earth and Planetary Sci. Letters, 4: 270-276.

Lee, W. H. K. and MacDonald, G. J. F., 1963. The global variation of terrestrial heat flow. J. Geophys. Res., 68: 6481-6492.

Lee, W. H. K. and Uyeda, S., 1965. Review of heat flow data. In: W. H. K. Lee (Editor), Terrestrial Heat Flow. Am. Geophys. Union, Washington, pp. 87-190.

Liebermann, R. C. and Schreiber, E., 1969. Critical thermal gradients in the mantle. Earth and Planetary Sci. Letters, 7: 77-81.

Lubimova, E. A., 1969. Heat flow patterns in Baikal and other rift zones. Tectonophysics, 8: 457-467.

Lubimova, E. A. and Polyak, B. G., 1969. Heat flow map

of Eurasia. In: P. J. Hart (Editor), The Earth's Crust and Upper Mantle. Am. Geophys. Union, Washington, pp. 82–88.

Lubimova, E. A. and Feldman, I. S., 1970. Heat flow, temperature, and electrical conductivity of the crust and upper mantle in the U. S. S. R. Tectonophysics, 10: 245–281.

Luyendyk, B. P., 1971. Comments on paper by J. W. Minear and M. N. Toksöz, "Thermal regime of a downgoing slab and new global tectonics". J. Geophys. Res., 76: 605–606.

Magnitsky, V. A., 1971. Geothermal gradients and temperatures in the mantle and the problem of fusion. J. Geophys. Res., 76: 1391–1396.

Makarenko, F. A., Polyak, B. G., and Smirnov, Y. A., 1969. Thermal regime of the upper parts of the lithosphere. Bull. Volcanol., 33: 281–291.

McKenzie, D. P., 1967. Some remarks on heat flow and gravity anomalies. J. Geophys. Res., 72: 6261–6273.

McKenzie, D. P., 1971. Comments on paper by J. W. Minear and M. N. Toksöz, "Thermal regime of a downgoing slab and new global tectonics". J. Geophys. Res., 76: 607–609.

McKenzie, D. P. and Sclater, J. G., 1968. Heat flow inside the island arcs of the northwestern Pacific. J. Geophys. Res., 73: 3173–3179.

Merritt, F. S., 1962. Mathematics Manual. McGraw-Hill, New York, 378 pp.

Minear, J. W. and Toksöz, M. N., 1970a. Thermal regime of a downgoing slab and new global tectonics. J. Geophys. Res., 75: 1397–1419.

Minear, J. W. and Toksöz, M. N., 1970b. Thermal regime of a downgoing slab. Tectonophysics, 10: 367–390.

Mizutani, H., Baba, K., Kobayashi, N., Chang, C. C., Lee, C. H., and Kang, Y. S., 1970. Heat flow in Korea. Tectonophysics, 10: 183–203.

Oliver, J. and Isacks, B., 1967. Deep earthquake zones, anomalous structures in the upper mantle, and the lithosphere. J. Geophys. Res., 72: 4259–4275.

Polyak, B. G., and Smirnov, Y. A., 1966. Heat flow on the continents. Dokl. Akad. Nauk SSSR, 168: 170–172.

Rao, R. U. M., Verma, R. K., Rao, G. V., Hamsa, V. M., Panda, P. K., and Gupta, M. L., 1970. Heat flow studies in the Godavari Valley. Tectonophysics, 10: 165–181.

Simmons, G. and Roy, R. F., 1969. Heat flow in North America. In: P. J. Hart (Editor), The Earth's Crust and Upper Mantle. Am. Geophys. Union, Washington, pp. 78–81.

Sugimura, A., 1968. Spatial relation of basaltic magmas in island arcs. In: H. H. Hess and A. Poldervaart (Editors), Basalts -- The Poldervaart Treatise on Rocks of Basaltic Composition, volume 2. Interscience, New York, pp. 537–571.

Sykes, L. R., Isacks, B. L., and Oliver, J., 1969. Spatial

distribution of deep and shallow earthquakes of small magnitude in the Fiji-Tonga region. Bull. Seismol. Soc. Am., 59: 1093-1113.

Sykes, L. R., Kay, R., and Anderson, D. L., 1970. Mechanical properties and processes in the mantle. Trans. Am. Geophys. Union, 51: 874-879.

Tatsch, J. H., 1963. Certain volcanological implications of applying a dual primeval planet model to the Earth. Trans. Am. Geophys. Union, 44: 892.

Tatsch, J. H., 1964. Distribution of active volcanoes: summary of preliminary results of three-dimensional least-squares analysis. Geol. Soc. Am. Bull., 75: 751-752.

Tatsch, J. H., 1967. Global geomagnetic evidence supporting the existence of a geophysical equator predicted as a consequence of applying a dual primeval planet model to the Earth. Trans. Am. Geophys. Union, 48: 59.

Tatsch, J. H., 1969. Sea-floor spreading, continental drift, and plate tectonics unified into a single global concept by the application of a dual primeval planet hypothesis to the Earth. Trans. Am. Geophys. Union, 50: 672.

Tatsch, J. H., 1970. Global seismicity patterns as interpreted in accordance with a dual primeval planet hypothesis. Geol. Soc. Am. Abs. w. Programs, 2: 153.

Tikhonov, A. N., Lubimova, Y. A., and Vsalov, V. K., 1969.

Heat flow from the Earth's interior depending on inner parameters variations. Bull. Volcanol., 33: 261-280.

Toksöz, M. N., Minear, J. W., and Julian, B. R., 1971. Temperature field and geophysical effects in a downgoing slab. J. Geophys. Res., 76: 1113-1138.

Torrance, K. E. and Turcotte, D. L., 1971. Structure of convection cells in the mantle. J. Geophys. Res., 76: 1154-1161.

Utsu, T., 1967. Anomalies in seismic wave velocity and attenuation associated with a deep earthquake zone. Fac. Sci. Hokkaido Univ. Ser. 7, 3: 1-25.

Utsu, T. and Okada, H., 1968. Anomalies in seismic wave velocity and attenuation associated with a deep earthquake zone, 2. J. Fac. Sci. Hokkaido Univ., Ser. 7, 3: 65-84.

Uyeda, S. and Horai, K., 1964. Terrestrial heat flow in Japan. J. Geophys. Res., 69: 2121-2141.

Uyeda, S. and Vacquier, V., 1968. Geothermal and geomagnetic data in and around the island arc of Japan. In: L. Knopoff, C. Drake, and P. Hart (Editors), The Crust and Upper Mantle of the Pacific Area. Am. Geophys. Union, Washington, pp. 349-366.

Verma, R. K., Hamsa, V. M., and Panda, P. K., 1970. Further study of correlation of heat flow with age of basement rocks. Tectonophysics, 10: 301-320.

Vine, F. J., 1966. Spreading of the ocean floor: New evidence. Science, 154: 1405-1415.

Von Herzen, R. P., 1959. Heat flow values from the south-eastern Pacific. Nature, 183: 882-883.

Von Herzen, R. P., 1967. Surface heat flow and some implications of the mantle. In: T. F. Gaskell (Editor), The Earth's Mantle. Academic, London and New York, pp. 197-230.

Von Herzen, R. P. and Uyeda, S., 1963. Heat flow through the eastern Pacific Ocean floor. J. Geophys. Res., 68: 4219-4250.

Von Herzen, R. P. and Langseth, M. G., 1965. Present status of oceanic heat flow measurements. Phys. Chem. Earth, 6: 367-407.

Von Herzen, R. P. and Vacquier, V., 1966. Heat flow and magnetic profiles on the Mid-Indian Ocean ridge. Phil. Trans. Roy. Soc. London, A, 259: 262-270.

Von Herzen, R. P. and Lee, W. H. K., 1969. Heat flow in oceanic regions. In: P. J. Hart (Editor), The Earth's Crust and Upper Mantle. Am. Geophys. Union, Washington, pp. 88-95.

Von Herzen, R. P., Simmons, G., and Folinsbee, A., 1970. Heat flow between the Caribbean Sea and the Mid-Atlantic Ridge. J. Geophys. Res., 75: 1973-1984.

Chapter 10
INTRUSIVE AND EXTRUSIVE ACTIVITY

Intrusive and extrusive activity has played an important role in the evolution of the Earth's tectonosphere during the past 4.6 b. y. This intrusive and extrusive activity includes volcanic activity, the creation of basaltic traps, both in continental and in oceanic areas, mantle upwelling at crustal rifts, and the emplacement of batholiths, laccoliths, and other intrusive bodies during the past 4.6 b. y.

This book discusses intrusive and extrusive eruptives as a single entity in order to emphasize the concept that a single global deep-seated long-lived mechanism is responsible for the behavior of both these types of eruptives. Also, the term eruptive is used in its general sense (Challinor, 1964), i. e., including both intrusive and extrusive activities involved in breaking through from one level to another within the tectonosphere. Thus, an eruptive that "breaks through" to a subsurface level is an intrusive; one that breaks through to the surface, an extrusive. An intrusive during one orogeny may become an extrusive through the action of a subsequent orogeny.

Arthur Holmes pointed out nearly 40 years ago that the 3 major problems of igneous petrology are: (1) the nature and source of primary magmas; (2) the mechanisms of magmatic intrusion and emplacement; and (3) the processes that have brought about the manifold diversity of igneous rocks. This chapter reviews these problems and analyzes possible solutions for them. In addition, it considers one additional problem area: (4) the nature and source of the energy for the ascensive mechanism.

Types of Intrusive and Extrusive Activity.

An intrusion may be defined as a body of igneous rock that invades older rock. The invading rock may be a plastic solid or magma that pushes its way into the older basement. Some magmas may be emplaced by magmatic stoping. Intrusion includes bodies of dimensions ranging from cm (veins and dykes) to thousands of km^2 (massive outcrops). Other typical forms of intrusions include sills, laccoliths, phacoliths, lopoliths, batholiths, stocks, and bosses.

An extrusion may be defined as the thrusting, pushing out, or emission of magmatic material at the Earth's surface. Extrusion includes lava flows, volcanic domes, and certain pyroclastic rocks. These may measure cm to many m in thickness. Volcanic complexes resulting from fissure eruptions may cover hundreds of thousands of km^2 to depths of hundreds of meters.

For an analysis of the evolution of the Earth's tectonosphere, the distinction between intrusion and extrusion is not as important as understanding the driving mechanism that causes the magmatic material to move upward either as an intrusion or as an extrusion. In the last analysis, an intrusive body behaves much like an extrusive one insofar as the basic geometrical, mechanical, chemical, and thermal problems within the tectonosphere are concerned. In fact, according to the tectonospheric Earth model, it is the geometry and mechanics of tectonospheric behavior in a given region that determines whether a magmatic body in that region during that period of geologic time is intrusive or extrusive.

459

Representative Extrusive Activities.

The most common type of extrusive activity is the volcano. Volcanoes range from the explosive type to the quiescent type and come in all sizes and shapes. How a volcano forms depends not only on the subsurface structure but also on the physical and chemical nature of the erupting lava, or magma.

One of the most important factors determining the shape and activity of a volcano is the magmatic viscosity. Some magmas are so viscous that their movement it imperceptible; others move more than 30 km/hr. Generally, the less viscous the magma, the more extensive the flow of lava, the flatter the resulting structure, and the fewer and weaker the explosive eruptions. A volcano formed primarily by quiescent effusions is saucer shaped. With increase of viscosity the structure becomes convex upward and approaches a conical shape.

In the temporal frame, early flows in the life of a volcano are usually less viscous than are later flows. Explosive activity becomes more frequent and violent during the later stages of volcano evolution. In the last stages, the viscosity approaches zero with the result that the magma is thrust upward from the "feeding pipe" in an almost-solid obelisk, such as was produced to a height of over 300 m atop the dome of Mt. Pelée in 1902.

Composite volcanoes are built up by a combination of flowing lava plus the explosive discharge of rock fragments. In these volcanoes, the slopes are concave and steepest near the summit. Examples are: Shasta, Rainier, Mayon, Orizaba, Popocatepetl, and Fuji.

An explosive type volcano is composed entirely of explosion debris, which causes the volcano to grow very rapidly into a symmetrical cone with equal slopes. Such a volcano may grow to a height of more than a km within a few years.

In the plateau type of volcanic activity, unlike the cylindrico-conical type thus far discussed, the magma issues from

long fissures or fractures in the crust and produces a plateau
composed of the issuing lava and ash.

The exact nature of the plateau depends upon the magmatic
composition. If the magma is basaltic (i. e., dark, viscous,
silica-poor, and rich in Fe, Ca, and Mg) it probably comes more-
or-less directly from a basaltic layer deep within the Earth
beneath both oceanic and continental areas. If the magma is
rhyolitic or acidic (i. e., rich in alkalis and silica) it probably comes in a less-direct manner from a shallower depth.

The Earth contains many examples of both basaltic and rhyolitic plateaus. As recent as 20 m. y. ago, large basaltic eruptions occurred from fissures in the Pacific Northwest. The
eruptions were not continuous but cyclical with a quiescent
period between eruptions, such that soils and forests developed
on each flow before being buried by the next. In these eruptions, lasting about 10 m. y., a volume of 450,000 km^3 of lava
covered an area of 500,000 km^2 to depths of as much as several
km, thereby completely burying a mountainous terrain. Examples
of plateau-type basaltic flows may be found in all the continents
and oceans.

Plateau-type rhyolitic flows have also occurred in all the
continents and oceans. In the rhyolitic fissure eruptions
in the Valley of Ten Thousand Smokes, swarms of fractures suddenly developed on the valley floor and produced a gas-charged
effervescent magma. Highly mobile and loaded with droplets of
incandescent liquid, the rhyolithic effusion traveled rapidly
and extremely extensively across the valley in the form of glowing avalanches. Notable examples of plateau-type rhyolitic
flows include those in Nevada, Utah, Arizona, California, New
Zealand, Sumatra, and El Salvador.

Closer scrutiny of the various types of extrusive activity
indicates that the differences in many cases are virtual rather
than real. Much evidence indicates that the lineaments of all

461

volcanoes (and not just of the fissure type) are based on crustal
fractures (See, e. g., Tija, 1969). Instead of two types of
magma (basaltic and rhyolitic) there seems to be a complete
spectrum of eruptives extending from ultrabasic to highly acidic.
Whether an effusive is basaltic or rhyolitic, the characteristics
of its emplacement are determined largely by the gases and volatiles it contains. These are considered in the following section.

Gases within Extrusive Magmas.

The gases present in parental magmas seem to be, in the order of importance, H, CO, and N, with lesser amounts of S, F, Cl, and other vapors. In the cloud of gas that emerges from a volcano, over 90% is water vapor, with CO_2 next in abundance. It is not known how much of this water vapor is due to oxidation of hydrogen in the magma, how much is ground water, and how much is derived from water-bearing rocks surrounding the magma reservoirs and pipes.

How enormous a quantity of gas that can evolve during volcanic eruptions may be appreciated from the fact that long after the glowing avalanches covered the Valley of Ten Thousand Smokes, in 1912, the deposits of pumice continued to produce steam at the rate of 6 million gal/sec and discharged into the atmosphere more than a million tons of HCl and 200,000 tons of HF during the first year.

The exact function of gases within magmas is not known. Speculations indicate that hot gases are instrumental in maintaining high temperatures in magmas, in keeping volcanoes active, and in awakening dormant volcanoes. Thus the activity of a given volcano depends upon hot gases from below combined with the combustion of these gases and with other heat-yielding reactions within the tectonosphere (e. g., friction, phase changes, exothermic reactions, and radioactivity).

The sudden release of gases produces the violent explosive phases of volcanic activity. The gases may be held in solution in a viscous magma until heat-yielding reactions near the surface produce cataclysmic boiling, such as occurred in Mt. Pelée in 1902.

To what extent magmatic gases have contributed to the evolution of the Earth's atmosphere and hydrosphere is not known. Some hypotheses (e. g., Rubey, 1951, 1955) maintain that such contribution was virtually 100%. Others (See, e. g., Ringwood, 1960, 1966a, 1966b, 1970, 1971) disagree. The basic differences

in these hypotheses depend upon what assumptions are made regarding the Earth's primordial atmosphere and hydrosphere (See,
e. g., Ringwood, 1971). Basically, if it is assumed that the
primordial Earth had no atmosphere nor hydrosphere, then the
interior of the Earth was the source of the atmosphere and hydrosphere, and they most likely evolved via magmatic extrusions.
If it is assumed, on the other hand, that the primordial Earth
had an atmosphere or a hydrosphere, or both, then there is less
requirement that they evolved from magmatic emanations deep
within the Earth during the past 4.6 b. y.

Whatever assumption is made about the primordial Earth,
it is clear that intrusive and extrusive activity has played
a significant role in reshaping the primordial Earth to produce its present surficial features.

<u>Temporal</u> <u>Sequences</u> <u>in</u> <u>Extrusive</u> <u>Activity</u>.

Most extrusive activity, when viewed as an entity separate from intrusive activity and over a long period of time, shows a variable sequence within the geometrical, mechanical, thermal, and chemical characteristics of the applicable subsurface. Thus, a typical eruptive sequence for a composite andesitic volcano arising from an earlier dacite volcanic center may be summarized:

1. <u>Early-cone-building</u> phase: growth of one or more dacite domes plus eruption of andesite lava flows and nuées ardentes culminating in aa-flows of olivine basalt (at time t= 0).

2. <u>Summit-dome-phase</u>: vulcanian eruption with airfall pumice (lower-layer) from summit vent (t = 50 m. y.); growth of large dacite summit dome accompanied by hot avalanches (t= 50 to 100 m. y.); subsequent flank eruptions of andesite as lava flows with precursory nuées ardentes.

3. <u>Post-dome</u> phase: vulcanian eruption with airfall pumice (upper-layer) from flank vents; growth of dacite dome from flank vents; flank eruptions of andesite from sides of cone (intermittently following t = 100 m. y.).

During this latter post-dome phase, many eruptive cycles may occur. Each cycle may vary from 50 to 100 m. y. in length and typically follows a progression from strong pyroclastic eruptions (volatile-rich magma) to viscous protrusions and glowing avalanches to blocky lava flows (volatile-poor magma), and from hornblende-pyroxene dacite to olivine-pyroxene andesite.

In this type of extrusive activity, the parent magma probably was andesite and magmatic differentiation probably occurred at a shallow depth, perhaps in the conduit. The derivatives were dacite and basalt. The exact depth and nature of differentiation, and therefore of the composition of the derivatives, depend upon the composition of the parent magma as well as upon the exact mechanical, geometrical, thermal, and chemical environment through which the feeder conduits pass.

This type of extrusive activity from an andesitic parent magma has been repeated hundreds of times during the past 4.6 b. y. These have been interspersed, during that time, with activity emanating from more-basaltic parent magmas.

Ultramafic Intrusions and Extrusions.

Ultramafic rocks that are contained in layered, stratiform, and other intrusions involving gabbro and dibase, together with accumulations or concentrations of mafic minerals, were formed from mantle-derived magmas (Wyllie, 1969a, 1969b). All other ultramafites appear to be associated with major tectonic features of the Earth's crust with the specific distribution patterns being controlled by deep-seated tectonics with linear trends.

Alpinotype ultramafic rocks are presently found along deformed mountain chains and along island arcs usually with associated gabbro or basic volcanic rocks. The occurrence of serpentinites along mid-ocean ridges suggests a third type of ultramafic belt, but its relationship to the other two is not known. There are reasons to believe that all ultramafic rocks are derived ultimately from a mantle source although the petrogenesis is sometimes extremely complicated by possible coupling between mantle-derived and crustal material.

The petrogenesis of Alpinotype intrusions in orogenic regions is complicated by several post-intrusion destructive metamorphic episodes. It appears that, when allowances are made for these metamorphic events, the original intrusions represented parts of a solid, or partially-fused, mantle that flowed into or through the crust along an unstable linear orogenic belt. Vast composite sheets of Mediterranean and Himalayan areas appear to represent voluminous extrusions of mantle material breaking through the floor of the Tethyan sea floor to depths of 8 and 10 km (Maxwell, 1968).

Genetic relationships between oceanic ultramafic rocks and Alpine ultramafic belts are suggested by the observation that

466

Puerto Rican serpentinites represent uplifted oceanic crust composed of altered mantle material exposed at the surface, as well as by the observation that a "spreading" ocean floor could cause tectonic incorporation of the serpentinites of the oceanic crust into the underlying sediments of the continental rise, thus producing ultramafic rocks of the Alpine type upon metamorphism of the sedimentary pile (Wyllie, 1969a, 1969b). Similarities between the basic pillow lavas and peridotites that appear to characterize the mid-ocean ridges may be the loci of ultramafic belts just as extensive as the Alpine ultramafic belts (Wyllie, 1969a, 1969b).

From the above and related analyses (See, e. g., Drake, 1969, 1970; Drake and Kosminskaya, 1969; Green, 1969; and Simonen, 1969), it may be seen that intrusive and extrusive activity of both the mid-ocean-ridge type and the Alpine-ultramafic-belt type, when viewed on a global scale, is motivated by a single long-lived global mechanism characterized by cyclical-but-aperiodic activity along extensive linear belts of deep-seated control. Closer scrutiny reveals that intrusive and extrusive activity forms a complete spatio-temporal spectrum within each cycle of orogenic activity, whether of the mid-ocean-ridge type or of the Alpine-ultramafic-belt type. Before discussing the causal mechanism for these cyclical spectrums of eruptive activity, it is well to review some of the other observational data that have a bearing on this subject: Precambrian intrusive and extrusive activity; the emplacement of large intrusive bodies; and the occasional "downward intrusion" of magmatic bodies.

Precambrian Intrusive and Extrusive Activity.

The earliest Precambrian crustal evolution, insofar as can be determined today, occurred in the oldest shields. The trends were from thin-crustal, unstable protocontinents to a thick-crustal, relatively stable continental mass, the respective units representing the pregeosynclinal and early geosynclinal stages of crustal development.

The sites of accumulation of the sediments and the volcanics progressed from randomly and sub-linearly distributed Archean basins and trenches to relatively long, continuous Proterozoic troughs, stable platforms, and primitive geosynclines. In North America, easterly and northerly trends of Archean time gave way to northeasterly and northwesterly trends of Proterozoic time (Goodwin, 1968). Other continents showed similar changes in direction of surficial-activity trends. Immature, flysch-type Archean sediments were succeeded by mature, shelf-type Proterozoic sediments. Volcanic activity changed from simple and sequential to varied and non-sequential. In addition, a major increase in the crustal content of the main heat-producing elements (U, Th, K, and Pb) is apparent in late Archean and Proterozoic rocks, an event which in all likelihood influenced all major aspects of crustal history: lithologic, orogenic, tectonic, and biologic.

Many of the largest intrusive bodies of the Earth were emplaced during the Precambrian. Some of these large intrusives are discussed in the next section.

468

The Emplacement of Large Intrusive Bodies.

Primary granite magmas and the material causing the "granitization" appear to be genetically related to the upward migration of granite elements from a deep source (See, e. g., Simonen, 1969). These granite elements form intrusive bodies of various sizes, the largest of which, batholiths, lie within the folded belts of the Earth's crust where interaction between the crust and upper mantle has been most active during the past 4.6 b. y. Because of the close relationship between batholithic emplacement and orogenic folding, batholiths are classified as preorogenic, synorogenic, late orogenic, postorogenic, and anorogenic.

Large plutonic batholiths are particularly abundant in the basement complexes representing deeply eroded sections of Precambrian folded areas dating back 3 b. y. and more. In some folded areas, silicic plutonic rocks occupy as much as 70% of the area. Enormous batholithic masses occur also in all younger folded belts, thereby suggesting: (1) that the granite crust of the continents was already present during the oldest geologic times; (2) that the granitoids are most abundant in the deepest sections of the folded areas; and (3) that plutonic activity might have been more effective during earliest Precambrian times.

Preorogenic plutons, of gneissose and cataclastic composition, belong to an ancient substratum of the folded belt remobilized and rejuvenated during subsequent orogenic activity. The operating factors in the emplacement of these preorogenic intrusives are orogenic thrust movements and the vertical movements of old granitoids.

Synorogenic plutons were emplaced during the main period of folding, and their structures are in harmony with that of the adjoining country rocks. Because the orogenic deformation and folding continued after the emplacement, the synorogenic plutonic rocks have a marked gneissose texture. They form elongated and ovoid masses, concordant with the adjacent country rocks, and occur especially in the anticlinal zones of the folded

469

areas.

Late orogenic plutons were emplaced after the main period
of folding but while some folding was still in progress. Like
in the case of synorogenic plutons, the structures of late orogenic plutons conform with that of the contiguous country rocks.
Their structural elements (foliation and lineation), however,
do not coincide with those of the principal or regional fold
belt, suggesting some twisting of the surficial layer since
the main period of folding or some arching of the substructure
during that period. Late orogenic intrusives are mainly granites
and are more massive than the synorogenic gneissose rocks.

Postorogenic plutons are significantly discordant and cut
sharply across the structures of the adjoining folded belt or
plutonic masses contemporaneous with the folding. Both spatially and temporally, they belong to a period of mountain folding. Anorogenic plutons are entirely disharmonious and are
usually considered to be more related to deep faults than they
are to the folding itself.

Although plutonic bodies are often separated into these
five groups, they actually form a continuous spectrum of types
(or a single family) ranging from preorogenic to anorogenic.
This spectrum of types is considered again in the last section
of this chapter.

Downward Intrusive Activity.

The direction of intrusive activity is normally vertically upward with a few cases in which the activity is obliquely upward, depending upon the exact direction of the unbalanced force vector acting on the intrusive body at the time of its emplacement. In those rare cases in which the unbalanced force vector is vertically or obliquely downward (e. g., pure gravity), the body may be said to "intrude" downward. Examples are: (1) the advanced stages of the subduction of lithospheric plates during which blocks of granitic material may be carried (by the action of the unbalanced-force vector) into the basaltic mantle (See, e. g., Watkins and Huggett, 1970); and (2) the "oceanization" of continental material by downward displacement of the lighter material (See, e. g., Beloussov, 1968).

In modern plate-tectonic concepts, some forms of downward intrusion are referred to as "subduction" of downgoing lithospheric plates (Chapter 18). The more general term "downward intrusion" is preferred in this book, particularly when discussing intrusive and extrusive activity on a global scale, over long periods of time, and including all directions of intrusion.

The Nature and Source of Parental Magmas.

The nature and source of parental magmas is indeterminate because present observational evidence does not clearly define the nature of primary mantle material, considered to be the source of parental magmas for all intrusive and extrusive activity. Certain speculations may be made, however, on the basis of what evidence is available.

Because the parental magmas are considered to be basaltic, speculations usually attempt to identify the basaltic magmas that can exist as liquids at depth within the mantle and are modified in composition by low-pressure fractionation during their ascent. This identification is difficult because most basalts observed at the surface have unique chemical compositions and exhibit crystallization sequences that have been determined by low-pressure crystal fractionation processes (O'Hara, 1965; Murata and Richter, 1966). There are plausible arguments, however, for considering that the mean composition of the 1959-1960 Kilauea eruptions (of olivine and quartz tholeiite) represented quite closely the magma originally existing at a 60 km depth (Green and Ringwood, 1967).

An extremely useful indication of parental magmas, unmodified by near-surface fractionation processes, is provided by an analysis of basalts containing xenoliths and xenocrysts of high-pressure origin and of density considerably higher than that of the enclosing magma. The most common of these inclusions are lherzolite (olivine + aluminous pyroxenes + aluminous spinel) restricted to undersaturated rocks ranging from alkali olivine basalts through olivine melilite nephelinite to alnoitic and kimberlitic magmas.

The presence of the lherzolite inclusions in a magma shows that the host magma traveled very rapidly (i. e., rapidly enough to transport the inclusions) from depths greater than 30 or 35 km. Although the host magma may have suffered some contamination during this rapid passage through the crust, the

472

compositions cannot have been modified by crystal fractionation above the level at which the lherzolite inclusions entered the host magma (Green, 1969).

Similar analyses in other parts of the globe will eventually define clearly the nature and source of parental magmas. But such is not yet possible. Some of the difficulties involved may be appreciated from a review of several other hypotheses related to parental magmas.

In Scheinmann's (1962, 1968) magma-producing mechanism, all oceanic magmas are postulated to originate below the M-discontinuity. Ultrabasic magmas are produced by melting of the entire material of the mantle; basaltic magma, by its partial melting. Magma, according to Scheinmann's hypothesis, is generated by deep fractures around which the pressure decreases and along which melts ascend preferentially. At higher temperatures this mechanism generates alkaline ultrabasic magmas by melting of the entire material of the mantle. An increase in the abundance of fractures and lengthening of their life permit the escape of the entire basaltic component.

In the mantle a "barren" layer melts at a particularly high temperature and consequently with a maximum tectonic activity (as a rule, under geosynclines). In oceanic areas, such magma is rarely found, or is entirely absent. Oceanic magmas thus originate, according to Scheinmann's hypothesis, mainly within the low-velocity layer and are directly connected with deep tectonic mechanisms. Scheinmann's hypothesis does not define the geometry and mechanics of the causative "deep tectonic mechanisms".

Other Hypotheses Regarding Parent Magmas.

The "primitive" lavas that constitute the major portions of the Hawaiian shield volcanoes are all members of the tholeiitic suite. No consistent differences exist between the tholeitic rocks of volcanoes that later produced hawaiite and those that produced mugearite. Intergradations in chemical composition between the tholeiite rocks and the later, far-less abundant rocks of the alkaline suite, and interbedding of the two types in the contact zone, suggest that the alkalic rocks have been derived from a tholeiitic magma. Crystal differentiation is probably the dominant process but such factors as volatile transfer and thermodiffusion may play a part (Macdonald and Katsura, 1962).

Whether all Hawaiian lavas are derived from a single parent magma type is undetermined. Observational evidence indicates, however, that such could be the case with the postulated single parent magma of more-or-less uniform composition. Volume and time-sequence analyses indicate that the hypothetical parent is tholeiitic. Powers (1935) refuted crystal differentiation as a means of deriving undersaturated alkalic basalt from saturated tholeiitic magma. Murata (1960), on the other hand, showed that abstraction of olivine and pyroxene crystals from an undersaturated tholeiitic magma yields alkalic olivine basalt. Yoder and Tilley (1957), however, showed that a thermal barrier exists in basaltic systems, making conversion from tholeiitic to alkalic magma, and vice versa, very unlikely through crystal differentiation.

Kuno and his associates (Kuno et al., 1957) suggested wholly independent parent magmas for the tholeiitic and alkalic suites. Kuno (1960) suggested a third independent parent magma for the high-alumina basalts.

In the Hawaiian volcanoes, the existence of a small ratio of alkalic-to-tholeiitic rock argues against a separate parent magma for the alkalic series. So do all the chemical gradations

474

between the two series, as well as the interbedding of tholeiitic, alkalic, and transitional rocks, suggesting that any basalt there might be represented as a linear combination, $aA + bT + c$, where $a + b = 1$, A is the alkalic fraction, and T is the tholeiitic fraction. This idea was furthered by the differentiation of undersaturated tholeiitic magma that accumulated in Kilauea Iki crater, during the 1959 eruption, to yield a basalt of decidely alkalic affinity, thereby suggesting that alkalic basalts can be formed from tholeiitic magma.

Crystal differentiation is not necessarily the only process for producing variations within each of the principal rock suites. Other processes, such as thermo-diffusion and migration of volatiles with attendant solutes, including alkalis, also occur within magmas. The degree to which each of these processes contributes to the magmatic modification is not known. All processes must therefore be considered in each analysis of possible conversion of tholeiitic to alkalic magma.

The seemingly greater gas content of alkalic magmas, as evidenced by their more explosive eruptions and other indications of compressive confinement of the volatiles, correlates well with a greater effect of volatile transfer in altering the magmatic composition. Other things being equal, alkalic magmas appear to be associated with greater pressure and flow of volatiles than are tholeiitic magmas. If a system is characterized by an environment having high pressure and a good source of volatiles, the magmatic products are more likely to be alkalic that tholeiitic.

Global and Regional Uniformity of the Magma-Generating Process.

Many cases of intrusive and extrusive activity suggest regional and global uniformity in the process of magma generation. One such example is the unweathered fresh chilled dibase from the Triassic dykes, sheets, and flows of southeastern Pennsylvania. This dibase consists of two very distinct populations distinguished by (1) relatively higher Al and S in one (the Roseville type, which verges on high-alumina basalt) and (2) higher Mg, Ti, and Cu in the other (the York Haven type, a typical quartz tholeiite).

Both sheets and dykes are represented in both populations, and there appear to be no chemical differences between chilled facies of the sheets and dykes within a type (Smith and Ross, 1970). The chilled contacts show no signs of local assimilation, and compositions are very uniform over the entire area spanning at least 230 km.

The compositional uniformity within each type over this relatively large distance implies that: (1) a well-homogenized magma migrated laterally for 100 km from a restricted body of magma; or (2) the mantle and magma-generating process were uniform within the entire 230 km of the Triassic province under study.

A composition similar to oceanic tholeiite, coupled with Triassic age, suggests that the Triassic sedimentary basin and the magma generation have constituted an aborted episode of continental rifting, which was actually accomplished (subsequently) farther east when North America separated from Europe (and/or from Africa).

Other studies indicate that there is evidence to support a hypothesis for worldwide uniformity in the magma-generating process.

The Ascensive Mechanism.

It has been determined that the ultimate source of primary magmas lies deep within the tectonosphere; but the nature of the mechanism that causes the magmas to rise to the surface is unknown. An analysis of some of the characteristics of the ascensive mechanism may be made, however, by examining certain phenomena related thereto: (1) the types of magma as a function of depth of formation; (2) the consanguinity of extrusive flows over large areas; (3) the temporal variations in mafic intrusions; (4) the stages in upward intrusion of mantle material; (5) the pene-contemporaneous emplacement of ultramafic bodies and their enclosing volcanic rocks; and (6) the sources of energy for intrusive and extrusive activity.

Types of Magma as a Function of Depth of Formation.

A spectrum of magma types may be produced as a function of their depth of formation, alpine ultramafic (i. e., low O^{18}) being associated with the shallowest parts of the upper mantle; alkaline ultramafics (i. e., high O^{18}), with the deeper regions underlying stable continental crust. This relationship suggests that magma types are controlled by a deep-seated global mechanism. The exact nature of the deep-seated mechanism can be determined only by further analyses of the ascensive mechanism.

One of the factors bearing on the ascensive mechanism is the consanguinity of extrusive flows over large areas. This may be seen in the close relationship of widely distributed Miocene basalts along the West Coast of North America. Snavely and his associates (Snavely et al., 1971) have suggested two possible mechanisms based on plate-tectonics concepts: (1) partial melting of a subducted lithospheric plate and (2) partial melting along a nearly horizontal shear zone at the base of the American plate.

Other causal mechanisms for the observed consanguinity of magmas are possible. One of these mechanisms is described

in the last section of this chapter where the tectonospheric
Earth model is described.

Temporal Variations in Mafic Intrusions.

A systematic temporal variation seems to be shown by the composition of mafic intrusions (Mueller, 1970). In the Beartooth
Mountains of Montana and Wyoming, for example, it appears that
separate intrusions occurred at 2.6, 2.0, 1.8-1.5, 1.5-1.2, and
0.7 b. y.

These intrusions, mainly dykes, can be separated into groups
based on their TiO_2 content. TiO_2, like K_2O and Na_2O, has a
negative correlation with age. The rocks of these dykes are
continental tholeiites with low Al_2O_3 content and low K/Rb
ratios (generally less than 230). Dykes of each TiO_2 group
show the Sr-depletion trend similar to other mafic rocks of the
Wyoming Province of the Canadian Shield, presumably because
of low-pressure fractional crystallization of plagioclase.

Mueller (1970) concluded that the secular variation of
TiO_2, K_2O, Na_2O, and MgO content might be caused by an increase
in the depth of melting with time (based on recent experimental
work regarding the effects of pressure on the partitioning of
TiO_2 during partial melting). He felt that these changes may
record a steady increase in thickness of the continental plate
or a decrease in the thermal gradient with time.

If this increase in TiO_2 content with time is characteristic of basaltic volcanism in continental areas during the late
Precambrian (i. e., during the period 2.6 to 0.7 b. y. ago),
then the contribution of basaltic material to continental growth
may have decreased steadily with time during that period.

Stages in the Upward Intrusion of Mantle Material.

The ascension of mantle material, in some cases, consists of a single stage (See, e. g., Wilson, 1969). In most cases, however, two or more stages are involved in the ascensive mechanism. Proshchenko and Nanashev (1969) have analyzed the complex structure of a large massif that was emplaced in two spatially associated non-synchronous stages each lasting about 20 m. y.

In the first stage (c 359 - 338 m. y.), ultramafic and mafic rocks were emplaced; in the second (c 238 - 223 m. y.), a group of alkaline granitoid rocks. Within each of these two main stages, numerous smaller phases are detectable indicating successive emplacements and repeated rejuvenations associated with later orogenies.

Three distinct intrusive stages are detectable in the emplacement of St. Paul's Rocks above a segment of the Mid-Atlantic Ridge: (1) at the greatest depths within this intrusive complex, i. e., at about 70 km, the peridotite below the peridotite-water solidus encloses highly alkaline liquid; (2) between 70 km and the hornblende-mylonite solidus at about 50 km, small amounts of interstitial silicate liquid are prominent; and (3) at depths shallower than 50 km, solid-state intrusives were emplaced through fluids evolved by the crystallization of the liquid phase (Millhollen and Wyllie, 1970).

Closer scrutiny of St. Paul's Rocks reveals that this complex

479

includes various types of mantle and non-mantle material. The mantle-type material is represented by spinel peridotite mylonite; the non-mantle material, by brown hornblende mylonite in bands measuring mm to m thick within spinel peridotite. This "non-mantle" material is ultrabasic (SiO_2 = 36.6%) but too high in Al_2O_3 (17.2%), in CaO (13.3%), and in alkalis (4.2%) to represent mantle material. It contains 16.5% normative nephiline and appears to represent a liquid formed by partial melting of the mantle in the presence of a pore liquid.

In some complex intrusive emplacement systems, the ultramafic material in the upper mantle appears to have been pulverized and displaced into the upper crustal layers in a solid state, along deep zones of compression, to form alpinotype peridotite (Knipper, 1969). Other systems involve solid, liquid, and plastic intrusions. In still others, there is evidence of pene-contemporaneous emplacement of ultramafic bodies and their enclosing host rocks. Some of the older of these dual emplacements are considered in the next section.

The Pene-Contemporaneous Emplacement of Ultramafic Bodies and
Their Enclosing Volcanic Rocks.

An analysis of Archean and Proterozoic eugeosynclinal belts
in Canada and of Archean belts in South Africa and West Australia
indicates that the emplacement of each of two distinctive groups
of ultramafic bodies was pene-contemporaneous with that of their
respective enclosing volcanic rocks (Naldrett, 1970).

These two groups are: (1) differentiated sills, up to
1.2 km thick, that have formed from olivine tholeiite magma,
in which gravity-stratified layers parallel stratification in
the country rocks, suggesting that crystallization preceded
folding; and (2) ultramafic lenses, commonly 15 to 150 m thick,
that are characterized by an olivine-rich core and a pyroxene-
rich magma.

The marginal zones of some lenses contain large skeletal
crystals of olivine, pyroxene, and chrome spinel very similar
to experimental quench products. The skeletal textures are
evidence that the lenses were emplaced in cool rocks, close
to the surface of the volcanic pile. Some lenses are probably
extrusive; interbonding between the lenses themselves, as well
as with basalts and chert, supports an extrusive origin.

Most sills and lenses are distinct from both Alpine-type
peridotites and ophiolite complexes in many respects, including
their autochthonous nature, and warrant recognition as a separate

481

class of syn-volcanic ultramafic bodies.

Closer scrutiny of these and other intrusive and extrusive systems reveals that, although the individual systems are complex and varied, their characteristics, when viewed on a global scale over a long period of time, form certain patterns that reveal the nature and source of the energy involved. Some of these characteristics are considered in the following section.

<u>The Nature and Source of the Energy Involved in Intrusive and Extrusive Activity.</u>

The nature and source of the energy involved in intrusive and extrusive activity may be analyzed by studying patterns of associated geophenomena and features on a global scale over a long period of time. Rift valleys and associated features and phenomena, for example, afford a fairly simple example of how this type of analysis may be used. Rift valleys, including those along the crests of oceanic ridges, are always associated with uplifted areas (See, e. g., Holmes, 1965). The converse is not true, because not all uplifted areas contain rifts. In addition to uplift, the production of surficial lava flows requires that there be crustal distension, fracturing, and the ascent of gases. This leads to the questions: What causes the uplift, crustal fracturing, and the ascent of gases? Whence comes the magma and why does it occur in localized pockets or "hot spots" that form linear patterns?

It has long been known that tectonomagmatic activity has been practically continuous during the past half-billion years, with maximum activity occurring during the Caledonian, Hercynian, and Alpine orogenies. What causes magmatic activity to be cyclical but aperiodic? The next section attempts to answer this question by examining the complex nature of a single global causal mechanism for producing all magmatic activity.

483

The Manifold Diversity of Igneous Rocks.

If all igneous rocks are derived from the mantle as an
ultimate source, why is there such a large diversity in the
types of igneous rocks? And why do oceanic rocks consist of
far fewer types than do continental rocks? To answer these
questions it is necessary to take a brief look at basaltic and
granitic rock types which comprise the ingeous rocks.

The average composition of the continental crust approaches
a mixture of basaltic and granitic rocks in a 1:1 ratio; that
for the crust of the entire Earth, in a 3:2 ratio. All rocks
fall into one of four fundamental categories: acidic, intermediate, basic, and ultrabasic. Within each of these categories
the diversity is almost unlimited. Some of the hypotheses developed to explain these differences may be examined briefly.

Fine-grained granites and coarse-grained granites within
a given intrusive episode have different ages, the fine-grained
granites being the most recently emplaced. In many areas, extrusives become less alkalic near the areas of maximum extrusion (See, e. g., Harris, 1969), suggesting petrogenetic control
by the geothermal environment.

The surface of the Earth, viewed on a global scale over
the past 4.6 b. y., may be described as consisting of 5 types
of material: (1) ancient granite material older than about
3.6 b. y.; (2) middle-age granitic material between 3.6 and

0.6 b. y. old; (3) young granitic material less than 0.6 b. y. old; (4) young basaltic material less than 0.6 b. y. old; and (5) various combinations of these. Below the surface of the Earth, the material consists of these five types plus one other: (6) basaltic material older than 0.6 b. y. This suggests that the rocks at the surface of the Earth differ from subsurface rocks primarily as a result of intrusive and extrusive activity during the past 4.6 b. y. Some of the features of this intrusive and extrusive activity that influenced the diversity of surface rocks will be discussed after a brief description of basaltic and granitic rocks.

The Nature of Basalt.

Basalt is a dark-colored, very fine-grained rock of wide-spread occurrence as lava flows of all geological ages as well as the most abundant type of lava erupted from present volcanoes. A high percentage of iron in basalt produces the characteristic dark color. Non-crystalline basalts result from sudden chilling thereby producing glass. Crystals may form in a magma before its eruption as a lava, producing porphyritic basalt with relatively large crystals in a fine-grained or glassy ground mass. (See, e. g., Hart, 1969).

The upper surface of a basaltic lava flow may be blown into a cinder-like froth by the expansion of escaping gases. Even in some compact basalts, gas-blown cavities of various sizes may occur. These may be empty, crystal-lined, or mineral-filled. The latter sometimes look like almonds thereby being designated as amygdaloidal basalts. The occurrence of silica minerals in some basalts suggests that they formed in a hyper-silica environment. Olivine basalts formed in a silica-deficient environment because olivine implies silica deficiency. Other details of the emplacement environment may be determined in a similar manner from the physical aspect of the basalt.

A contrasting class of rocks, the granites, are considered in the next section.

486

The Nature of Granite.

Granite is a medium—coarse-grained rock composed essentially of quartz, feldspar, and mica. Some granites have a distinctive pattern, or porphyritic texture, due to the development of ortho-clase (primarily feldspar constituents) as conspicuous, isolated crystals much larger than those of the granular groundmass in which embedded.

The problem of whether granite crystallized from an aqueous solution or from a hot molten state led to the Neptunist-Plutonist controversy. Werner, leader of the Neptunists, reasoned that granite must constitute the "oldest" rocks, having been the first "precipitate" from the "universal ocean". Hutton, leader of the Plutonists, as well as an astute observer and researcher, rejected the Neptunist doctrine on the basis of the relative insolubility of quartz and other granitic minerals in water. In view of the limited knowledge of his day, the only alternative was that granite might have "risen in a fused condition from subterranean regions". He thereby inferred that the invaded strata would be broken, distorted, and veined, an inference subsequently confirmed by observation showing veins of granite penetrating rocks apparently there prior to the granite. Hutton also recognized many of the phenomena of metamorphism often associated with the emplacement of granitic rocks.

Although Hutton showed that granite was not the oldest

of rocks, as Werner had supposed, he failed to show that granite must necessarily have consolidated from a molten state. What was proved was simply that the up-arching granite was in a hot mobile state at the time it was breaking through and veining the surrounding rocks which were brittle and forced to break. Hutton sensed that granite might not have to be molten to flow and used thermal fluidity with the reservation that his conclusion would be invalid "if there were any other cause for fluidity besides the operation of heat".

Almost two centuries have elapsed since Hutton's day, but man still does not know exactly how granite is formed, and no one has yet been able to produce granite from its known constituents. Evidence indicates that many granites are the product of an "ultra-metamorphism" far short of actual fusion, but the details of such a postulated process are elusive (See, e. g., Read, 1955, 1957; Hart, 1969).

Many granite-like rocks have consolidated from rhyolite magma, but this does not necessarily mean that granite itself must also have consolidated from a melt. Rocks of the granite-diorite series normally are categorized in terms of their present constituents rather than on the basis of their origin or evolution (See, e. g., Holmes, 1965).

Variations in Intrusive and Extrusive Activity as the Cause for the Diversity of Surficial Rock Types.

Some of the temporal and spatial variations in intrusive and extrusive activity may be summarized to show how these variations have contributed to the diversity of surficial rock types. Whether viewed temporally or spatially, surficial rock types display definite variations. Rock series in the circum-Pacific regions, for example, display a gradual transition from less-alkalic and more-siliceous types on the oceanic side to more-alkalic and less-siliceous types on the continental side (Kuno, 1966).

Petrologic zonations are not limited to the circum-Pacific regions. A similar zonality is evident in rock series associated with intracontinental geosyncline regions (Ivanov, 1968) and other similar series may be found in all parts of the globe and in all geologic ages.

Lateral variation in the composition of volcanic rocks across island arcs and their correlation with depth to the Benioff zone has been established by a number of researchers (e. g., Kuno, 1966; Dickinson and Hatherton, 1967; Sugimura, 1968). Tholeiites occur on the oceanic side of island arcs, followed toward the continent by low, medium, and high-potassium calc-alkaline rocks, and finally be shoshonites (Jakes and White, 1969) and/or alkaline rocks. This sequence is valid only for

489

regions with a seismic zone dipping toward the continent.

Bass (1970) found that the order of extreme magma type, as a function of the spreading rate, correlates with the order of increasing magma-source depth and suggests that maximum source depth is greater under ridges with slower spreading rates. Both spatial and temporal variations are found in the compositions of basaltic liquids forming at ocean ridges (Peterman et al., 1970). Variations occur, at a given ridge, over a period of time. Variations occur, at a given time, as one moves along that ridge or to another ridge.

Because of stoping and related contaminative processes affecting ascending intrusives and extrusives, the ultimate magmatic parents of these bodies are indeterminate in most cases as are the compositions of the various levels of the tectonosphere through which the magma has ascended.

Fractional crystallization causes a gradual enrichment of Si, K, Rb, and Na with concurrent depletion of Ca, Al, Sr, Zr, Ti, Fe, and Mg as the crystallization proceeds. It is not known, however, to what extent fractional crystallization is more important than other mechanisms tending to diversify the surficial rock types because the extent of partial melting cannot be determined accurately. All that can be said is that a multitude of causes has been operating over a period of many years to produce the presently observed diversity in surficial rock types. There have been many causes and all have left their marks (See, e. g., Hart, 1969; Benes, 1968; Stemprok, 1968; Misik, 1968; Wynne-Edwards, 1969).

The next section attempts to show how the tectonospheric Earth model provides a single global deep-seated long-lived mechanism for producing the parent magmas for all rocks.

Intrusive and Extrusive Rocks: A Summary of Their Evolution.

Intrusive and extrusive rocks have much in common. Both result from surficial manifestations of the Earth's internal behavior and possibly from the same deep-seated global mechanism. Both share a long history deep within the Earth's tectonosphere.

The driving mechanism for both intrusive and extrusive magmas may be expressed in general terms as $aA + bB + C$, where a and b are constants such that $a + b = 1$, A is the energy from ascending heat and possibly of volatiles, B is the energy from local manifestations of the Earth's equilibrating energy (Chapter 6), and C is "local" energy inherent in the local environment (local stress, local radioactivity, etc.). Local environments in areas of relative horizontal tension will normally have b greater than a, with resulting basaltic flows; areas of relative horizontal compression will normally have a greater than b, with resulting andesitic flows.

The nature of the actual effect of a given driving mechanism, $aA + bB + C$, depends upon the environment in which it operates. The same driving mechanism placed in two different environments will produce entirely different effects: in one, the result might be a combination of intrusion and metamorphism; in another, a combination of extrusion and metamorphism. In most cases, however, part of the energy of the driving mechanism will be expended in transforming rocks, i. e., metamorphism.

492

In the simplest case, part of the energy of the local manifestation C of the driving mechanism, aA + bB + C, will be expended in contact metamorphism, or transformation of rocks by direct thermal contact. In most real-world cases, the metamorphism is a linear combination of contact or thermal metamorphism plus metamorphism due to folding and recrystallization.

In summary, the history of intrusive and extrusive rocks is closely associated with: (1) Earth movements (uplift and depression of oceanic and continental areas; mountain building by folding and overthrusting of rocks; earthquakes); (2) metamorphism (transformation of pre-existing rocks into new types by the action of heat, pressure, and stress; and of hot, chemically-active, migrating fluids); and (3) igneous activity (emplacement of intrusions; emissions of lavas and gases and other volcanic products). In each case, the history of any rock is the story of its reaction to the particular driving mechanism, aA + bB + C, which happened to be effective within each part of the Earth during each period of the past 3.6 b. y.

A review of the geometrico-mechanical aspects of the basic tectonospheric Earth model indicates how widely the 5 parameters, a, A, b, B, and C have varied during the past 3.6 b. y. Each rock, whether intrusive or extrusive, is a record of these long-lived cyclical but aperiodic variations of the 5 parameters responding to the ceaseless attempt of the Earth to equilibrate

493

itself. Until eventual equilibration might be reached within the Earth, intrusive and extrusive rocks will continue to record their respective responses to an ever-changing environment as they are pursued by the ever-changing local manifestations of the driving mechanism, $aA + bB + C$. In a sense, the history of each rock is a record of the variations of the 5 parameters, a, A, b, B, and C comprising its particular driving mechanism during the existence of the rock.

<u>Intrusive</u> and <u>Extrusive</u> <u>Activity</u> <u>as</u> <u>Interpreted</u> <u>by</u> <u>the</u> <u>Tectono-</u>
<u>spheric</u> <u>Earth</u> <u>Model</u>.

Intrusive and extrusive activity, according to the tectonospheric Earth model, is associated with the "cones of activity"
and "wedge-belts of activity" of the model (Chapter 6). That
is, intrusive and extrusive activity due to a given energy source
(existing at the surface of the subtectonosphere) would be expected to occur somewhere within a cone of activity centered
on the "energy source" as a vertex and intersecting the surface
of the Earth in a circle with a radius of approximately 785
km in the idealized case (Fig. 6-2e-2, Chapter 6).

When local cones of activity are formed for every point
on the subtectonospheric fracture system, the surficial trace
of their vertexes consists of 3 mutually-orthogonal great circles,
aggregating about 102,000 km in length. The resulting cones
of activity form an infinity of cones whose "envelope" may be
described as 3 global tectonospheric "wedge-belts of activity".
At the surface of the Earth, where the radius is 1.18 times
that at the base of the tectonosphere, the centerlines of these
cones form 3 mutually-orthogonal great circles aggregating 102,000
x 1.18 = 120,000 km.

All intrusive and extrusive activity, according to the
model, is motivated by the Earth's driving mechanism. Intrusive and extrusive activity resulting <u>directly</u> from the Earth's

driving mechanism would normally be expected to occur within the present position of an active wedge-belt of activity; that resulting _indirectly_ from the Earth's driving mechanism may occur anywhere within the Earth without regard to present position of the "active" wedge-belts of activity.

Because, at any given time, active or effective energy sources may be expected along about 50% of the subtectonospheric fracture system (i. e., along about 51,000 km), intrusive and extrusive activity would be expected to occur within about half of the global wedge-belt system (i. e., along "belts" aggregating about 60,000 km on the surface of the Earth) at any one time. Intrusive and extrusive activity would not necessarily occur simultaneously in all active segments of the global tectonospheric wedge-belts of activity. Also certain types of magmatic activity normally would not accompany certain types of tectonic activity. For example, ultrabasic effusives would not normally accompany compressive tectonics. Nor would more than one specific type of magmatism be expected at a given location at a given time.

496

Intrusive and Extrusive Activity Related to Evolutionary Modes
Postulated by the Tectonospheric Earth Model.

Depending upon what basic assumptions are made, the Earth's
tectonosphere developed, according to the tectonospheric Earth
model, through any one of 3 basic modes of accretionary evolution: octantal-fragment, multiple-fragment, and composite-
fragment (Chapter 3). The type and behavior of magmatic activity
during the past 4.6 b. y. would have been different in each
of these 3 accretionary modes of tectonospheric evolution.

In the octantal-fragment mode for the evolution of the
Earth's tectonosphere, the 5 octants of Earth Prime are not
completely fragmented prior to accretion onto the basic 5400-
km primordial Earth (Chapter 3). In such case, parts of each
of the 5 octants of Earth Prime might still exist today as unmodified portions of the Earth's present tectonosphere, perhaps
even as subsurficial portions of presently-existing continental
shields.

If the Earth's tectonosphere evolved through the octantal-
fragment mode, the intrusive and extrusive activity during the
past 4.6 b. y. would have been somewhat different from that
in the multiple-fragment and composite-fragment modes for the
evolution of the Earth's tectonosphere. We need not consider
the details of such differences.

The multiple-fragment mode differs from the octantal-fragment mode primarily in the degree to which the 5 "terrestrial" octants of Earth Prime were fragmentized prior to accretion onto the basic 5400-km primordial Earth. If the degree of pre-accretion fragmentation was fairly complete, then it is not very likely that any of the actual surficial rocks of present continental shields would now be identifiable with unmodified portions of the original octants of Earth Prime because intrusive and extrusive activity during the past 4.6 b. y. would have modified these multiply-fragmentized primordial rocks.

The composite-fragment mode may be considered as an intermediate mode lying between the octantal-fragment mode and the multiple-fragment mode. Depending upon the specific degree of fragmentation assumed for the octants of Earth Prime prior to accretion onto the basic 5400-km primordial Earth, this mode occupies any one of a myriad of intermediate positions within the entire spectrum, or envelope, bounded by the octantal-fragment and multiple-fragment modes as extremes. A similar statement may be made about the modification of these fragments by intrusive and extrusive activity during the past 4.6 b. y. That is, the amount and degree of modification by intrusive and extrusive activity in the composite-fragment mode would be intermediate between that in the octantal-fragment and that in the multiple-fragment mode of tectonospheric evolution.

<u>Intrusive</u> and <u>Extrusive</u> <u>Activity</u> <u>as</u> <u>a</u> <u>Function</u> <u>of</u> <u>Paths</u> <u>of</u> <u>Pre-</u>
<u>ferential</u> <u>Heat</u> <u>Flow</u> <u>within</u> <u>the</u> <u>Tectonosphere</u>.

Within the tectonospheric wedge-belts of activity, the
nature and location of intrusive and extrusive activity depends
largely upon: (1) the nature of the local stress field and (2)
the paths of preferential heat flow.

If an energy source at some point along the subtectonospheric
fracture system consisted of heat, and if the tectonosphere
above that point were homogeneous, then it might be a relatively
simple matter to determine the most probable paths that the
heat might follow through the tectonosphere to the surface of
the Earth. Because the tectonosphere (and specifically the
wedge-belt of activity) is not homogeneous, the preferential
heat-flow path (from the surface of the subtectonosphere to
the surface of the Earth) might be quite circuitous in order
to circumvent heterogeneities of low thermal conductivity. The
exact path followed by the heat would influence the nature and
location of the resulting magmatic activity, whether intrusive
or extrusive. This, according to the model, accounts for the
circuitous routes taken by some "feeders" or conduits to volcanoes and intrusive bodies.

The Relationship between Intrusive and Extrusive Magma and the Low-Velocity Zone as Interpreted by the Tectonospheric Earth Model.

The low-velocity zone is related to intrusive and extrusive activity, according to the tectonospheric Earth model, through a common mechanism, i. e., the low-velocity zone, produced by incipient melting within the upper mantle (See, e. g., Lambert and Wyllie, 1970), and magmatic activity are motivated by the same driving mechanism. The incipient melting is caused by the same mechanism that melts the magma which ascends to produce intrusive and extrusive activity. Thus, the low-velocity zone may be considered a complex global network of horizontal "intrusions" between asthenospheric plates. In these regions, the unbalanced force vector on some protomagma is horizontal (i. e., there are insufficient vertical conduits to convey all the protomagma upward). Magma unable to proceed upward will proceed sideward, if there is a horizontal path, or remain in situ if there are no paths.

Intrusive and extrusive activity is, therefore, closely related to both the low-velocity zone and the paths of preferential heat flow.

<u>Intrusive and Extrusive Magmatism and Tectonic Activity as Inter-
preted by the Tectonospheric Earth Model</u>.

Basaltic and rhyolitic magmatism is associated with crustal
thinning which in turn may be caused by (1) crustal tension,
(2) heating from below, or (3) a combination of these. Because
the amount of crustal tension and heating in a given area at a
given time is dependent upon contributions from the geometrical,
mechanical, thermal, and chemical factors of the subsurface
environment, the composition of intrusive and extrusive magmatism
forms a temporal spectrum, ranging from ultra-basicity to ultra-
acidity. This compositional spectrum repeats in a cyclical
but aperiodic manner, according to the tectonospheric Earth
model, and accounts for the great variety of basic-to-acid rocks
that have been emplaced on the surface of the Earth during the
past 3.6 b. y.

Intrusive and Extrusive Activity Associated with Uplifts and
Thrusts as Interpreted by the Tectonospheric Earth Model.

Many mountain ranges are the product primarily of vertical
uplift rather than of horizontal crustal compression as was
once thought (See, e. g., Eardley, 1963). The Laramide Rockies,
for example, east of the Paleozoic miogeosyncline, were formed
primarily by intrusive activity on a large scale consisting
essentially of oval or irregularly broad shapes having structural relief of a few hundred m (Bowdoin Dome) to 12 km (Wind
River Uplift). Later Tertiary faulting modified considerably
some of these Laramide "intrusives". Younger Laramides were
modified by erosion, sedimentation, and subsequent intrusive
and extrusive activity. When the thrust faults of the Laramide
province are charted, they prove to be, for the most part, marginal to the uplifts or intrusives. There is a general correlation between the amount of uplift and the type and prominence
of border thrusts, suggesting a common causal mechanism. This
association is very pronounced in cases of large uplifts (i. e.,
6 km or more) but may be almost undetectable in intrusives of
low or intermediate uplift.

These and similar relationships between uplifts and thrusts
suggest that vertical uplift (or intrusion from below) was the
primary deformation and that thrusting was a secondary lateral
deformation caused by unbalanced force vectors acting vertically

and obliquely downward (i. e., gravity sliding, flowing, and large-scale erosion).

In summary, the tectonospheric Earth model considers mountain uplifts as well as related features of the Earth's surface to be due to megasills and megalaccoliths deep within the silica (granitic) layer. The size, shape, and depth of intrusion, necessary to produce the various surface structures, and the nature of the border faults required to produce mountain ranges are discussed in Chapter 12.

Diapiric Intrusions as Positive and Negative Surficial Features
According to the Tectonospheric Earth Model.

Diapiric intrusives usually are evidenced by positive surficial features such as salt domes. In some cases, however,
a negative surficial feature may result when an intrusive loses
its upward unbalanced force vector and drops vertically downward.

One such negative feature is the Carswell circular structure
of Saskatchewan, a nearly circular downdropped block 40 km in
diameter, set in undisturbed clastic rocks of the Athabasca
Formation (Currie, 1969). Inside this large block, Carswell
dolomite is exposed in a marginal ring syncline, folded and
deformed. The Athabasca, undeformed, is exposed in a ring between the Carswell and a central core of basement complex characterized by deformation lamellae on quartz and by local potash
metasomatism and iron depletion. Both features are associated
with fault zones containing lenticular masses of Cluff breccia.
Currie (1969) feels that the rising diapir dragged up the peripheral Athabasca as the intrusion proceeded upward, causing
the Carswell to dome over the diapir and then slide down the
flanks, the last stage being marked by the escape of the volatiles.

Diapiric intrusions are motivated, according to the tectonospheric Earth model, by the upward unbalanced force vector

504

present in wedge-belts of activity. Thus, the Carswell circular
structure is associated with an early position of the Aleutians-
to-Galapagos wedge-belt of activity. Many other similar circular structures of western North America are associated with
this wedge-belt of activity; similar features in other parts
of the globe are identifiable with the other wedge-belts of
activity of the model.

REFERENCES

Bass, M. N., 1970. Variation of ocean basalts with spreading rate. _Trans. Am. Geophys. Union_, 51: 762.

Beloussov, V. V., 1968. Some general aspects of development of the tectonosphere. In: K. Benes (Editor), _Upper Mantle (Geological Processes)_, Report of 23rd Session, IGC, pp. 9-17.

Beloussov, V. V., 1970. Against the hypothesis of ocean floor spreading. _Tectonophysics_, 9: 489-511.

Benes, K. (Editor), 1968. _Upper Mantle (Geological Processes)_. Academia, Prague, 260 pp.

Challinor, J., 1964. _Dictionary of Geology_, 2d ed. Oxford University Press, New York, 289 pp.

Christensen, N. I., 1970. Composition and evolution of oceanic crust. _Marine Geology_, 8: 139-154.

Currie, K. L., 1969. Geological notes on the Carswell circular structure, Saskatchewan. _Can. Geol. Surv. Paper 67-32_, 6 pp.

Dewey, J. F. and Bird, J. M., 1970. Mountain belts and the new global tectonics. _J. Geophys. Res._, 75: 2625-2647.

Dickinson, W. R. and Hatherton, T., 1967. Andesitic volcanism and seismicity around the Pacific. _Science_, 157: 801-803.

Dietz, R. S. and Holden, J. C., 1970. East Indian basin (Wharton basin) as pre-Mesozoic ocean crust. _Geol. Soc. Am. Abs. w. Programs_, 2: 537.

Drake, C. L., 1969. Continental margins. In: P. J. Hart (Editor), The Earth's Crust and Upper Mantle. Am. Geophys. Union, Washington, pp. 549-556.

Drake, C. L., 1970. A long-range program of solid Earth studies. Trans. Am. Geophys. Union, 51: 152-159.

Drake, C. L. and Kosminskaya, I. P., 1969. The transition from continental to oceanic crust. Tectonophysics, 7: 363-384.

Eardley, A. J., 1963. Relation of uplifts to thrusts in Rocky Mountains. In: The Backbone of the Americas: Tectonic History from Pole to Pole. Am. Assoc. Petr. Geol. Memoir 2.

Glikson, A. Y., 1970. Geosynclinal evolution and geochemical affinities of early Precambrian systems. Tectonophysics, 9: 397-433.

Gnibidenko, H. S. and Shashkin, K. S., 1970. Basic principles of the geosynclinal theory. Tectonophysics, 9: 5-13.

Goodwin, A. M., 1968. Archean protocontinental growth and early crustal history of the Canadian shield. In: K. Benes (Editor), Upper Mantle (Geological Processes). Academia, Prague, pp. 69-89.

Green, D. H., 1969. The origin of basaltic and nephelinitic magmas in the Earth's mantle. Tectonophysics, 7: 409-422.

Green, D. H. and Ringwood, A. E., 1967. The genesis of basaltic magmas. Contr. Mineral. Petr., 15: 103-190.

Harris, P. G., 1969. Basalt magma type in African rift tectonism. Tectonophysics, 8: 427-436.

Hart, P. J. (Editor), 1969. The Earth's Crust and Upper Mantle. Am. Geophys. Union, Washington, 735 pp.

Holmes, A., 1965. Principles of Physical Geology. Ronald, New York, 1288 pp.

Ivanov, R., 1968. Zonal arrangement of rock series with respect to deep-seated masses. In: K. Benes (Editor), Upper Mantle (Geological Processes). Academia, Prague, pp. 43-56.

Jacoby, W. R., 1970. Instability in the upper mantle and global plate movements. J. Geophys. Res., 75: 5671-5680.

Jakes, P. and White, A. J. R., 1970. K/Rb ratios of rocks from island arcs. Geochim. Cosmochim. Acta, 34: 849-856.

Johnson, H. and Smith, B. L. (Editors), 1970. The Mega-tectonics of Continents and Oceans. Rutgers, New Brunswick (N. J.), 284 pp.

Julian, B. R., 1970. Regional variations in upper mantle structure in North America. Trans. Am. Geophys. Union, 51: 359.

Knipper, A. L., 1969. Mantle rocks at the Earth's surface. Priroda, 7: 41-48.

Kuno, H., 1960. High-alumina basalt. J. Petrol., 1: 121-145.

Kuno, H., 1966. Lateral variation of basalt magma type across continental margins and island arcs. Bull. Volcanol., 29: 195-222.

Kuno, H., Yamasaki, K., Iida, C., and Nagashima, K., 1957.

Differentiation of Hawaiian magmas. Japan J. Geol. Geophys., 28: 179-218.

Lambert, I. B. and Wyllie, P. J., 1970. Low-velocity zone of the Earth's mantle: incipient melting caused by water. Science, 196: 764-766.

Macdonald, G. A. and Katsura, T., 1962. Relationship of petrographic suites in Hawaii. In: G. A. Macdonald and H. Kuno (Editors), The Crust of the Pacific Basin. Am. Geophys. Union, Washington, pp. 187-195.

MacGillavry, H. J., 1970. Turbidite detritus and geosyncline history. Tectonophysics, 9: 365-393.

Maxwell, J. C., 1968. Continental drift and a dynamic Earth. Am. Sci., 56: 35-51.

McBirney, A. R., 1970. Cenozoic igneous events of the Circum-Pacific. Geol. Soc. Am. Abs. w. Programs, 2: 749-751.

Millhollen, G. L. and Wyllie, P. J., 1970. Relationship of brown hornblende mylonite to spinal peridotite mylonite at St. Paul's Rocks: experimental melting study at mantle pressures. Geol. Soc. Am. Abs. w. Programs, 2: 625.

Misik, M. (Editor), 1968. Orogenic Belts. Academia, Prague, 327 pp.

Morgan, W. J., 1970. Plate motions and deep mantle convection. Trans. Am. Geophys. Union, 51: 822.

Mueller, P. A., 1970. Secular variations in the mafic rocks of the southern Beartooth Mountains, Montana and Wyoming.

Geol. Soc. Am. Abs. w. Programs, 2: 632.

Murata, K. J., 1960. A new method of plotting chemical analyses of basaltic rocks. Am. J. Sci., 258-A: 247-252.

Murata, K. J. and Richter, D. H., 1966. The settling of olivine in Kilauean magma as shown by lavas of the 1959 eruption. Am. J. Sci., 264: 194-203.

Naldrett, A. J., 1970. Synvolcanic ultramafic bodies: a new class. Geol. Soc. Am. Abs. w. Programs, 2: 633-634.

Ocola, L. C. and Meyer, R. P., 1970. Regional upper crustal structure of mid-continent of U. S. A. Geol. Soc. Am. Abs. w. Programs, 2: 638.

O'Hara, M. J., 1965. Primary magmas and the origin of basalts. Scot. J. Geol., 1: 19-40.

Ollier, C., 1970. Volcanoes. MIT Press, Cambridge, 177 pp.

Peterman, Z. E., Coleman, R. G., and Hildreth, R. A., 1970. Sr^{87}/Sr^{86} in mafic rocks of the Troodos Massif, Cyprus. Geol. Soc. Am. Abs. w. Programs, 2: 650.

Powers, H. A., 1935. Differentiation of Hawaiian lavas. Am. J. Sci., (5) 30: 57-71.

Proshchenko, Y. G. and Nenashev, N. I., 1969. Geologic interpretation of the radiogenic age of the rocks of the Tommotsk complex gabbro-syenite intrusion. Akad. Nauk SSSR Izv. Ser. Geol., 12: 26-32.

Read, H. H., 1955. Granite series in mobile belts. In:

A. Poldervaart (Editor), Crust of the Earth. Geol. Soc. Am. Spec. Paper 62, pp. 409-430.

Read, H. H., 1957. The Granite Controversy. Murby and Co., London, 430 pp.

Ringwood, A. E., 1960. Some aspects of the thermal evolution of the Earth. Geochim. Cosmochim. Acta, 20: 241-259.

Ringwood, A. E., 1966a. Chemical evolution of the terrestrial planets. Geochim. Cosmochim. Acta, 30: 41-51.

Ringwood, A. E., 1966b. The chemical composition and origin of the Earth. In: P. M. Hurley (Editor), Advances in Earth Science. MIT Press, Cambridge, pp. 287-356.

Ringwood, A. E., 1970. Phase transformation and the constitution of the mantle. Phys. Earth Planet Interiors, 3: 109-155.

Ringwood, A. E., 1971. Core-mantle equilibrium. Geochim. Cosmochim. Acta, 35: 223-230.

Rubey, W. W., 1951. Geologic history of sea water: An attempt to state the problem. Bull. Geol. Soc. Am., 62: 1111-1147.

Rubey, W. W., 1955. Development of the hydrosphere and atmosphere with special reference to probable composition of the early atmosphere. Geol. Soc. Am. Spec. Paper 62: 631-650.

Salop, L. I. and Scheinmann, Y. M., 1969. Tectonic history and structures of platforms and shields. Tectonophysics, 7: 565-597.

Sammis, C., Jordan, T., and Anderson, D. L., 1970. Inhomogeneity in the upper mantle. Trans. Am. Geophys. Union, 51: 828.

Scheinmann, Y. M., 1962. A mechanism of formation of oceanic magma. In: G. A. Macdonald and H. Kuno (Editors), The Crust of the Pacific Basin. Am. Geophys. Union, Washington, pp. 181-186.

Scheinmann, Y. M., 1968. Magma and tectonic processes of depths. Tectonophysics, 5: 427-439.

Simonen, A., 1969. Batholiths and their orogenic setting. In: P. J. Hart (Editor), The Earth's Crust and Upper Mantle. Am. Geophys. Union, Washington, pp. 483-489.

Smith, R. C. and Rose, A. W., 1970. The occurrence and chemical composition of two distinct types of Triassic dibase in Pennsylvania. Geol. Soc. Am. Abs. w. Programs, 2: 688.

Snavely, P. D., MacLeod, N. S., and Wagner, H. C., 1971. Consanguinity of Miocene basalts in coastal Oregon and Washington and on the Columbia Plateau. Geol. Soc. Am. Abs. w. Programs, 3: 197-198.

Stemprok, M. (Editor), 1968. Endogenous Ore Deposits. Academia, Prague, 425 pp.

Sugimura, A., 1968. Spatial relation of basaltic magmas in island arcs. In: H. H. Hess and A. Poldervaart (Editors), Basalts -- The Poldervaart Treatise on Rocks of Basaltic Composition, vol. 2. Interscience, New York, pp. 537-571.

Tija, H. D., 1969. Fracture pattern on Lamongan volcano, East Java. Bull. Volcanol., 33: 594-599.

Watkins, J. S. and Huggett, T., 1970. Evidence of middle Paleozoic sea-floor spreading in the Southern Appalachians. Trans. Am. Geophys. Union, 51: 824.

Wilson, J. T., 1967. Theories of building of continents. In: T. F. Gaskell (Editor), The Earth's Mantle. Academic, London and New York, pp. 445-473.

Wilson, J. T., 1969. Aspects of the different mechanics of ocean floors and continents. Tectonophysics, 8: 281-284.

Wyllie, P. J., 1969a. The origin of ultramafic and ultra-basic rocks. Tectonophysics, 7: 437-455.

Wyllie, P. J., 1969b. The ultramafic belts. In: P. J. Hart (Editor), The Earth's Crust and Upper Mantle. Am. Geophys. Union, Washington, pp. 480-488.

Wynne-Edwards, H. R. (Editor), 1969. Age Relations in High-Grade Metamorphic Terrains. Geol. Assoc. Can., Toronto, 228 pp.

Yoder, H. S. and Tilley, C. E., 1957. Basalt magmas. Carnegie Inst. Washington Yearbook 56: 156-161.

Chapter 11

MORPHOLOGY OF THE EARTH

The expression "morphology of the Earth", as used in this book, is synonymous with "geomorphology" in its general sense: the science of the Earth's surficial features, their character, origin, and evolution (Challinor, 1964).

In the preceding chapters we have discussed how some of the disturbances operating deep within the Earth have evolved. We have discussed also how the tectonosphere behaves and how this behavior is related to its evolution during the past 4.6 b. y. In the following chapters we will discuss some of the surficial manifestations of the Earth's internal behavior: mountain building; gravity anomalies; geomagnetic variations including polarity reversals; ever-changing panoramas of oceans and continents; sea-floor spreading and other types of crustal rifting; continental drift and polar wandering; plate and block motions of slabs and pieces of the Earth's tectonosphere.

It is the premise of this book that these surficial manifestations are the result of the Earth's internal behavior. That is, the surface of the Earth has presented an ever-changing form in response to the ceaseless behavior of the Earth's interior during the past 4.6 b. y.

Prior to a discussion of the individual entities of the

514

Earth's morphology during the past 4.6 b. y., this chapter summarizes these morphological entities; attempts to place them
in their proper perspective with each other and with the Earth's
internal behavior; and proposes that this morphological behavior
results from a single global deep-seated long-lived mechanism.

The Morphological Behavior of the Primordial Earth.

In analyzing the morphological behavior of the Earth, it
is well to begin the analysis as far back as possible, i. e.,
to the beginning of the Earth, if possible. To do this, it is
necessary to make certain assumptions about the primordial Earth.
Those assumptions made in Chapter 4 will be used in this analysis.

In any analysis of the Earth's behavior during the past
4.6 b. y., it is necessary to consider the beginning, the present, and at least one intermediate time that appears to be
critical. Since the Earth's evolutionary development appears
to have been interrupted 3.6 b. y. ago, that time, therefore,
will be used as the critical intermediate time. The analysis
will consider 3 aspects of the Earth's morphological development: (1) the most probable morphological behavior of the proto-
Earth 4.6 b. y. ago; (2) the most probable morphological behavior
of the Earth 3.6 b. y. ago; and (3) the evolution of the Earth's
present morphological behavior during the past 3.6 b. y.

<u>The Most Probable Morphological Behavior of the Proto-Earth</u>
<u>4.6 Billion Years Ago.</u>

The most probable morphological behavior of the proto-Earth
4.6 b. y. ago may be approximated by speculating upon the configuration of the primordial Earth's most probable surficial
features and the nature of the Earth's internal behavior at
that time. This configuration, in turn, depends upon which
hypotheses are adopted to explore the evolution of the Earth
from the solar nebula (Chapter 1) and to explain the evolution
of the Earth-Moon system.

If it is assumed that the Earth accreted from the solar
nebula and that the Moon formed in the vicinity of (but not
part of) the Earth, then the Earth's surface would probably
have been devoid of any features other than those that would
be genetic to an accretionary process. Thus, the Earth's surface would have presented a more-or-less homogeneous aspect
4.6 b. y. ago, both from the standpoint of topography (smooth)
and composition (uniform) over the entire surface.

If, on the other hand, it is assumed that the Earth's surficial composition and/or topography were not homogeneous 4.6
b. y. ago, then such features most probably would not have survived until today without internal or external rejuvenation,
at least on one occasion since then. For the purpose of this
analysis, therefore, it may be assumed that the Earth's surface

4.6 b. y. ago was devoid of any features other than those that would have been genetic to an accretionary process of planetary formation from the solar nebula at that time.

The Earth's morphological behavior 4.6 b. y. ago would have been, therefore, primarily dependent upon the nature of the Earth's internal behavior at that time.

The Most Probable Morphological Behavior of the Earth 3.6 Billion Years Ago.

If it is assumed that the Earth had no appreciable surficial features 4.6 b. y. ago, when were the surficial features first acquired and what caused them? Did they result from an external cause of from an internal cause? If external, what was this cause? If internal, why did it occur when it did rather than sometime earlier or later?

If the surficial features resulted from external causes during the period 4.6 b. y. ago to 3.6 b. y. ago, such features would not now be active but would resemble the Precambrian shields and would have an age of about 3.6 b. y.

If the Earth acquired its surficial features from internal causes during the period 4.6 b. y. to 3.6 b. y. ago, such features would not now be active unless the same causative internal behavior has been effective ever since 3.6 b. y. ago, in which case it would be reasonable to assume that the internal causal mechanism has been operating more-or-less continuously during the past 4.6 b. y. (since nothing _internal_ to the Earth would have been available to activate such a mechanism _3.6_ b. y. ago).

In any case, it appears that the Precambrian shields existed 3.6 b. y. ago, but not necessarily in their present forms and locations, and that they have existed in one or more locations and forms without having been severely heated or shocked ever since then.

From the above it follows that the Earth's morphological behavior 3.6 b. y. ago would have been dependent primarily upon the nature of the Earth's internal behavior at that time.

The Evolution of the Earth's Present Morphological Behavior.

Whether it be assumed that the Earth's surficial features of 3.6 b. y. ago were externally induced or that they were internally induced, we must consider certain questions about what has happened to the Earth's surface since then. Have there been significant externally-induced catastrophes that might have altered the Earth's surficial features during the past 3.6 b. y.? If so, what were they, when did they occur, and what caused them? If no such external influences have operated during the past 3.6 b. y., is it reasonable to assume that all modifications to the Earth's surface since then were induced by the Earth's internal behavior?

518

To answer these and related questions about the evolution of the Earth's present morphological behavior, global-scale analyses must be made regarding both the present and the past internal behavior of the Earth. Many such analyses have been made during the past decade by individual investigators and under the supervision of international bodies such as the International Upper Mantle Commission and the Interunion Commission of Geodynamics (See, e. g., Delaney and Smith, 1969; Drake, 1970; Hart, 1969; Knopoff et al., 1968; Stacey, 1969; Takeuchi et al., 1967; Wynne-Edwards, 1969).

When the Earth is considered as a global entity, the results of the above analyses indicate that most past and present geophenomena are manifestations of the Earth's internal behavior during the past 4.6 b. y., but that the geometry and mechanics of the causative internal behavior and the nature of the ultimate driving mechanism remain to be defined. It is the writer's premise that the tectonospheric Earth model provides answers for the undefined geometry and mechanics and for the nature of the driving mechanism. Before considering how this driving mechanism explains the morphological behavior of the Earth during the past 4.6 b. y., it is well to review some of the energy and force considerations that are instrumental in the morphological behavior of the Earth.

Energy and Force Considerations in the Earth's Morphological
Behavior during the Past 4.6 Billion Years.

Regardless what assumptions are made about the form of
the Earth 4.6 b. y. ago, that form has been changing continually
ever since then. Evidence of crustal instability is undeniable.
No matter how fixed the Earth's form might have appeared at a
given time and place on the Earth, it was changing somewhere
ever since the Earth began -- and is doing so today. Parts
of the Earth are moving as a hot liquid; other parts are shifting
up or down or sideways as solid blocks or plates; and still
other parts are being washed, blown, or eroded down slopes.
Throughout geologic history this continued unrest has turned
inland seas into mountains and these, in turn, into lowlands.
It has turned valleys into mountains; mountains into valleys.

The varieties of crustal unrest during the past 4.6 b. y.
can conveniently be divided into three major groups under the
geologic processes of (1) gradation, (2) igneous activity, and
(3) diastrophism. Gradation includes weathering in the form
of rock breakup and erosion (i. e., the removal and eventual
dumping elsewhere of the debris from weathering). The agents
of erosion are surface water, wind, underground water, ocean
waves and currents, and glaciers, all aided by the force of
gravity that pulls loose fragments down any available slopes
and cliffs or into holes and fractures. Igneous activity

520

includes the extrusion of lava flows and volcanoes and the intrusion of liquid and plastic rock into the crust. <u>Diastrophism</u> includes faulting, folding, uplifting, and downwarping, i. e., in general any movement of solid parts of the Earth with respect to each other.

The major geomorphological features (e. g., mountains, plateaus, and other surficial features lifted above the geoid) are surficial manifestations of the Earth's internal behavior (i. e., diastrophism and igneous activity). It is gradation that destroys these uplifted regions and, in so doing, produces the finer details of the geomorphological panorama: cliffs, rapids, waterfalls, hill slopes, and the great diversity of rock shapes and surfaces.

Before taking a closer look at the geomorphological panorama of the Earth's surface during the past 4.6 b. y., it is well to consider briefly the energy and forces that have produced these changes continuously since the Earth began. A general (but hopelessly undefinitive) answer is that the Sun has supplied, and continues to supply, the energy for the Earth's morphological changes. Unfortunately, energy <u>alone</u> cannot produce morphological changes on the surface of a sphere. No matter how much energy has been supplied, nor for how long, the morphological changes occurred not as a result of that energy but as a result of that energy's disequilibrated state.

This chapter examines the nature, source, and consequences of the disequilibrated state of the Earth's energy during the past 4.6 b. y. Generally, morphological changes occur only when an unbalanced force is acting. Gradation, diastrophism, and igneous activity operate only in response to an unbalanced force; and the direction of the morphological change at any time and place can be only in the direction of the unbalanced force. If, for example, the unbalanced force on a geoblock or on a plate is upward, uplift occurs. If the unbalanced force on an uplifted region is downward, gradation occurs. If the unbalanced force on a magmatic body is upward, we have intrusion or extrusion upward. If the unbalanced force on a magmatic body is downward, we have downward intrusion, subduction, or collapse. If the unbalanced force on a magmatic body is horizontal, a sill (if intrusive) or a plateau (if extrusive) is formed.

Before considering the nature, source, and consequences of the unbalanced forces in the Earth's interior during the past 4.6 b. y., it is well to review the nature of the driving mechanism for the Earth's morphological behavior.

<u>The</u> <u>Driving</u> <u>Mechanism</u> <u>for</u> <u>the</u> <u>Earth's</u> <u>Morphological</u> <u>Behavior</u>
<u>during</u> <u>the</u> <u>Past</u> <u>4.6</u> <u>Billion</u> <u>Years</u>.

An approach to the definition of a driving mechanism for
the Earth's morphological behavior during the past 4.6 b. y.
has been made by the International Upper Mantle Commission and
by the Interunion Commission of Geodynamics. These bodies concluded that significant lateral inhomogeneities within at least
the upper 700 km of the Earth may be evidence for the driving
mechanism for large-scale motions of the upper portions of the
Earth. These motions, in turn, may provide, they suggest, the
process whereby the inhomogeneities are produced (Hart, 1969;
Drake, 1970). Although these investigative bodies failed to
clearly identify the exact geometrico-mechanical processes involved in the above conclusion, this and other findings left
little doubt that the driving mechanism for the Earth's morphological behavior is deep-seated and worldwide.

Beloussov (1969) concluded that the history of the development of the Earth's surficial features is fundamentally dependent upon processes operating at depths to 1000 km beneath the
surface. Other researchers concluded that it is undetermined
yet whether a process of convection is required to explain the
Earth's morphological behavior (Knopoff, 1969; Lyustikh, 1969).
More specifically, the following details regarding the possibility of mantle convection remain undetermined: (1) whether

the convection pattern can extend over the entire mantle or only over a part of it; (2) whether the pattern is continuous or intermittent; (3) what the most probable number of convection cells is; (4) whether the cells are arranged in one or several layers; and (5) what causes changes in the convection pattern.

Most researchers agree that simple models of mantle convection are unsatisfactory and must be extended to embody more details regarding the known structure and inhomogeneities of the real-world Earth (Lyustikh, 1969). Runcorn (1969) concluded that any successful hypothesis must incorporate both tensional and compressional tectonic features within a single driving mechanism. In one such hypothesis regional metamorphism may be set in motion by the flow of heat from below (Beloussov, 1969; Mehnert, 1969).

Geosynclines, among the largest snd most numerous of the Earth's morphological features, present many enigmas. It is not even known whether geosynclinal deformation is due to horizontal or to vertical forces, or to a combination of both (Beloussov, 1969; Khain and Muratov, 1969). The relationship between the distribution of geosynclines within different tectonic cycles and the Earth's internal behavior remains to be determined.

According to the tectonospheric Earth model (chapter 6), the basic behavior of the primordial Earth (i. e., the present

subtectonosphere) is assumed to be such that it tends to equilibrate itself to a state of minimum energy. The driving mechanism of the system may be expressed as a function of the disequilibration energy inherent in its initial state. In simplest
terms, then, the driving mechanism for the Earth's morphological
behavior during the past 4.6 b. y. consists essentially of the
resultant of two factors: (1) the potential energy inherent
in the geogenetically disequilibrated shape of the basic 5400-km
primordial Earth; and (2) the selectively-channeled energy from
the preferential flow of heat, and possibly of volatiles, outward from the subtectonospheric fracture system.

This driving mechanism has been responsible for all of the
Earth's morphological behavior during the past 4.6 b. y. It has,
in short, been responsible for mountain building, gravity anomalies,
geomagnetic variations including polarity reversals, facets of
the ever-changing panorama of oceans and continents, sea-floor
spreading and other types of crustal rifting, continental drift
and polar wandering, and plate and block motions of slabs and
pieces of the Earth's tectonosphere that have served to change
the Earth's morphology.

Geosynclines, Mountains, and Related Rifting and Quasi-linear Uplift: The Elongate Linear Features of the Earth's Morphology during the Past 4.6 Billion Years.

Geosynclines, mountains, and related rifting and quasi-linear uplift represent over 100,000 km of elongate linear features of the Earth's morphology. The relationship that exists between the elements of this family of elongate geomorphological features is not completely understood. It appears, however, that there is an antithetical, Phoenix-like association between them. The nature of this association becomes more obvious when mountains and geosynclines are analyzed simultaneously as morphological features resulting from the Earth's internal behavior during the past 4.6 b. y.

The most obvious feature common to mountains and geosynclines is their linearity when viewed on a global scale. Also, both appear to date far back into the history of the Earth, i. e., at least 3.6 b. y. A geosyncline, generally, appears to be an alongated downwarp in the Earth's crust, the bottom apparently subsiding deeply beneath accumulating sediments. The Alleghany mountain range evolved from a long-lived linear subsidence, followed by an accumulation of sediments during the entire Paleozoic, i. e., a span of over 300 m. y. This was culminated by great breakings, faultings, and foldings of strata along with other morphological changes that appeared to have a deep-seated causal mechanism.

Because the deposition of geosynclinal sediments probably was in shallow water, some observers (See, e. g., Hall, 1857) concluded that the geosynclinal depression and long-lived subsidence was caused by the weight of the accumulating sediments. But others (See, e. g., Dana, 1873) felt that the accumulation was a consequence, rather than a cause, of the subsidence.

Some, but not all, mountains and geosynclines appear to form along the edges of continents. Some appear to form entirely within continental areas. Others seem to form partly in and partly out. If crustal rifting may be considered as the earliest stage of the mountain-building cycle, then some mountains and geosynclines may form completely within oceanic areas. Should the paired-belt concept of miogeosynclines (non-volcanic) and eugeosynclinal (volcanic) be used (See, e. g., Stille, 1940), the identification of geosynclines along continental margins becomes confusing and even meaningless because the separation between the mio and eu elements of a geosynclinal pair may be more than a thousand km in many cases.

Closer scrutiny reveals that many eugeosynclinal elements (e. g., the American Cordilleran and Appalachian) were analogous to present island arcs and that the sediments were derived from active volcanoes within the associated sedimentary basin (Kay, 1951). Both present and past geosynclines form in an orderly

527

manner in all parts of the globe, suggesting the possibility
of orderly control on a global scale (Drake et al., 1959; Dietz,
1963; Dietz and Holden, 1966). Thus, the structure and development of the Alpine geosyncline seem to have been similar to those
of the Indonesian arcs, both having consisted of a complex system
of troughs and ridges throughout most of their histories (Aubouin, 1965).

From these and other analyses (See, e. g., Argand, 1916;
Kuenen, 1967; Laubscher, 1969; Schuchert, 1923; Sylvester-Bradley,
1968), it appears that the elongate linear features of the Earth's
morphology are controlled by a single global deep-seated long-
lived mechanism.

According to the dual primeval planet hypothesis, the elongate
linear morphological features of the Earth are associated with
the wedge-belts of activity of the tectonospheric Earth model
(Chapter 6).

<u>Geoidal</u> <u>Undulations</u>, <u>Geomagnetic</u> <u>Anomalies</u>, <u>and</u> <u>the</u> <u>Ocean-</u>
<u>Continent</u> <u>Panorama</u>: <u>The</u> <u>Quadrate</u> <u>Features</u> <u>of</u> <u>the</u> <u>Earth's</u> <u>Mor-</u>
<u>phology</u> <u>during</u> <u>the</u> <u>Past</u> <u>4.6</u> <u>Billion</u> <u>Years.</u>

The evolution of the Earth's continents during the past
4.6 b. y. may be described as an accretionary process whereby
more-or-less quadrate central bodies, or cratons, grew along
their peripheries by the action of linear elongate features,
or orogens (Chapter 7). It is not surprising, therefore, that
geoidal undulations, geomagnetic anomalies, and the ocean-
continent panorama display quadrate patterns when viewed globally.
If, for example, the Earth's gravity field and the shape of
the geoid are morphological reflections of the Earth's internal
behavior during the past 4.6 b. y., then they should have certain characteristics correlatable with such internal behavior
on a global scale during that period of time.

Under ideal conditions of isostatic equilibrium, all crustal columns exert equal pressure at some depth, and the column
extending above has a mass exactly equal to that of the compensating mass at depth (See, e. g., Heiskanen and Vening Meinesz,
1958; Heiskanen and Moritz, 1967; Caputo, 1967; Woollard, 1969a,
1969b, 1969c). Although the Earth does not behave as an ideal
isostatic system, an understanding of the effect of such behavior on the morphology of the Earth may be gained by a brief
analysis of the nature and extent to which the Earth's internal

behavior departs from that of an ideal isostatic system when viewed on a global scale over a long period of time. In making such an analysis, it is necessary to make appropriate spatial and temporal corrections to compensate for the fact that the Earth is a dynamic body and that its morphological behavior may lag behind its deep internal behavior.

When these and other appropriate adjustments are made, there is still a lack of correlation between the oceanic-continental features and the shape of the geoid. This independence of geoidal and gross morphological features suggests that the geoidal features are due to either: (1) density differences deep within the tectonosphere (deeper than the so-called "low-velocity" layer, by virtue of which isostatic balance might be feasible); or (2) density differences maintained by convection, or by a similar mechanism, in which case the density differences are more apt to exist in the upper mantle. Geoidal features could also be caused by a combination of these two types of density differences.

On the basis of present evidence, it appears that the Earth's gravity field and the shape of the geoid reflect the morphological behavior of the Earth during the past 4.6 b. y. However, it appears that mass redistributions lag behind the forces causing the Earth's internal behavior. Thus the shape of the geoid, as a geomorphological feature, correlates with the tectonospheric driving mechanism rather than with the lagging surficial manifestations of that driving mechanism.

Similar analyses and conclusions may be made in connection with the Earth's magnetic field. These and other effects of the Earth's gross quadrate features are discussed more fully in later chapters.

According to the dual primeval planet hypothesis, the large quadrate morphological features of the Earth are associated with the eight geoblocks of the tectonospheric Earth model (Chapter 6).

<u>Sea-Floor Spreading, Crustal Plate Motions, and Continental</u>
<u>Drift as Surficial Adjustments to Compensate for the Disequili-</u>
<u>brated and Ever-Changing Morphology of the Earth's Interior.</u>

When viewed on a global scale and over a long period of
time, sea-floor spreading and other forms of crustal rifting
appear to represent the morphological behavior of the Earth in
those areas where the Earth's internal behavior causes crustal
tension. Crustal plate motion, subsurface plate and block motion,
and continental drift, in a similar manner, represent the morphological behavior of the Earth in those areas where the resultant of the Earth's internal behavior is representable as
a tangential unbalanced force vector acting on a plate, on a
block, or on a continent to produce motion of that block, plate,
or continent relative to the bulk of the Earth or to a datum
therein.

In short, sea-floor spreading, crustal plate motion, subsurface plate and block motion, and continental drift, according to the tectonospheric Earth model, are surficial adjustments
to compensate for the disequilibrated and ever-changing morphology of the Earth's interior. The Earth's surficial morphology is
a temporarily delayed attempt of the Earth's surface to adjust
itself to the ceaseless equilibrating behavior of its interior.
In some cases a block may be moved to effect equilibration;
in others, a plate may be moved. In still other cases, a rift

532

may be formed to effect the equilibration. Sometimes, world-wide mountain ranges grow from these rifts in a single orogeny. At other times, an earthquake occurs when a block or plate adjusts itself to a more nearly equilibrated position in the never-ending attempt of the Earth to assume a morphology of minimum energy.

The following chapter considers some of the details of one of the largest morphological changes that the Earth has undergone: geosynclinal formation and mountain building.

REFERENCES

Argand, E., 1916. Sur l'arc des Alpes occidentales. _Eclogae Geologicae Helvetiae_, 16: 179–182.

Aubouin, J., 1965. _Geosynclines_. Elsevier, Amsterdam, 335 pp.

Beloussov, V. V., 1969. Interrelations between the Earth's crust and upper mantle. In: P. J. Hart (Editor), _The Earth's Crust and Upper Mantle_. Am. Geophys. Union, Washington, pp. 698–712.

Bird, J. F., 1970. General concepts of orogenesis in terms of lithosphere plate tectonics. _Geol. Soc. Am. Abs. w. Programs_, 2: 733–734.

Caputo, M., 1967. _The Gravity Field of the Earth_. Academic, New York, 202 pp.

Challinor, J., 1964. _A Dictionary of Geology_, 2d. ed. Oxford Univ. Press, New York, 289 pp.

Dana, J. D., 1873. On some results of the Earth's contraction from cooling, including a discussion of the origin of mountains and the nature of the Earth's interior. _Am. J. Sci. Ser._ 3, 5: 423–443; 6: 6–14, 104–115, 161–172.

Delany, P. and Smith, C. H. (Editors), 1969. _Deep-Seated Foundations of Geological Phenomena._ Special issue _Tectonophysics_, 7: 359–610.

Dewey, J. F. and Bird, J. M., 1970a. Mountain belts and the new global tectonics. _J. Geophys. Res._, 75: 2625–2647.

534

Dewey, J. F. and Bird, J. M., 1970b. Plate tectonics and geosynclines. Tectonophysics, 10: 625-638.

Dietz, R., 1963. Collapsing continental rises: an actualistic concept of geosynclines and mountain building. J. Geol., 71: 314-333.

Dietz, R. and Holden, J. C., 1966. Miogeosynclines in space and time. J. Geol., 74: 566-583.

Drake, C. L., 1970. A long-range program of solid Earth studies. Trans. Am. Geophys. Union, 51: 152-159.

Drake, C. L., Ewing, M., and Sutton, G. H., 1959. Continental margins and geosynclines: the east coast of North America north of Cape Hatteras. In: L. H. Ahrens (Editor), Physics and Chemistry of the Earth, vol. 3. Pergamon, London, pp. 110-198.

Griffin, V. S., Jr., 1970. Relevancy of the Dewey-Bird hypothesis of cordilleran-type mountain belts and the Wegmann stockwork concept. J. Geophys. Res., 75: 7504-7507.

Hall, J., 1857. Direction of the currents of deposition and source of the materials of the older Paleozoic rocks. Can. Naturalist and Geologist, 2: 284-286.

Hart, P. J. (Editor), 1969. The Earth's Crust and Upper Mantle. Am. Geophys. Union, Washington, 735 pp.

Heiskanen, W. A. and Moritz, H., 1967. Physical Geodesy. Freeman, San Francisco, 364 pp.

535

Heiskanen, W. A. and Vening Meinesz, F. A., 1958. *The Earth and Its Gravity Field*. McGraw Hill, New York, 470 pp.

Hsü, K, J., 1968. Principles of mélanges and their bearing on the Franciscan-Knoxville problem. *Bull. Geol. Soc. Am.*, *79*: 1063.

Kay, M., 1951. *North American Geosynclines*. Geol. Soc. Am. Mem. 48, 143 pp.

Khain, V. E. and Muratov, M. V., 1969. Crustal movements and tectonic structure of continents. In: P. J. Hart (Editor), *The Earth's Crust and Upper Mantle*. Am. Geophys. Union, Washington, pp. 523-538.

Knopoff, L., 1969. Continental drift and convection. In: P. J. Hart (Editor), *The Earth's Crust and Upper Mantle*. Am. Geophys. Union, Washington, pp. 683-689.

Knopoff, L., Drake, C. L., and Hart, P. J. (Editors), 1968. *The Crust and Upper Mantle of the Pacific Area*. Am. Geophys. Union, Washington, 522 pp.

Kuenen, P. H., 1967. Geosynclinal sedimentation. *Geologische Rundschau*, *56*: 1-19.

Laubscher, H., 1969. Mountain building. *Tectonophysics*, *7*: 551-563.

Lyustikh, E. N., 1969. Problem of convection in the Earth's mantle. In: P. J. Hart (Editor), *The Earth's Crust and Upper Mantle*. Am. Geophys. Union, Washington, pp. 689-692.

536

Mehnert, K. R., 1969. Petrology of the Precambrian basement complex. In: P. J. Hart (Editor), The Earth's Crust and Upper Mantle. Am. Geophys. Union, Washington, pp. 513-518.

Runcorn, S. K., 1969. Convection in the mantle. In: P. J. Hart (Editor), The Earth's Crust and Upper Mantle. Am. Geophys. Union, Washington, pp. 692-698.

Schuchert, C., 1923. Sites and natures of North American geosynclines. Geol. Soc. Am. Bull., 34: 151-260.

Stacey, F. D., 1969. Physics of the Earth. Wiley, New York, 324 pp.

Stille, H., 1940. Einführung in den Bau Nordamerikas. Borntraeger Verlagsbuchhandlung, Berlin, 717 pp.

Sylvester-Bradley, P. C., 1968. Tethys, the lost ocean. Sci. J., 4: 47-53.

Takeuchi, H., Uyeda, S., and Kanamori, H., 1967. Debate about the Earth. Freeman-Cooper, San Francisco, 253 pp.

Woollard, G. P., 1969a. Tectonic activity in North America as indicated by earthquakes. In: P. J. Hart (Editor), The Earth's Crust and Upper Mantle. Am. Geophys. Union, Washington, pp. 125-133.

Woollard, G. P., 1969b. Standardization of gravity measurements. In: P. J. Hart (Editor), The Earth's Crust and Upper Mantle. Am. Geophys. Union, Washington, pp. 283-293.

Woollard, G. P., 1969c. Regional variations in gravity.

In: P. J. Hart (Editor), The Earth's Crust and Upper Mantle.
Am. Geophys. Union, Washington, pp. 320-341.

Wynne-Edwards, H. R. (Editor), 1969. Age Relations in
High-Grade Metamorphic Terrains. Geol. Assoc. Can., Toronto,
228 pp.

Chapter 12

MOUNTAIN BUILDING

The processes that raise mountains and that crumple the
Earth's surface have traditionally posed some of the more baffling
and least-understood problems in the Earth sciences.

One of the better-developed explanations for mountain building
has been the principle of isostasy, i. e., that areas of the
crust being loaded with sediments tend to sink and that areas
relieved of some of their material by erosion tend to rise.
Another proposal is that thermal convection in the mantle beneath the continents produces subsurface movements that push the
mountains upward. Still another proposal is that mountains are
built principally by horizontal compression. Lateral movements
from the compression tend to form the sediments into large folds
of mountain ranges.

Recent observation has shown that none of these earlier
ideas provides an adequate explanation for mountain building.
Any analysis of the evolution of the Earth's tectonosphere must
consider those mountain-building processes that are related to
the internal behavior of the Earth during the 3 or 4 b. y. that
mountains have been growing on the surface of the Earth. One
of the surficial manifestations that appear to be related to
mountain building is geosynclinal development.

Geosynclines.

Since mountain building appears to be closely related to geosynclines, an analysis of geosynclinal evolution should provide some of the details regarding the exact nature of this association.

Why, for example, beginning about 350 m. y. ago and lasting for 100 m. y., was the crust in an area such as the Rocky Mountains slowly depressed to form a long narrow geosyncline? Why was this geosyncline then crushed and uplifted into mountains during a relatively shorter period of intense orogeny? Why should that region then remain quiescent while the mountains were slowly eroded away, and then, finally, rise gently as a broad arch without further crushing?

In order to answer these and related questions, it is necessary to consider the salient elements of geosynclinal evolution. One of these elements involves lithogic compositional variations within geosynclines.

Compositional Variations within Geosynclines.

Geosynclinal composition, until recently, were considered diagnostic attributes of geosynclinal assemblages, i. e., some rock compositions were thought to reflect a eugeosynclinal genesis, others a miogeosynclinal genesis. Recent observations have shown, however, that this simple relationship does not exist between rock type and geosyncline assemblage.

Schwab (1970) found from a study of Applachian and Cordilleran geosynclinal sequences that lithologic compositional variations within geosynclines are attributable to variations in geosynclinal activity. Thus, if the same type of activity occurs in eugeosynclines and in miogeosynclines, the resulting rocks will be similar in the two cases. Although any sedimentary rock type may be found in either type of geosynclinal assemblage, some rocks such as evaporites are rare in both assemblages. Other rocks such as carbonates are particularly common to miogeosynclines while others such as graywackes are particularly common to eugeosynclines. Still other rocks such as shales are common throughout both eugeosynclines and miogeosynclines.

These similarities and differences, according to Schwab, reflect different types of geosynclinal activity. When an analysis is made regarding miogeosynclines and present-day continental shelf areas genetically related thereto, the contrasting compositions between them may be shown to reflect long-term

541

secular changes. These are reflected in: (1) the mode and locus of sedimentation as a result of organic evolution; (2) the volume, location, and relief of continental blocks; and (3) the development and growth of ocean basins. Thus, variations in geosynclinal rocks reflect variations in the tectonic environment in which they evolved. Specific similarities and differences in geosynclinal rock types reflect corresponding similarities and differences in tectonic environments.

Similar analyses show that eugeosynclinal assemblages are compositionally unlike sediments covering continental rises and deep-ocean basins. Allowances must be made in the analyses of eugeosynclinal assemblages for differences in "pre-drift" oceanic sedimentation, if it is assumed that eugeosynclinal assemblages were produced by laterally accreting deep-ocean sediments onto continental cratonic peripheries (Schwab, 1970).

Geosynclinal Activity.

Geosynclinal activity appears to be related to two deep-seated processes: (1) a "sagging" of the whole crust in the process of sedimentation; and (2) "absorption" by the upper mantle of the basal horizons of the crust and subsequent upward migration of the M discontinuity. Both these processes produce the same overall result, i. e., gradual submergence of crustal material into the mantle.

Three types of tectonic structure can be identified in continental areas as being derived from crust—mantle relationships inherent in these two deep-seated geosynclinal processes: (1) platforms, where crustal absorption into the mantle is not intensive; (2) miogeosynclinal belts, where crustal absorption is intensive; and (3) eugeosynclinal belts, where some crustal material is being absorbed by the mantle through sedimentary subsidence while some of the mantle material is absorbed into the crust throught volcanism and related magmatic activity.

Rezanov (1970) concluded from a global analysis of the cyclical "absorption" of crustal material into the mantle and of mantle material into the crust that the average chemical composition of crustal material submerged in the mantle is approximately the same as that of the volcanic products derived from the mantle.

To understand the tectonic processes associated with the

cyclical interchange between crustal and mantle material, it is well to consider certain fundamental questions regarding the role and source of the forces that drive the processes inherent in geosynclinal activity.

Is the deformation within a geosynclinal belt produced by the same force that moves crustal material into the mantle and mantle material into the crust? Or is the crushing deformation observed within a geosyncline produced by an unrelated horizontally-directed force tangential to a spherical radius? If so, does this horizontally-directed force originate outside the geosyncline? Or is it possible to explain geosynclinal deformation without invoking horizontal forces?

On the assumption that a generalized global tangential force is required to produce geosynclinal deformation, geologists, for many years, used a general contraction hypothesis, postulated on cooling of the Earth, shrinking of its surface, and consequent compression of geosynclines between the hard masses of platforms. They saw the results of this compression in the folded formations associated with geosynclines and mountains.

Although the contraction hypothesis is no longer generally accepted, an alternate explanation for folding in geosynclines has not been satisfactorily detailed. Many scientists are convinced that geosynclinal folding is related to compressive forces

resulting from either: (1) the relative horizontal motions of
continents (Chapter 17); or (2) the drag of the descending branch
of a mantle convection system (Chapter 14).

Some kinds of folding (block and injection) connot readily
be related to horizontal compression. In fact, only one kind
of folding (geosynclinal or holomorphic) requires horizontal
compression of layers. Furthermore, the limited distribution
of this type of folding (usually surrounded by other types of
long, linear folding associated with geosynclines and mountains)
suggests that the required horizontal compression is localized
rather than global. In almost all cases, holomorphic folding
is attributed to one of two causes (Beloussov, 1969): (1) the
pressure of masses of rocks rising from depth through the upper
layers of the crust, or (2) the pressure of masses of rocks
sliding from elevated positions to lower levels under the influence of gravity. Over a long time period, these two causes
for holomorphic folding are not mutually exclusive; i. e., if
cause (1) did not operate, then cause (2) could not operate
over any appreciable length of time, at least not over the billions of years during which the crust has been evolving.

In any case, this point of view does not require major horizontal motions in the mantle (Beloussov, 1969); it may be satis-
fied by vertical circulation of material combined with a limited
horizontal motion sufficient to permit passage of 2 oppositely

545

directed vertical currents. Resolution of this point of view must await more-detailed observational evidence.

Another point of view derived from a long-term global analysis of geosynclinal activity postulates that all geosynclinal behavior may be categorized as either low-pressure (Hercynian) or high-pressure (Alpine). Furthermore, an entire spectrum of geosynclinal activity lies between the low-pressure and high-pressure extremes represented by the Hercynian and Alpine categories of geosynclinal activity.

At one extreme of this low-to-high pressure spectrum, we find a set of distinct characteristics associated with low-pressure geosynclinal activity: high temperature, shallow metamorphic depth, thin metamorphic zone, steep geothermal gradient, many granites and migmatites, few magmatic basics and ultrabasics, broad extent of the geosynclinal belt with relatively little associated uplift and folding. Earlier examples of low-pressure (Hercynian) geosynclinal activity include the Sveconorwegian, Karelian, Namaqualand, Ketilidian, Hudsonian, Ukrainian, West African, South American, Saamidian, Pre-Ketilidian, Kenoran.

At the opposite extreme of the spectrum of geosynclinal behavior, we find a contrasting set of characteristics associated with high-pressure (Alpine) geosynclinal activity: low temperature, very deep metamorphic depth, thick metamorphic zone, low geothermal gradient, few granites and migmatites, many magmatic

546

basics and ultrabasics, narrow extent of geosynclinal belt with much uplift and folding. Other examples of high-pressure (Alpine) geosynclinal activity include global systems evolved during the Tertiary and the Mesozoic.

Lying roughly between the extremes of Hercynian and Alpine types of geosynclinal activity are intermediate types represented by the Grenville, Assyntian, Uralian, Mozambique, and Damara.

Specific details of the diversity of geosynclinal activity inherent in the Hercynian-to-Alpine spectrum are discussed in the following section.

The Diversity of Geosynclinal Behavior.

Classically, all geosynclines fall into one of two categories: eugeosynclines and miogeosynclines. Actually, each geosyncline is a combination of both, aA + bB, where A represents the eugeosynclinal component, B represents the miogeosyncline component, a and b represent constants, either of which may be zero provided a + b = 1. Thus, if a geosyncline is purely eugeosynclinal, b is zero; if purely miogeosynclinal, a is zero. In most actual geosynclines, neither a nor b is zero, i. e., the geosyncline is partly eugeosynclinal and partly miogeosynclinal. Paticularly if viewed over a span of time, every actual geosyncline shows features of both A and B, with a and b varying with time in a cyclical but aperiodic manner.

Also the A and B elements are not necessarily single elements but multiplicities. For example, the composite Caledonian geosyncline that crosses the British Isles can be regarded, judged by the criterion of vulcanism, as consisting of at least three eugeosynclines which differ widely amongst themselves in style of tectonic deformation and in the proportion of granite rocks exposed at the present surface. At best, miogeosynclines in the Caledonian system are represented only on a very minor and temporary scale, but characteristically B-type sediments occur on the foreland platforms of the Northwest Highlands and Welsh Borderlands. On the basis of the sedimentation cri-

548

terion, however, the Welsh and Moffat geosynclines are classic examples of A-type; the Highland geosyncline, in which the Dalradians were deposited, is an almost ideal B-type with many quartzites and limestones. But the volcanic criterion, which conventionally takes priority, makes the Dalradian an A-type.

A similar diversity exists in the Alps, where the inner Pennine belt has been regarded as A-type from its Lower Cretaceous ophiolites and the extreme mobility of its subsequent development, whereas the outer belts (Helvetic of Switzerland, Dauphine of France, and Dolomites of Italy) were regarded as marginal geosynclines from the nature of their sediments. However, ophiolites are now known to exist in all three of the B-type outer belts, thereby placing them in the A-category, at least during the earlier stages of their geosynclinal histories. Other cases exist in which early A-type geosynclines are gradually transformed into B-type, suggesting that local geography may be of less importance than is a global mechanism in determining whether a given geosyncline shall evolve as an A or B type.

Another classical generalization that has been over-emphasized in older texts is that eugeosynclines characteristically develop on the oceanic side of rising geanticlines. This and related concepts are based on the observation that the orogenic belts of the Asiatic island-arc type are marginal to the Pacific and Indian oceans and that the Appalachians, like the Caledonians of Norway, East Greenland, and Spitzenbergen, are marginal to the North Atlantic Ocean. But there is no evidence that the Atlantic existed while these Paleozoic geosynclines were developing into orogenic belts. All of them were partly filled with sediments from what is now the oceanic side, i. e., from land areas that are no longer there. Because there is also good evidence that Europe, Greenland, and North America lay much closer together during the Paleozoic (Chapter 17) and that the development of the North Atlantic accompanied their separation, it has become more meaningful to interpret the Caledonian and older Appalachian geosynclines as seaways bordered on both sides by continental land.

Other examples of ancient intracontinental geosynclines exist in practically all continents (Holmes, 1965; Aubouin, 1965). Here again, this suggests that the characteristics of a given geosyncline and of its evolutionary behavior are not determined by local ephemeral oceanic-continental geography but by a long-lived deep-seated global mechanism.

Summarizing, most geosynclines are partly A-type (eugeosynclinal) and partly B-type (miogeosynclinal) particularly when viewed on a global scale and over a long span of time. In some cases, the dynamic behavior of geosynclines in a given area erases specific A and B characteristics to the extent that the geosyncline appears to belong exclusively to neither type. There are so many temporal and spatial variables in the evolution of any geosyncline that simple A and B type categorizations are not only difficult but meaningless in many cases. Perhaps it might be better to to say that there is an infinity of geosynclinal types of the form $aA + bB + c$, where $\underline{a}$ and $\underline{b}$ may be zero but $\underline{c}$, a special parameter embodying local conditions, never is.

Hypotheses Regarding Geosynclinal Evolution.

Before considering how geosynclinal development and mountain building are related to a single global deep-seated long-lived mechanism it is well to review briefly some of the hypotheses that have been developed to explain geosynclinal evolution. Most modern geosynclinal models involve one or more stages embodying subsidence, volcanism, and tensional forces, followed by a late compressional phase of turbidite deposition and cessation of volcanism.

MacGillavry (1970) used a model embodying a 2-phase interaction between 2 global crustal plates. In the first phase, the plates move in the same direction with attenuation through tension; in the second phase, the plates undergo "sideswiping compression". MacGillavry did not specify the nature or source of the force vectors driving these plates in the highly-constrained manner required by his hypothesis.

Mikhaylov (1970) invoked high plasticity of crustal matter, granitization, and recrystallization to provide initial geosynclinal folding, followed by a main stage of geosynclinal development during which folding is produced by horizontal cruatal movements to form linear folds in the sedimentary and volcanic strata. In the final stage, vertical movements produce block and diapiric folding. Like other geosynclinal evolutionary hypotheses, Mikhaylov's did not specify the nature and source of the driving mechanism required to provide the characteristic sequence of events embodied in his hypothesis.

Glikson (1970) based his hypothesis on a type section of the Archean Kalgoorlie system of western Australia. This system displays a trend of evolution from the eugeosynclinal ophiolite stage into the turbidite-deposition stage that grades into a molasse-like conglomerate stage representing the termination of geosynclinal deposition. In Glikson's hypothesis, the increase in the energy of sedimentation went hand-in-hand with a weakening of the associated igneous activity.

Trends of geosynclinal evolution similar to those observed by Glikson are detectable in other areas such as Canada, South America, and India. The commonly isochemical nature of low-grade regional metamorphism affecting these systems is normally indicated by the consistent geochemical character of the metabasalts, as well as by the systematic variations shown within the associated volcanic sequences. Glikson concluded that the observed geochemical trends reflect an evolution from a thin primordial "oceanic" crust to a geosynclinal pile. He felt that the crustal subsidence associated with this development resulted in magma generation at progressively greater depths and pressures. He showed that this trend is represented by the predominance of oceanic tholeiites at low stratigraphic levels, as well as by the occurrence of hi-alumina basalts and alkaline volcanics at intermediate and high stratigraphic levels. Glikson's

hypothesis did not include the nature nor source of the driving
mechanism that can duplicate the geosynclinal evolutionary sequence embodied in his hypothesis.

Ernst (1970) concluded from an analysis of late Mesozoic
Franciscan rocks of the California Coast ranges that a 4-stage
model explains the geosynclinal evolutionary behavior associated
with the Great Valley sequence: (1) rapid late Jurassic relative northeastward or eastward spreading of the Pacific Ocean
floor, coupled with westward or southwestward encroachment by
the continental lithospheric plate; (2) mid-Cretaceous buoyant
uplift of portions of the eugeosynclinal prism during a period
of less-intense spreading; (3) accelerated post-Cretaceous con-
vergence between oceanic and continental lithospheric plates;
and (4) Miocene-Pleistocene diapiric uplift of the Franciscan
melange related to northwestward sea-floor spreading. Ernst,
like earlier researchers, failed to provide the geometrical and
mechanical details of a driving mechanism for his model.

Hsu (1971) studied this same melange and concluded that
the evidence does not support the traditional view of a eugeosyncline as an elongated trough having active volcanism. His
conclusion suggests that the mio and eu elements of a geosyncline are functions of geometry and mechanics rather than of
heat and chemistry. The last section of this chapter discusses
how the geometrico-mechanical as well as the thermo-chemical

554

elements of a generalized geosyncline are derived from the tectonospheric Earth model.

Other geosynclinal evolutionary hypotheses have been proposed by Beloussov (1968, 1970), Davies (1968), Goodwin (1968), Laubscher (1969), Khain and Muratov (1969), Gnibidenko and Shashkin (1970), Fisher et al. (1970), to name only a few of the more recent. Earlier geosynclinal evolutionary hypotheses are summarized in standard works such as Aubouin (1965), Holmes (1965), and other standard textbooks.

Wynne-Edwards (1969) has assembled 22 papers on geosynclinal evolutionary behavior in high-grade metamorphic belts during the past 3.6 b. y. in various parts of the globe. These papers, like many other in-depth analyses, emphasize, among other things, that rejuvenated segments of older metamorphic terrains exist in even the oldest orogenic belts.

Summarizing, most geosynclinal evolutionary belts have followed parallel lines of development in various parts of the Earth and have been subjected to repeated orogenic activity during the past 3.6 b. y. This suggests (1) cyclical but aperiodic geosynclinal migration; (2) a well-ordered global causative mechanism operating deep within the Earth over a long period of time, perhaps since the Earth began; and (3) a repeating geosynclinal evolutionary pattern, extending back to the beginning of the Earth, such that each younger geosyncline, such as the Alpine, is but a well—ordered modification from a long succession of earlier geosynclines ending with the Hercynian and Caledonian but extending back far into the Precambrian.

Before considering the nature and source of the global deep—seated long—lived causative mechanism that governs geosynclinal evolutionary processes, it is well to define the exact relationship that exists between the geosynclinal evolutionary sequences and corresponding mountain—building evolutionary processes.

The Relationship between Geosynclines and Mountain Building.

Geological evidence in all ages and in all continents shows that mountain building is closely related to geosynclinal development. Considered as elements of a single family, geosynclines, mountains, and related rifting and quasi-linear uplift represent over 100,000 km of elongated linear features on the surface of the Earth. The relationship that exists between the elements of this family of elongated surficial features is not completely understood. It appears, however, that there is an antithetical Phoenix-like association between them, suggesting that they are motivated by a single global deep-seated long-lived mechanism.

The most obvious feature common to mountains and geosynclines is their linearity when viewed on a global scale. Also, both appear to date far back into the history of the Earth, i. e., at least 3.6 b. y. A geosyncline, generally, appears to be an elongated downwarp in the Earth's crust, the bottom apparently subsiding deeply beneath accumulating sediments. Mountain ranges appear to be uplifted from these long-lived elongated subsidences, following accumulations of sediments to depths of over 20 km spanning, in some cases, several hundred m. y. The uplifts are culminated by great breakings, faultings, and foldings of strata along with other surficial changes that appear to be motivated by a deep-seated causal mechanism on a global scale and in a cyclical but aperiodic manner.

Because the deposition of geosynclinal sediments probably was in shallow water, some observers (See, e. g., Hall, 1857) concluded that the geosynclinal depression and long-lived subsidence was caused by the weight of the accumulating sediments. But others (See, e. g., Dana 1873) felt that the accumulation of sediments was a consequence, rather than a cause, of the subsidence.

Some mountains and geosynclines appear to form along the edges of continents. Some appear to form entirely within continents. Others seem to form partly in and partly out of continental areas. If crustal rifting may be considered as the earliest stage of the mountain-building cycle, then some mountains and geosynclines may be said to form completely within oceanic areas. Should the paired-belt concept of miogeosynclines (non-volcanic) and eugeosynclines (volcanic) be used (Stille, 1940; Kay, 1951; Aubouin, 1965), then the identification of geosynclines along continental margins becomes confusing and even meaningless because the separation between the <u>mio</u> and <u>eu</u> elements of a geosynclinal pair may be more than a thousand km in many cases.

Closer scrutiny reveals that many eugeosynclinal elements (e. g., the American Cordilleran and Appalachian) were analogous to present island arcs and that the sediments were derived from active volcanoes within the associated sedimentary basin (Kay, 1951). Both present and past geosynclines appear to form in an orderly manner in all parts of the globe, suggesting the possibility of well-behaved control on a global scale (Drake et al., 1959; Dietz, 1963; Dietz and Holden, 1966). The structure and development of the Alpine geosyncline seems to have been similar to those of the Indonesian arcs, both having consisted of a complex system of troughs and ridges throughout most of their history (Aubouin, 1965).

From these and other analyses (See, e. g., Argand, 1916; Kuenen, 1967; Laubscher, 1969; Schuchert, 1923; Sylvester-Bradley, 1968) it appears that geosynclines and mountains are controlled by the same mechanism. This association is shown schematically in Fig. 12-1-5-1, representing the 4 stages in the hypothesized development of a geosyncline and a mountain range from a single global deep-seated long-lived causal mechanism.

The _first_ _phase_ shows two land masses, L and H, separated by a shallow sea S several hundred km wide (See Chapter 15 for a discussion of the genesis of the shallow sea and of the adjacent land masses).

The _second_ _phase_ shows the erosion of the highland area H and the filling of the adjacent sea S with sediments. The sea-bed S sinks partly because of the sedimentary load but mainly

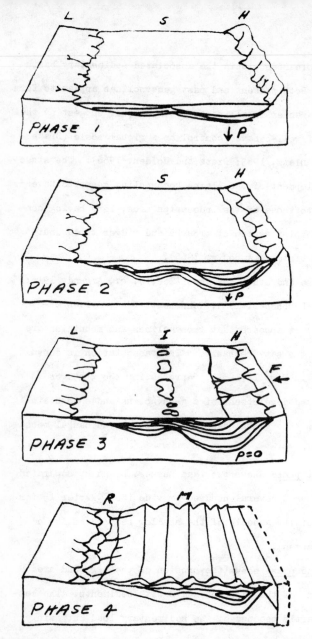

Figure 12-1-5-1. Schematic representation of the 4-stage hypothetical development of a geosyncline and a mountain range. See text for details.

560

due to a progressive downward unbalanced force vector P, "pulling" the seabed downward as much as 20 km over a period of time and in a well-ordered manner.

The third phase shows the cessation of the sinking as the downward unbalanced force vector P ceases to "pull" downward. The right-hand "jaw" H of the geosyncline begins closing upon the left L in a well-ordered manner in response to an unbalanced force vector F acting on the right-hand jaw. The sedimentary strata are forced upward, and a narrow line of islands I emerges above the surface of the sea. If F is a couple (i. e., embodying rotation as well as translation) I will assume a curvature during the operation of the couple (Chapter 15).

The fourth phase shows the cessation of the uplifting, leaving a range of parallel folded mountains M having curvature appropriate to any couple operating during the third and fourth phases. The old highlands H have been worn down to their roots. At the left, a new river R carries away the sediments eroded from the young range M.

Mountain-Building Episodes.

Mountain building begins in a long narrow belt that is sinking as a result of an undefined downward-acting force. The subsiding belt, or geosyncline, gradually fills with sediments and sinks far below the level expected from isostatic equilibrium.

Eventually, for reasons unknown, the downward-acting force ceases to act and the geosynclinal sediments begin to rise. But isostatic equilibrium is only a partial explanation; the bulk of the uplift is due to an unbalanced force vector acting upward with sufficient energy to create a mountain many km high. The uplift is accompanied by three processes: (1) intense folding and fracturing of the rocks; (2) the transformation of the deeper strata into metamorphic or igneous material such as granite; and (3) the addition of material from the depths due to igneous activity.

Many geologists regard folding as primarily the result of compression, i. e., the forcing together of the two sides of the geosyncline. Usually one side moves while the other remains fixed. As one side closes upon the other, rocks within the geosyncline form folds and ridges. As the two sides meet, the folds are bent over and heaped one atop the other to form nappes, some of which measure 50 km from their roots to their sharply folded tips and therefore represent 100 km of rock strata doubled back upon themselves. Tips of nappes often become detached to form klippes. The great folds of large mountains represent at least one-third of the original geosynclinal width.

Nappes and klippes are not formed by being _pushed_ over the underlying rocks. Rather they are formed by gliding to their present positions under the influence of gravity. Heat and possibly volatiles rising from below perform an important function in mountain building, since hot rocks are folded and contorted easier than cold rocks.

Brief History of Mountain Building.

Mountain-building episodes, or orogenies, have occurred many times since the Earth began, but the remnants of Precambrian mountains have been eroded into shields, or low-lying areas of old, and often complex, rock that forms the nuclei of the continents (Chapter 7). Although some of the margins of these shields have been buried by shallow accumulations of younger rocks from later peripheral orogenies, the shields themselves have not been invaded orogenically since the Archean.

The exact number of mountain-building episodes that have occurred on various shields is not known, but five distinct Precambrian orogenies can be detected in the Canadian shield. The oldest of these occurred about 2 b. y. ago, in northern Manitoba, to the west of Hudson Bay. Others occurred about 1.35 b. y. ago (the Great Bear Lake); about 1 b. y. ago (the Grenville Province); about 800 m. y. ago (Huronian); and about 600 m. y. ago (Appalachian). Mountain-building is other shields occurred at about these same times.

Since Cambrian times, there have been 3 major mountain-building episodes: in the Paleozoic, in the Mesozoic, and in the Cenozoic. The Lower Paleozoic ended about 400 m. y. ago in the folding of the Devonian Period that formed a great mountain range stretching from northwest Ireland to Northern Norway in the European continent and similar ranges in the other

564

continents: Taconic-Acadian in North America, Caledonian in Greenland, and Tasman in Australia.

The Upper Paleozoic ended about 225 m. y. ago in the building of ranges that stretched eastward across central Europe, from France to Czechoslovakia, and similar ranges in other continents: Appalachian in North America, Hercynian in the British Isles, the Cape Folds in Africa, late Tasman in Australia.

The Alpine orogeny ended the virtually continuous accumulation of sediments during the Triassic, Jurassic, and Cretaceous that ended about 65 m. y. ago. Similar orogenies occurred in other continents: Laramide in North America, Andean in South America, Atlas in Africa, and Alpine in India.

From the global contemporaneity of orogenies in all continents, it is clear that mountain-building and geosynclinal developments have a deep-seated long-lived global causal mechanism. The cyclical but aperiodic nature of that mechanism is evident from the fact that each period of orogeny is followed by a long intervening period of relative stability during which erosion and sedimentation continue almost undisturbed by folding or uplift.

Closer scrutiny reveals that some type of surficial change is occurring somewhere at all times, i. e., volcanic activity, crustal rifting, plate-block motion, and other indications of the Earth's internal behavior never cease. These manifestations of the Earth's internal activity shift from place to place and vary their intensities but they never cease. Uplift and subsidence continue. Parts of the Tethyan zone are rising; others are sinking. The folding of the Tethyan geosyncline is far from complete. In other parts of the world, geosynclines are just beginning to form (circum-Pacific, Mediterranean, Caribbean).

Even in the past, geosynclinal development was cyclical. Subsidence was occurring in some geosynclinal segments while uplift or folding was occurring in other segments of the same geosyncline. The earliest Mesozoic geosynclinal folding occurred near the end of the Jurassic Period, about 140 m. y. ago, when the Urals and part of the Alps were formed. The Rockies and Andes were formed, about 70 m. y. later, toward the end of the Cretaceous. The main folding of the Alps and Himalayas did not occur until about 30 m. y. later. Along the same belt (the Tethyan), geosynclinal development is in its infancy (Mediterranean, Indonesian, West Indian). Thus within a given geosynclinal belt, such as the circum-Pacific (or the Tethyan extended to a complete circle) mountain-building at any one time may be seen in various stages of development from subsidence, to uplift, to folding, to erosion and pleneplanation.

With the passing of time, the geosynclinal development appears to "migrate" in a dual manner: (1) the types of orogenic activity migrate linearly along the belt; and (2) the belt itself migrates parallel to its erstwhile position. Off the east coast of the United States, a typical double-geosyncline belt lies parallel to the Appalachian Mountains which were built from a similar double sedimentary belt about 500 m. y. ago. In both these systems, the sequences were the same: (1) shallow-water deposits in the inner basin and (2) deeper-water and more-mixed deposits in the outer basin. At the time of folding, it was the outer basin that was subject to the greater amount of metamorphism and intrusion of igneous rock.

Other areas of the Earth in all geologic ages show similar migrations of geosynclinal activity. The last section of this chapter discusses the causal mechanism for such geosynclinal migration. There it will be shown also: (1) why all geosynclines do not form along the edges of continents but may form within continents, within oceans, or partly within each; (2) why geosynclines have a myriad of forms (i. e., some with only one trough, others with two); (3) why some geosynclines are symmetrical, with sediments arriving from both sides, while others are highly asymmetrical; (4) why the rate of geosynclinal deposition can exceed the rate at which the geosynclinal floor subsides; and (5) why the geosyncline need not develop in water (e. g., the Indo-Gangetic plain of northern India, which is a geosyncline that has been kept full of sediments by the many rivers flowing into it from the Himalayas).

The Relationship between Metamorphism and Mountain Building.

Metamorphism is closely related to mountain building and geosynclinal development but the exact nature of the relationship is not known. Metamorphism refers to the mineralogical and structural changes of solid rocks due to conditions (especially temperature and pressure) differing from those under which the rocks originated. Contact metamorphism is caused by the thermal effect of an intrusive igneous body on the surrounding rocks. Regional metamorphism, the most important type, is caused by the thermal effects and pressures during orogeny. The mechanism causing the orogeny, therefore, causes metamorphism.

In the simplest case, a thick pile of sedimentary rocks accumulates in a geosynclinal area during the early stages of orogeny. The pile is not only folded and faulted, but also heated, thereby producing metamorphic recrystallization. Individual parts of the geosynclinal pile are transformed from soft sediments to hard mineral assemblages corresponding to the various temperatures and pressures within the pile.

Regional metamorphic rocks are primarily schists and gneisses derived from argillaceous and arenaceous sediments. Complexes of this type are seen in all dissected axial zones of all orogenic belts dating back to the earliest Precambrian. Such belts, in many cases, are traceable for more than 1000 km. The temperature and pressure of metamorphic recrystallization in most

568

cases appears to increase toward the centerline of the zone, suggesting that the source of heat was linear and centered deep within the zone.

Many metamorphic belts are accompanied by abundant granitic and basaltic igneous rocks. Basaltic volcanism usually occurs during the geosynclinal stage; granitic emplacement, during or after the main phase of regional metamorphism. Closer scrutiny reveals that the metamorphic belts contribute to the orogenic-cratonic evolution of the continents (Chapter 7).

Older concepts of regional metamorphism (that the temperature and pressure within geosynclinal sediments are due to the weight of the sediments) are no longer tenable (See, e. g., Miyashiro, 1969). Depending upon the temperature and pressure, an entire spectrum of rocks is produced by regional metamorphism, ranging from andalusite, through kyanite, to glaucophane, lawsonite, and jadeite-quartz as the pressure and temperature increase.

Mountain-Building Forces.

If mountain building were better understood, it would be easier to identify the forces that have created mountains ever since the Earth began. In the absence of an explanation of how mountains grow, man has formulated numerous hypotheses for mountain building (See, e. g., any standard text in geology). When these hypotheses are analyzed, it is found that none of them is completely satisfactory to explain all mountain-building activity that has occurred since the Earth began. None of them, in fact, is completely satisfactory to explain even the mountain-building activity that is occurring today.

Closer scrutiny of the tenable hypotheses for mountain building reveals that these hypotheses have one thing in common: each hypothesis, in the last analysis, attempts to provide the one element without which it is impossible to build a mountain, i. e., an unbalanced force. More specifically, a successful mountain-building hypothesis must provide an unbalanced force field that has a circular spatio-temporal distribution such that (1) it "begins" as a downward-acting unbalanced force vector (to cause the original geosynclinal subsidence); (2) it ceases to be a downward-acting force vector and becomes an upward-acting unbalanced force vector (to cause the geosynclinal pile to be uplifted); and (3) it provides horizontally-acting unbalanced force vectors (to cause the geosynclinal pile to be compressed).

The second element (uplift) is absent from some hypotheses; the third (compression) is absent from others. All hypotheses embody the first element (initial subsidence).

A complete analysis of all mountain-building hypotheses is not required in an analysis of the evolution of the Earth's tectonosphere. Suffice it here to analyze representative of these hypotheses in terms of the unbalanced-force-vector concept. All mountain-building hypotheses embody the requirement for the creation of a downward-acting unbalanced force vector. In the last analysis, this means that a trough must be formed so that gravity can act to place sediments into the trough. Gravity acts at all times, but gravity alone cannot become an unbalanced force until a trough or rift is created in the crust of the primordial planar Earth.

Rifts or troughs in a sphere may be created by any one of several methods: (1) extraterrestrial influences; (2) expansion of the sphere; (3) compression of the sphere; (4) the upward flow of heat, and possibly of volatiles, through the surface; (5) the movement of plates and/or blocks within the sphere; and (6) a combination of two or more of these.

Earlier mountain-building hypotheses (See, e. g., Holmes, 1965) used the first three of these mechanisms (extraterrestrial influences, global expansion, and global compression) to create the initial geosynclinal subsidence, rift, or trough. More recent hypotheses (See, e. g., Dewey and Bird, 1970) use the last three mechanism to create the initial geosynclinal subsidence, rift, or trough.

The last section of this chapter analyzes some of the difficulties involved in producing and maintaining mechanisms (4) and (5) in a spherical system such as the Earth over a period of 4.6 b. y.

Mountain-Building Hypotheses.

Ever since man first saw mountains, he has tried to explain them. An analysis of the evolution of the Earth's tectonosphere need not consider all the hypotheses that have grown up to explain mountains. We need consider only a few of the more-recent ones in order to determine to what extent these hypotheses correlate with the salient observable characteristics of mountains:

1. They are long linear or arcuate features.

2. They embody distinctive zones of sedimentary, deformational, and thermal patterns roughly parallel to the mountain ranges.

3. They contain complex internal geometry, with extensive thrusting and mass transport juxtaposing very dissimilar rock sequences with the result that many former features have been destroyed or obscured.

4. They exhibit extreme stratal shortening features and, often, extensive crustal shortening features.

5. They have asymmetric deformational and metamorphic patterns.

6. They constitute marked sedimentary composition and thickness changes normal to the trend of the belt.

7. They comprise mainly marine sediments.

8. Their underlying basement is dominantly continental, but the belts themselves contain zones in which basic and ultra-

573

basic (ophiolite suite) rocks occur as basement and as upthrust slivers.

9. They contain some sedimentary sequences deposited during very long intervals without volcanicity within the sedimentary regions.

10. Their duration of intense deformation and metamorphism is comparatively short-lived compared with the duration of the sediments now constituting the bulk of the mountains.

One of the first comprehensive hypotheses that embodied most of these ten observable characteristics of mountains was proposed by Bucher.

The Bucher Hypothesis.

Bucher (1933, 1950, 1955) concluded from a study of orogenic belts that the deformation within these belts begins with a "down-bending" of the crust over constrictions in an actively shrinking subcrustal shell beneath which lie the zones of deep-focus earthquakes. He felt that the weight of the sedimentation within the subsiding geosyncline would tend to accentuate the process, but his analyses had clearly shown that such weight was not the primary cause of the crustal subsidence. His observations showed that, during the process of subsidence, there will be an eventual failure of the crust, as revealed by a shearing through low-angle thrusts. Simultaneously, superheated water and volatiles escape, in Bucher's model, from the subcrust to produce regional metamorphism and plutonism.

Compression eventually weakens the zone sufficiently to shape the yielding blocks of the geosyncline into crustal folds. These, lengthening upward, become either recumbent folds or thrust blocks. Then, in Bucher's model, these advancing recumbent folds or thrust masses cause the sedimentary cover to "peel off" far beyond the borders of the active zone, thereby producing marginal deformation and related phenomena.

Bucher felt that "pre-existing" fracture zones are instrumental in mountain-building processes. Depending upon their location with respect to the direction of movement of the mountain,

these fractures may determine basin-formation phenomena within
the original geosyncline and the location of anomalies in the
outermost skin folds. Because of these local influences, the
actual structures in cross-section and ground plan in Bucher's
model present ever-changing views as the process evolves through
the various mountain-building stages. These changes are accen-
tuated by vast quantities of magma extruded as the orogenic
belt develops. In such cases, the magma replaces the major
anticlines or thrust blocks. These and other complexities tend
to modify or even destroy the "normal" structural configuration
of a "typical" orogenic belt.

Fig. 12-4-1 shows the three separate cases visualized by
Bucher in different segments of an orogenic belt at a given
time or in a given segment of a belt at three different times.
The first case shows a "typical" orogenic belt with a major
anticline; the second case, an orogenic belt with thrust blocks;
and the third case, an orogenic belt with extrusion of vast
quantities of magma.

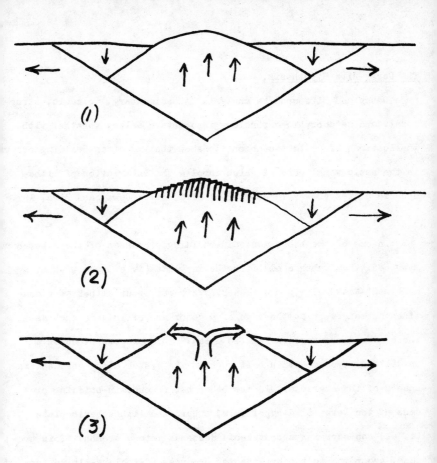

Figure 12-4-1. Schematic representation of Bucher's three orogenic belt configurations: (1) with major anticline; (2) with
thrust blocks; and (3) with magmatic extrusion. See text for
details. Superimposed on an idealized wedge-belt of activity
of the tectonospheric Earth model (Chapter 6).

<u>The Dewey-Bird Hypothesis.</u>

Dewey and Bird used an analysis of sedimentary, volcanic, structural, and metamorphic chronology in mountain belts, together with implications of plate tectonics, to show that mountain building occurs in two basic ways, both of which involve the interaction of lithospheric plates (Dewey 1969a, 1969b; Bird, 1970; Dewey and Bird, 1970a, 1970b).

In one of the basic mountain-building processes of the Dewey-Bird hypothesis, an unburdened lithospheric plate (i. e., one without a continent) descends into a trench. This can occur either at a continental margin or at the edge of a major arc of islands such as those around the western Pacific Ocean. This process produces a cordilleran-type mountain belt such as the system along the western coasts of the Americas. In the other basic mountain-building process of the Dewey-Bird hypothesis, a burdened lithospheric plate (i. e., one carrying a continent) descends into a trench. This process, according to the hypothesis, produces a Himalaya-Alpine type of mountain belt.

In the formation of the cordilleran-type mountain belts, the dominant mechanism is thermal. This thermal mechanism, according to the Dewey-Bird concept, arises from a complex series of events derivable from the geometry and mechanics of moving lithospheric plates. Basically, in their hypothesis for cordilleran-type mountain belts, wedges of oceanic crust and mantle are driven oceanward

578

as a lithospheric plate begins to sink at a continental margin. They postulate that the rate of descent of the plate is related to the thickness and density of the plate, both of which factors increase as the plate "grows" away from a mid-ocean ridge.

As the plate descends to depths of 100 km or more, the high-load pressures and shear stresses of oceanic crust on the descending plate produce partial melting. This melting, in turn, produces basaltic and calc-alkalic magma. The heat generated by the ascent of the magma creates a subsurface dome of rock with magma in the center. As the dome expands and grows toward the continent, the high-temperature deformation and metamorphism begin to affect the sediments of the lower continental rise.

The ascending welt soon rises above sea level, forming a trough between it and the continent. As this trough fills with sediments, this material is slowly driven toward and thrust onto the continent, thereby completing the Dewey-Bird cordilleran-type mountain range.

Whenever this process begins adjacent to a continental margin, the resulting mountain belts form along the edge of the continents. From there, Dewey and Bird postulate that the deformations can migrate into the continental interior to produce intracontinental mountain ranges.

When, on the other hand, this process begins at a trench considerably removed from a continent, an island arc grows within the ocean, and a small ocean basin forms between the arc and the continent. This, according to Dewey and Bird, is the situation around the western Pacific.

Himalaya—Alpine type mountains form in a less subtle process according to this hypothesis; i. e., when two lithospheric plates carrying continents interact, a single trench zone of plate consumption is replaced by a cracking and splintering of lithosphere over a wide area. This collision—like interaction produces large mountains such as the Himalayas. If one of the plates is carrying a group of islands rather than a continent, the resulting mountains are much smaller, such as those of northern New Guinea.

The details of the Dewey—Bird hypothesis need not be considered here. Suffice it here to say merely that Dewey and Bird cite stratigraphic sequences of geosynclinal developments in various mountain belts to support their hypothesis in generally the same manner as Bucher did. Niether Bucher's hypothesis nor that of Dewey and Bird is completely consistent, however, with the observational evidence when such is analyzed on a global scale and extending back through even a few of the major orogenies that appear to have been instrumental in building mountains during the past 3.6 b. y.

What, for example, is the mechanism that opens and closes the oceans? Why does this mechanism operate at certain times

580

and places but not at other? Why are the mountain ranges narrow
and long? These and other questions concerning the long-lived
global geometrico-mechanical details of mountain-building mechanisms are considered in the last section of this chapter.

<u>The</u> <u>Griffin-Wegmann</u> <u>Hypothesis.</u>

Wegmann's stockwork concept of mountain building was based
upon studies of the Alps, the Variscan crystallines of middle
Europe, and the Caledonides of Norway and Scotland (See, e. g.,
Wegmann, 1935; Griffin, 1970). In Wegmann's basic concept the
infrastructure mushrooms upward against a roof of stiffer supra-
structure, separated by a zone of intermediate character, designated the "transitional zone" by Wegmann.

Griffin (1970) pointed out the striking similarity of this
stage of the Wegmann concept to a comparable stage of the Dewey-
Bird model (Dewey and Bird, 1970a, 1970b; Bird, 1970). The
infrastructure in both concepts is mushroom-shaped and symmetric
toward the associated craton. In both concepts, the infrastructure is preceded by a migmatite front that intrudes the overlying deposits. In both concepts tensional effects occur above
the rising mountain core along with subsidence and underthrusting
on the oceanic side of both the rising mobile core and the infrastructure.

In Wegmann's concept the infrastructure develops within
an older basement through "remobilization", whereas in the Dewey-
Bird concept the mobile core develops through "heating and deformation of the sediments on the lower continental rise". The
primary difference between the two concepts is that the Dewey-
Bird model lists a precise mechanism for the heat that initiates

582

orogenesis; i. e., the descending plate in the Dewey-Bird concept produces the heat that develops and drives the mobile core upward.

Minear and Toksöz (1970) provided theoretical support for the Dewey-Bird heat-producing mechanism. But their explanation has not been universally accepted (See, e. g., Hasebe et al., 1970; Jacob, 1970; Luyendyk, 1971; and MacKenzie, 1971).

Other Mountain-Building Hypotheses.

An analysis of the evolution of the Earth's tectonosphere need not analyze all mountain-building hypotheses. It is well, however, to review briefly the salient conclusions of some that bear directly on problems related to tectonospheric evolution.

One of these involves the creation of relatively short mountain ranges such as the Taurus Mountains. Girdler (1970) concluded that this range was created by the northward movement and counterclockwise rotation of the Arabian peninsula and its impingement against Eurasia. This hypothesis is incomplete because (1) it does not specify the nature and origin of the unbalanced force vector that produced the translations and (2) it does not identify the location, depth, and cause of the fracture zones between the lithospheric plates involved.

Artyushkov (1971) has shown that the sources of regional isostatic anomalies, such as those instrumental in some mountain-building, lie below the low-viscosity zone formerly thought to be the seat of isostatic adjustment.

Price and Mountjoy (1970) found that the entire Rocky Mountain-Cordillera complex evolved as an integrated system and not as separate unrelated orogenies as had been previously thought, thereby emphasizing the unity of the causal mechanism for all elements of mountain building.

Reesor (1970) found that the Paleozoic rocks in the metamorphic culminations of parts of Canada were deeply depressed in early Mesozoic time, were subsequently metamorphosed, and then were significantly uplifted in late Mesozoic time. His analysis emphasized the narrowness of the zones of heat rise essential to explain the observed metamorphism.

Woollard (1969) concluded that the Rocky Mountains represent a typical Mesozoic-type orogenic feature similar to the Alps, the Himalayas, and other ranges of the same approximate age. Even more importantly, all these mountains have many characteristics similar to those of ocean ridges, thereby emphasizing the global nature of the mountain-building mechanism in both continental and oceanic areas.

Much of the crust in the Rocky Mountain area appears to be about 3 km too thin for the surface elevation. This abnormally thin crust in this mountainous area is normally interpreted as a large variation from equilibrium conditions. Left unanswered is the question regarding the cause for this extremely large variation from equilibrium conditions.

Dorofeyef (1968) has interpreted the Ural Mountains as being a very complex structure formed under the influence of stresses from different directions. Other hypotheses have explained the Urals as a suture formed by the coalescing of two lithospheric plates (See, e. g., Wilson and Fairbridge, 1971). Neither of these hypotheses provides the details of a driving mechanism for the unbalanced forces required to produce the "stresses from different directions" or to drive large plates toward each other.

Wilson and Fairbridge (1971) analyzed Appalachian cycles of pleneplanations in terms of plate tectonics. They concluded that repeated uplift of the Appalachian peneplains, dating 200, 130, and 60 m. y. ago, can hardly be explained by revivals of Appalachian roots or by isostatic response to unloading, neither of which should have a period of 60 m. y. or more. Their analysis showed that the observed cyclicity shows some similarity to major episodes of plate tectonics. This need not necessarily mean that either is caused by the other; both the cyclical uplift and the plates might be driven by a common (third) mechanism. They concluded that the cyclical uplift could have been "in response to" high heat-flow epochs, which in turn could have been "in response to" the same mechanism that drives the plates. In any case, most recent mountain-building hypotheses embody some provision for uplift, or at least a requirement for such. But none of them specifies the geometrico-mechanical details of the causal mechanism to produce such.

De Jong (1971) used vertical uplift in the Apennines as the basis for his analysis. He concluded that an obvious relationship between mountain building and horizontal movement (such as sea-floor spreading or continental drift) is not obvious in that area. He felt that vertical movement of the sea floor is more evident there than is horizontal movement of the sea floor.

Sollogub (1970) found evidence for both horizontal and vertical movements connected with mountain-building episodes in southeast Europe, resulting in a block-layered structure with a multiplicity of horizontal discontinuities representing upper-mantle transitions of various ages. These transitions range from sharp seismic interfaces to smooth transition zones many km in thickness. In many areas these "supplementary M discontinuities" appear to be relics of (1) ancient discontinuities, or (2) those of an erstwhile surface that formed in the process of phase transformation and then migrated during orogenesis.

Similar conclusions had been drawn earlier by Beloussov (1969) from a study of the deep structures associated with various mountain-building episodes. He found that the thickness as well as the density of the crust involved in such episodes had changed many times during geologic time and that in most cases these changes extend through the M discontinuity into the mantle.

From his study Beloussov concluded also that most vertical movements of the crust are anti-isostatic, and that the causal mechanism for these vertical movements appears to be more complicated than that which might be expected from the influence of changes in density and volume directly under the observed region. He did not suggest the nature and source of a causal mechanism that could produce the observed behavior.

Summarizing, most mountain-building hypotheses postulate some form of vertical movement downward during initial stages and some form of uplift during latter stages, in addition to horizontal movement during the intermediate stages. Details regarding the nature and source of the driving force are not specified in any mountain-building hypotheses. The following section examines this driving force in terms of the tectonospheric Earth model.

<u>Mountain Building as Interpreted by the Tectonospheric Earth</u>
<u>Model.</u>

Mountain building, as postulated by the tectonospheric
Earth model, is inseparately associated with the orogenic-
cratonic evolution of the continents, with the Earth's deep
seismicity, with its geothermal activity, and with its intrusive and extrusive activity. This relationship of the Earth's
internal behavior to mountain-building and related activities
at the surface of the Earth will be discussed in this section.

Mountain building is one of the several surficial manifestations of the Earth's internal behavior that result from the
basic driving mechanism of the tectonospheric Earth model (Chapter 6). That is, all mountain-building and related activities,
since the Earth began, have been motivated by the same driving
mechanism that has been described in Chapter 11.

Geosynclinal development and associated mountain-building
activities may best be analyzed in terms of a wedge-belt of
activity of the model, an idealized form of which is shown in
Fig. 12-10-0-1. The wedge-belt in its idealized form is more-
or-less bilateral or symmetric with respect to the center of
the figure. Actually, this idealized bilateralism or symmetry
is seldom realized because it is modified by the individual
tectonics of the associated octantal blocks, G and G'.

If, for example, the surfaces of the octantal blocks, G
and G', are at equal levels, then the geosynclines developing
at the right (primed) side of the figure will be similar to

588

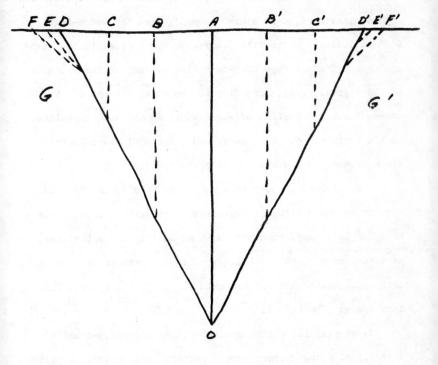

Figure 12-10-0-1. Schematic diagram showing a transverse cross-
sectional view of the predicted relationship of several tectonospheric features of a typical idealized cone of activity and
wedge-belt of activity of the tectonospheric Earth model. See
textual description of Fig. 6-2e-3.

those developing at the left (unprimed) side of the figure. If, on the other hand, G is higher than G', then the wedge-belt DOD' will tend   to tilt toward the right side, thereby producing a different type of geosyncline at each of the two sides of the figure. This, according to the model, is one of the causes for the formation of eugeosynclines and miogeosynclines in the evolution of most geosynclinal systems; other causes are discussed in the chapter on plate tectonics.

Fig. 12-10-0-1 shows that mountain-building and related activities occur within wedge-belts of activity. Thus, beginning 4.6 b. y. ago, mountain-building and related activities, as viewed on a global scale, have formed a pattern of 3 mutually-orthogonal belts identifiable with the subtectonospheric fracture system (Chapter 6).

As an example, all mountain-building and related activities 2.7 b. y. ago formed such a pattern passing through parts of each of the orogens associated with all cratons then in existence (Chapter 7): Yellowknife, Kenoran, Ketilidian, Scourian, Saamian, Brazilian, Shamvaian, Swaziland, Bundelkhand, Dharwar, Pilbara, and Kalgoorlie, to name some of the better known ones.

One hundred million years after these orogenies, the mountain-building activities also formed this pattern of 3 mutually-orthogonal belts identifiable with the subtectonospheric fracture system. But this pattern differed from the pattern 2.7

b. y. ago in a significant respect. During the 100-m. y. interval, some of the tectonospheric plates and blocks had shifted with respect to each other and with respect to the subtectonospheric fracture system (Chapter 4). At a rate of 1 cm/yr, motion on the surface of the Earth amounts of 1000 km in 100 m. y. Therefore, segments of the two mountain-building patterns at 2.6 and 2.7 b. y. ago could conceivably be separated by thousands of km.

The concept of a "pre-drift configuration" of continents, plates, and blocks is meaningless because, from the above analysis, an infinity of so-called pre-drift configurations could have existed during the past 3.6 b. y. Each of these hypothetical configurations at any time showed the same pattern of 3 mutually-orthogonal belts identifiable with the subtectonospheric fracture system of the model.

This characteristic pattern may be seen in all mountain-building and related activities during the past 3.6 b. y., including those of the more recent Charnian, Caledonian, Hercynian, and Alpine orogenies of the Phanerozoic. "Continental drift" and other causes since the Hercynian have obscured parts of some of the patterns of the Hercynina, Caledonian, Charnian, and Precambrian orogenies, but many remnants of the older belts are still detectable.

Mountain-building and related activities occuring today form, according to the model, the same characteristic pattern of 3 mutually-orthogonal belts identifiable with the subtectono-spheric fracture system of the model.

Global Patterns of Geosynclinal Evolution as Interpreted by the Tectonospheric Earth Model.

In their analysis of the evolution of geosynclinal succes-sions and continental margins in terms of ocean-floor spreading and lithospheric-plate underthrusting, Mitchell and Reading (1969) classified geosynclines into the following basic types: (1) Atlantic, (2) Andean, (3) island-arc, (4) Japan Sea, and (5) Mediterranean.

Atlantic-type geosynclines, in the Mitchell-Reading cate-gorization, occur on and oceanward of continental margins near where there is no differential movement between lithospheric plates. This type of geosyncline includes an ocean rise, some abyssal plains, a continental rise, a continental shelf, and a coastal plain. The geosynclinal deposition sequence is ap-proximately: (1) ultrabasics; (2) tholeiitic volcanics and pelagics; (3) compositionally-mature turbidites and pelagics; (4) geostrophic-contour-current deposits; and (5) shallow-water and paralic sediments.

Andean-type geosynclines, including a submarine trench and a mountain arc, occur on and beside continental margins beneath which spreading lithosphere descends. For the trench element, the geosynclinal deposition sequence is roughly: (1) pelagics; (2) compositionally-mature and immature turbidites; and (3) wildflysch. For the mountain-arc element, it is: (1) coarse-grained molasse and (2) calc-alkaline volcanics.

Island-arc type geosynclines, including a submarine trench and a volcanic island arc, occur where spreading lithosphere descends some distance from a continent. Geosynclinal deposits are (1) compositionally-immature turbidites and pelagics, (2) calc-alkaline volcanics around the island arc, and (3) occasional shallow-water carbonates.

Japan-Sea type geosynclines occur in small ocean basins between continental margins and bordering island arcs. Geosynclinal deposits are compositionally-mature turbidites, mass-flow material, tuffs, and pelagics grading into shallow-water sediments. Depending upon the nature of the basin floor, Japan-Sea type deposits may overlie either (1) continental crust or (2) tholeiites typical of Atlantic-type geosynclines.

Mediterranean-type geosynclines occur in small ocean basins within or between continents. Geosynclinal deposits are similar to those of the Japan-Sea type geosynclines.

When the tectonospheric Earth model is applied to an analysis of the global distribution pattern of the above Mitchell-Reading categorization of geosynclines, it is found that (1) geosynclines of the Atlantic type occur on and oceanward of

593

continental margins near where there is no current wedge-belt
of activity (Chapter 6); (2) geosynclines of the Andean type
occur on and beside continental margins near where there is a
current wedge-belt of activity; (3) geosynclines of the island-
arc type occur some distance from a continent but near where
there is a current wedge-belt of activity; (4) geosynclines
of the Japan-Sea type occur in small ocean basins between continental margins and bordering island arcs but near where there
is a current wedge-belt of activity; and (5) geosynclines of
the Mediterranean type occur in small ocean basins within or
between continents but near where there is a current wedge-belt
of activity.

Extending the concepts of the tectonospheric Earth model,
the 5 Mitchell-Reading geosynclinal categories may be reassem-
bled into two new categories: (1) those occurring near a current wedge-belt of activity of the model; and (2) those not
occuring near a current wedge-belt of activity. Thus, it follows that (1) because it does not occur near a current wedge-
belt of activity, the Atlantic type of geosyncline is the only
one that falls into the second category; and (2) because they
occur near a current wedge-belt of activity, the Andean, island-
arc, Japan-Sea, and Mediterranean types of geosynclines fall
into the first category postualted by the model.

From this, the following observations may be made: (1)
there is no current orogeny associated with the Atlantic type

594

geosyncline (because it is not near a current wedge-belt of activity) nor is there any differential movement between the lithospheric plates involved in the geosyncline (for the same reason); and (2) in all other types of geosynclines there is associated current orogeny (because the geosynclines are near a current wedge-belt of activity) and there may be differential movement between the lithospheric plates involved in the geosynclines (for the same reason). It may be concluded, therefore, that geosynclinal type is a function of its position with respect to a current wedge-belt of activity of the tectonospheric Earth model, as modified by the environment in which the wedge-belt of activity lies (i. e., whether oceanic, continental, or marginal).

Summarizing, present geosynclines and related features associated with crustal rifting and mountain building are identifiable with the present global positions of the current wedge-belts of activity (Fig. 6-2b-2, Chapter 6; Fig. 8-4d-1, Chapter 8). Similarly, fossil or ancient geosynclines and related features associated with ancient crustal rifting and mountain building are identifiable with the corresponding positions of the wedge-belts of activity that were effective at the time the "fossil" geosynclines evolved. Thus, fossil geosynclines of Cambrian age would be identifiable with the Cambrian positions of the three mutually-orthogonal wedge-belts of activity; fossil geosynclines of Silurian age, with Silurian positions of the wedge-belts; fossil geosynclines of Triassic age, with Triasic positions of the wedge-belts; and so on.

<u>The</u> <u>Migration</u> <u>of</u> Geosynclines <u>as</u> <u>Interpreted</u> <u>by</u> <u>the</u> <u>Tectonospheric</u>
<u>Earth</u> <u>Model</u>.

In the tectonospheric Earth model, the tectonospheric plates
and blocks move with respect to the subtectonosphere. Therefore, the behavior of the surface of the Earth is constrained
by the fracture system within the subtectonosphere. Thus, "inactive" or "fossil" geosynclines are identifiable with former
positions of the subtectonospheric fracture system. To an external observer, these fossil or inactive geosynclines appear
to have drifted, in time, with respect to presently active geosynclines. Such drifting or migration may involve both a translation and a rotation, where neither is necessarily a linear
function of time.

It might be more accurate, from a geometrico-mechanical
standpoint, to say that the subtectonospheric fracture system,
rather than the geosynclines, have migrated, or, perhaps, that
both the geosynclines and the subtectonospheric fracture system
have migrated, depending upon which part of the Earth is selected
as a fixed or as a mobile datum during an analysis. In either
case, the results will be identical: geosynclines appear to
undergo spatial migration when viewed in a temporal frame. The
migration is of the form aA + bB, where $\underline{a}$ and $\underline{b}$ represent constants, either of which may be zero so long as $a + b = 1$, and
where A and B represent rotation and translation, respectively.

597

If the translational and rotational migrations between the Earth's surficial features and the subtectonospheric fracture system were simple linear functions of time, it would be relatively simple to determine the age of fossil geosynclines from their positions with respect to active geosynclines (i. e., with respect to presently active wedge-belts of activity). Approximations, based on linear assumptions, may be made by measuring the gross distance a particular fossil geosyncline has drifted or migrated from the present position of its genetic wedge-belt of activity. Thus, it is not too surpsising that the Appalachian (Upper Paleozoic) geosyncline is geographically removed by about $30°$ from the Galapagos-Gibralatar wedge-belt of activity, a drift or migration of 1 cm/yr being equivalent roughly to about $30°$ per 300 m. y. In this case, there is very little rotational migration, the center line of the Appalachians lying almost exactly "parallel" to the present position of the genetically-associated wedge-belt of activity (i. e., Galapagos to Gibraltar).

Similar approximations may be made in analyzing other "fossil" geosynclines. More detailed analyses require that one consider the surface geometry of a sphere and the temporo-spatial constraints of moving a system of 3 mutually-orthogonal belts over such a surface. Specifically, one must estimate, as part of the analysis, the "limit-stop" effects of the cratons upon

598

the freedom of the wedge-belt system to "migrate" (Chapter 7).

In addition to this type of gross migration involving geo-synclines, it is well to consider the "small-scale" migration of certain plates and groups of plates lying between the subtec-tonospheric fracture system and the surface of the Earth. This type of migration may involve one or more plates of the 3 basic types comprising the tectonosphere, i. e., lithospheric plates, asthenospheric plates, and subasthenospheric plates (Chapter 6).

Each of these plates moves, at any particular time, in a direction parallel to that of the unbalanced force acting on it. Thus, a current wedge-belt of activity behaves, plate-wise, as an entity only when no unbalanced force is acting on any of its lithospheric, asthenospheric, or subasthenospheric plates. If an unbalanced force is acting on only the uppermost litho-spheric plate of the wedge-belt being analyzed, then the sur-face of the Earth in that wedge-belt is moving with respect to the 999 hypothetical plates lying beneath it (Chapter 6).

Motion of this type is considered in greater detail in sub-sequent chapters when plate tectonics and continental drift are discussed. Because, however, one aspect of this type of motion involves mountain building, it is well to consider here the extent to which geosynclinal evolution and mountain-building activities may be influenced by the motion of surficial plates moving with respect to the subsurficial plates lying beneath the uppermost plates, as represented by the behavior of some of the presently-active Circum-Pacific mountain ranges.

An Analysis of Circum-Pacific Mountain Building Activities Using the Tectonospheric Earth Model.

The Circum-Pacific area is an excellent "proving ground" for any mountain-building hypothesis, because this area covers approximately one half of the globe and is almost completely encircled by mountains (Fig. 12-10-3-1).

McBirney (1970) used the Circum-Pacific mountains to analyze mountain-building hypotheses based on sea-floor spreading concepts. Briefly, McBirney concluded that there is a fundamental discordance between the steady nature of the theoretical mechanism of sea-floor spreading and the observed episodic character of igneous and tectonic processes comprising present Circum-Pacific mountain-building activities. On the one hand, McBirney showed that there is a relatively constant rate of lateral motion of the oceanic crust; on the other, that there are distinct surges of volcanism, plutonism, and tectonism affecting wide regions of the Circum-Pacific belt.

Some of McBirney's questions may be listed here. Why, for example, he asked, has andesitic volcanism in the Japanese arc, 6500 km from the East Pacific Rise, been more prolonged and intense than it has been in South America, only 1600 km

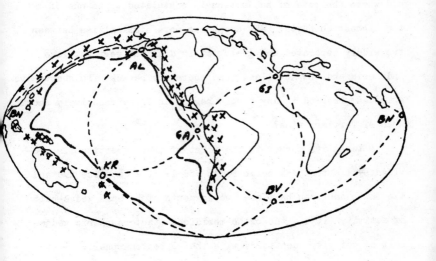

Figure 12-10-3-1. Mollweide projection of the Earth showing the
Circum-Pacific mountain-building activity (crosses) and the
"andesitic line" (dashes). The points AL (Aleutians), BV (Bouvet), GA (Galapagos), GI (Gibraltar), BN (Bengal) and KR (Kermadecs) represent radial surficial projections of the 6 subtectonospheric points of the tectonospheric Earth model. Schematic; not to scale.

from the hypothesized spreading center, where temperatures of the postulated underthrust oceanic plate should be much higher and where the rate of subduction is postulated to be one of the highest on Earth? Why are there more similarities between two widely separated and geologically distinct regions, like Indonesia and the Antilles, than there are between the successive episodes observed in any single region during a relatively short span of Tertiary time?

A tentative answer to the first of these questions is provided by the tectonospheric Earth model: the Japanese arc and the South American belt are two segments of the same wedge-belt of activity (i. e., the ortho-meridional tectonospheric wedge-belt of activity identified with the subtectonospheric points Aleutians, Galapagos, Bouvet, and Bengal (Chapter 6).

A tentative answer to the second question is provided by the fact that the Indonesian arc and the Antilles arc are two segments of the equatorial tectonospheric wedge-belt of activity identified with the subtectonopsheric points Galapagos, Gibraltar, Bengal, and Kermadecs.

From these and other considerations, it may be concluded that Circum-Pacific mountain-building and related activities behave more consistently with respect to the predictions of the tectonospheric Earth model than with respect to those of "sea-floor spreading" concepts (Chapter 16).

REFERENCES

Argand, E., 1916. Sur l'arc des Alpes occidentales. Eclogae Geologicae Helvetiae, 16: 179-182.

Artyushkov, E. V., 1971. Rheological properties of the crust and upper mantle according to data on isostatic movements. J. Geophys. Res., 76: 1376-1390.

Aubouin, J., 1965. Geosynclines. Elsevier, Amsterdam, 335 pp.

Beloussov, V. V., 1968. Some general aspects of development of the tectonosphere. In: K. Benes (Editor), Upper Mantle (Geological Processes), Report of 23rd Session, IGC, pp. 9-17.

Beloussov, V. V., 1969. Interrelations between the Earth's crust and upper mantle. In: P. J. Hart (Editor), The Earth's Crust and Upper Mantle. Am. Geophys. Union, Washington, pp. 698-712.

Beloussov, V. V., 1970. Against the hypothesis of ocean floor spreading. Tectonosphysics, 9: 489-511.

Bird, J. M., 1970. General concepts of orogenesis in terms of lithospheric plate tectonics. Geol. Soc. Am. Abs. with Programs, 2: 733-734.

Bucher, W. H., 1933. The Deformation of the Earth's Crust. Princeton Univ. Press, Princeton, 518 pp.

Bucher, W. H., 1950. Megatectonics and geophysics. Trans. Am. Geophys. Union, 31: 495-507.

603

Bucher, W. H., 1955. Deformation in orogenic belts. In: A. Poldervaart (Editor), The Crust of the Earth. Geol. Soc. Am., New York, pp. 343-368.

Cannon, R. T., Hopkins, D. A., Thatcher, E. C., Peters, E. R., Kemp, J. Gaskell, J. L., Ray, G. E., 1969. Polyphase deformation in the Mozambique Belt, northern Malawi. Geol. Soc. Am. Bull., 80: 2615-2622.

Dana, J. D., 1873. On some results of the Earth's contraction from cooling, including a discussion of the origin of mountains and the nature of the Earth's interior. Am. J. Sci. Ser. 3, 5: 423-443; 6: 6-14, 104-115, 161-172.

Davies, H. L., 1968. Papuan ultramafic belt. In: K. Benes (Editor), Upper Mantle (Geological Processes). Academia, Prague, pp. 209-220.

DeJong, K. A., 1971. Importance of vertical movements in mountain building: Northern Apennines. Trans. Am. Geophys. Union, 52: 356.

Dewey, J. F., 1969a. Evolution of the Appalachian/Caledonian orogen. Nature, 22: 124-129.

Dewey, J. F., 1969b. Continental margins: a model for the transition from Atlantic type to Andean type. Earth Planetary Sci. Letters, 6: 189-197.

Dewey, J. F. and Bird, J. M., 1970a. Mountain belts and the new global tectonics. J. Geophys. Res., 75: 2625-2647.

604

Dewey, J. F. and Bird, J. M., 1970b. Plate tectonics and geosynclines. Tectonophysics, 10: 625-638.

Dietz, R., 1963. Collapsing continental rises: an actualistic concept of geosynclines and mountain building. J. Geol., 71: 314-333.

Dietz, R. and Holden, J. C., 1966. Miogeosynclines in space and time. J. Geol., 74: 566-583.

Dorofeyev, B. V., 1968. Development of associated activated zones in the Urals. Issled. Zemnoy Kory i Verkhney Mantii, 1: 139-146.

Drake, C. L., Ewing, M., and Sutton, G. H., 1959. Continental margins and geosynclines: the east coast of North America north of Cape Hatteras. In: L. H. Ahrens et al. (Editors), Physics and Chemistry of the Earth., vol. 3. Pergamon, New York, pp. 110-198.

Ernst, G. W., 1970. Tectonic contact between the Franciscan Melange and the Great Valley sequence: crustal expression of a late Mesozoic Benioff zone. J. Geophys. Res., 75: 886-901.

Fisher, G. W., Pettijohn, F. J., Reed, J. C., and Weaver, K. N. (Editors), 1970. Studies of Appalachian Geology: Central and Southern. Wiley-Interscience, New York, 460 pp.

Girdler, R. W., 1970. The structure and evolution of the Red Sea and the nature of the Red Sea, Gulf of Aden, and Ethiopian Rift Junction. Tectonophysics, 10: 579-582.

605

Glikson, A. Y., 1970. Geosynclinal evolution and geochemical affinities of early Precambrian systems. Tectonophysics, 9: 397-433.

Gnibidenko, H. S. and Shashkin, K. S., 1970. Basic principles of the geosynclinal theory. Tectonophysics, 9: 5-13.

Goodwin, A. M., 1968. Archean protocontinental growth and early crustal history of the Canadian shield. In: K. Benes (Editor), Upper Mantle (Geological Processes). Academia, Prague, pp. 69-89.

Griffin, V. S., Jr., 1970. Relevancy of the Dewey-Bird hypothesis of cordilleran-type mountain belts and the Wegmann stockwork concept. J. Geophys. Res., 75: 7504-7507.

Hall, J., 1857. Direction of the currents of deposition and source of the material of the older Paleozoic rocks. Canadian Naturalist and Geologist, 2: 284-286.

Hanks, T. C. and Whitcomb, J. H., 1971. Comments on paper by J. W. Minear and M. N. Toksöz, "Thermal regime of a downgoing slab and new global tectonics". J. Geophys. Res., 76: 613-616.

Hasebe, R., Fujii, N., and Uyeda, S., 1970. Thermal processes under island arcs. Tectonophysics, 10: 335-355.

Holmes, A., 1965. Principles of Physical Geology. Ronald, New York, 1288 pp.

Hsu, K. J., 1971. Franciscan melanges as a model for eugeosynclinal sedimentation and underthrusting tectonics. J. Geophys. Res., 76: 1162-1170.

606

Jacob, K, H., 1970. P-residuals and global tectonic structures investigated by three-dimensional seismic ray tracings with emphasis on Longshot data. Trans. Am. Geophys. Union, 51: 359.

Kay, M., 1951. North American Geosynclines. Geol. Soc. Am. Mem. 48, 143 pp.

Khain, V. E., 1970. Present day concepts on the origin of geosynclinal folding. Geotectonics, 3: 3-29.

Khain, V. E. and Muratov, M. V., 1969. Crustal movements and tectonic structure of continents. In: P. J. Hart (Editor), The Earth's Crust and Upper Mantle. Am. Geophys. Union, Washington, pp. 523-538.

Kuenen, P. H., 1967. Geosynclinal sedimentation. Geologische Rundschau, 56: 1-19

Laubscher, H., 1969. Mountain building. Tectonophysics, 7: 551-563.

Luyendyk, B. P., 1971. Comments on paper by J. W. Minear and M. N. Toksöz, "Thermal regime of a downgoing slab and new global tectonics". J. Geophys. Res., 76: 605-606.

MacGillavry, H. J., 1970. Turbidite detritus and geosyncline history. Tectonophysics, 9: 365-393.

McBirney, A. R., 1970. Cenozoic igneous events of the Circum-Pacific. Geol. Soc. Am. Abs. with Programs, 2: 749-751.

McKenzie, D. P., 1971. Comments on paper by J. W. Minear

and M. N. Toksöz, "Thermal regime of a downgoing slab and new global tectonics". J. Geophys. Res., 76: 607–609.

Mikhaylov, A. Y., 1970. The development of geosynclines and folding. Sovetskaya Geol., 3: 18–25.

Minear, J. W. and Toksöz, M. N., 1970. Thermal regime of a downgoing slab and new global tectonics. J. Geophys. Res., 75: 1397–1419.

Mitchell, A. H. and Reading, H. G., 1969. Continental margins, geosynclines, and ocean floor spreading. J. Geol., 77: 629–646.

Miyashiro, A., 1969. Metamorphism and its relation to depth. In: P. J. Hart (Editor), The Earth's Crust and Upper Mantle. Am. Geophys. Union, Washington, pp. 520–522.

Price, R. A. and Mountjoy, E. W., 1970. Geologic structure of the Canadian Rocky Mountains between Bow and Athabasca Rivers: a progress report. Geol. Asc. Canada Spl Paper 6: 7–25.

Reesor, J. E., 1970. Some aspects of structural evolution and regional setting in part of the Shuswap metamorphic complex. Geol. Asc. Canada Spl. Paper 6: 73–86.

Rodgers, J., 1970. The Tectonics of the Appalachians. Wiley-Interscience, New York, 272 pp.

Rezanov, I. A., 1970. On the trend and rate of evolution of the Earth's crust. Vyssh. Ucheb. Zavedeniya Izv. Geol. i Rozved., 2: 3–15

Schuchert, C., 1923. Sites and natures of the North American geosynclines. Geol. Soc. Am. Bull., 34: 151-260.

Schwab, F. L., 1970. Geosynclinal compositions and the new global tectonics. Geol. Soc. Am. Abstracts with Programs, 2: 677.

Sollogub, V. B., 1970. On certain regularities of crustal structure associated with the major geologic features of southeastern Europe. Tectonophysics, 10: 549-559.

Stille, H., 1940. Einführung in den Bau Nordamerikas. Borntraeger Verlagsbuchhandlung, Berlin, 717 pp.

Sylvester-Bradley, P. C., 1968. Tethys, the lost ocean. Sci. J., 4: 47-53.

Wegmann, C. E., 1935. Zur Deutung der Migmatite. Geol. Rundsch., 26: 20-350

Wheeler, J. O. (Editor), 1970. Structure of the Southern Canadian Cordillera. Geol. Assoc. Canada Spl. Paper 6, 166 pp.

Wilson, J. T. and Fairbridge, R. W., 1971. Appalachian peneplains, paleosols, and plate tectonics. Trans. Am. Geophys. Union, 52: 350.

Woollard, G. P., 1969. Regional variations in gravity. In: P. J. Hart (Editor), The Crust and Upper Mantle of the Earth. Am. Geophys. Union, Washington, pp. 320-341.

Wynne-Edwards, H. R. (Editor), 1969. Age Relations in High-Grade Metamorphic Terrains. Geol. Assoc.Can., Toronto, 228 pp.

Chapter 13

THE EARTH'S GRAVITY FIELD

It is the premise of this book that the Earth's gravity field and the shape of the geoid (Challinor, 1964) have presented an ever-changing panorama in response to the restless behavior of the Earth's interior during the past 4.6 b. y. It is a further premise that the tectonospheric Earth model provides a geometrico-mechanical framework within which may be defined (1) the details of the driving mechanism for the Earth's internal behavior and (2) the relationship of this behavior to the geoidal panorama representing the Earth's ever-changing gravity field.

Before considering these premises, some of the observational evidence regarding the shape of the geoid and how this shape is related to other geophenomena and features will be summarized.

General Correlations between the Shape of the Geoid and Other
Static and Dynamic Geophenomena.

The shape of the geoid and other geophenomena can be correlated generally if certain spatio-temporal adjustments are
made. The eariest of these correlations showed that high heat
flow values correspond with low gravity if adjustments are made
for topography (Lee and MacDonald, 1963). Thus, specific geoidal
anomalies do not correlate with present locations of oceans and
continents but with more-fundamental deep-seated features of
the tectonosphere.

Certain geophenomena correlate with the age of the basement, the general correlation being that gravity increases with
increase in age of the basement (Lubimova, 1969; Fotiadi et
al., 1970; Verma et al., 1970). Consequently, the lack of exact
spatial correlation between gravity and heat flow represents
a temporal lag between the Earth's internal thermal behavior
and the causal mechanism producing changes in gravity. This
lag could conceivably amount to millions of years between the
measured heat flow and the measured gravity at the surface of
the Earth. Converted into spatial dimensions at a rate of a
few cm/yr, a temporal lag of 10 m. y. represents several hundred km on the surface of the Earth.

Most geophenomena, including gravity, when viewed globally,
are not symmetric with respect to the axis of rotation of the

611

Earth. This makes correlations difficult, particularly when a shortage of data is involved. This problem, as other problems involving a paucity of data, may be partially overcome if the low "signal" of the observations (i. e., the paucity of observational data over a large area) can be offset by decreasing the "noise" of the system. This may be done by transforming the coordinates from geographical (polar-oriented) coordinates to the geotectonospheric coordinates described in Chapter 3. When this is done, geoidal patterns correlate generally with patterns of other geophenomena and features viewed on a global scale (Tatsch, 1963, 1964, 1967, 1969, 1970).

The above and other analyses showed that the shape of the geoid does not correlate with the Earth's surficial features. At what depth, then, does the shape of the geoid correlate with geophenomena and features? To answer this question, Kaula (1967) compared the correlation between the potential coefficients of a combination solution from gravity and satellite data with those from a harmonic development of the Earth's topography. For degrees greater than 5, he found a positive correlation. This, combined with the observed decrease in potential coefficients for higher degrees, suggested that the origin of long-wavelength geoidal undulations should lie within the upper 400 km of the tectonosphere.

Toksöz et al. (1967, 1969) used geoidal-variation analyses

to show that density variations within the mantle may lie anywhere between depths of 100 km and the core-mantle interface (approximately 2900 km). Similar analyses by Woollard and Khan (1970) showed that anomalous masses may lie at depths ranging from 150 km to the base of the tectonosphere (1000 km). It is likely that the depth limits within which these anomalies lie are different for different parts of the Earth.

Other analyses have defined the association between geoidal undulations and actual mass excesses and deficiencies represented by salient surficial features such as ocean rises, trenches, island arcs, ocean basins, and areas of recent glaciation (See, e. g., Orlin, 1966; Kaula, 1969, 1970; Gaposchkin and Lambeck, 1970).

Hide and Malin found that there is a significant correlation between the shape of the geoid and the non-dipole part of the geomagnetic field, provided the latter is displaced in longitude (Hide and Malin, 1970; Hide, 1970).

In some parts of the Earth, the relationship between the shape of the geoid and other geophenomena presents an enigma. The gravity field in Afghanistan, for example, is characterized by large negative mean free-air anomalies in the eastern Hindu Kush and by essentially zero anomalies in the western Hindu Kush, suggesting that the western part of this mountain range is in isostatic equilibrium whereas the eastern part is associated with a large mass deficiency. Other parts of the Earth show similar complexities within single surficial features in more-or-less small areas.

Even when local anomalies are ignored the geoid contains approximately eight large-scale variations that distinguish it from that of a simple figure of rotation (See, e. g., Uotila, 1962; Guier and Newton, 1965; Izsack, 1965; Strange, 1966; Woollard, 1969a; Khan, 1969). Some of these large variations are negative; others are positive. Collectively, they may be spoken of as the "poles" of the geoid.

When the above and other analyses are considered on a global scale, it is found that generally regions of oldest basements correlate with highest gravity; regions of modern orogenesis, with low gravity. This suggests that the same causal mechanism that controls the Earth's orogenic behavior on a long-lived global scale also controls the shape of the geoid.

<u>The Shape of the Geoid as a Function of the Earth's Internal</u>
<u>Behavior</u>.

If the Earth's gravity field and the shape of the geoid
are surficial manifestations of the Earth's internal behavior
during the past 4.6 b. y., then they should have certain characteristics correlatable with the Earth's internal behavior during
that period of time. Under ideal conditions of isostatic equilibrium, all crustal columns exert equal pressures at some depth,
and the column extending upward from that depth has a mass exactly equal to that of the compensating mass below that depth
(See, e. g., Heiskanen and Vening Meinesz, 1958; Heiskanen and
Moritz, 1967; Caputo, 1967; Woollard, 1969a, 1969b, 1969c). Although the Earth does not behave as an ideal isostatic system,
much can be gained from a brief analysis of the nature and extent to which the Earth's internal behavior departs from that
of an ideal isostatic system when viewed on a global scale over
a long period of time.

Particularly germane to an analysis of the Earth's surficial manifestations in this connection are the following observations related to the isostatic behavior of the Earth: (1)
significant changes (25 to 50 mgal) in isostatic anomalies are
observed where there are changes in crustal composition (i. e.,
in regions of horizontal and vertical heterogeneities); (2)
there is a strong correlation between gravity values and topographic relief, changes in geology, and areas of change in crustal composition and thickness, particularly when the areas are

615

narrow; (3) there appears to be a slight elevation dependence, governed more by the wavelength (width) of a particular topographic feature than by its amplitude (elevation); (4) major changes appear to result from variations in crustal and upper mantle parameters including deep mass anomalies; (5) in cases of major changes, sea-level-intercept values may vary by as much as ± 50 mgal; and (6) generally gravity-anomaly gradients exceed normal values in the case of graben-like surficial features and are subnormal in the case of horst-like surficial features.

Thus, viewed on a global scale, it appears that the Earth's gravity field and the shape of the geoid may reflect the Earth's internal behavior during the past 4.6 b. y. provided approximate spatial and temporal adjustments are made for the fact that the Earth is a dynamic body and that the development of its surficial features may lag behind its deep internal behavior. Adjustments must be made, for example, for the fact that ideal conditions of isostatic equilibrium can exist only if the unbalanced force acting on a vertical column is zero; in all other cases the column tends to move in a direction parallel to that of the unbalanced force vector. It is extremely difficult to visualize a case of absolute ideal isostatic equilibrium within the real dynamic Earth at any time during the past 4.6 b. y. In fact, actual gravity surveys confirm this to be the case at the present time. However, approximate isostatic balance of subglobal or regional features would be expected, and such is observed, but not in agreement with present oceanic—continental configurations.

Numerous attempts have been made to interpret the shape of the geoid in terms of geometrico—mechanical concepts. Artemjev and Artyushkov (1971) concluded that certain facets of the shape of the geoid can be understood only if it is postulated that some geoidal features are maintained by "gravity convection" rather than by thermal convection. In their model, large—scale mantle movements created by the hypothesized gravity convection cause crustal extension which, in turn, leads to rifting. But this model does not include the details of a suitable mechanism to sustain the postulated gravity—convection system over a period of 4.6 b. y. essential to an explanation of the internal behavior of the Earth during this period of time.

The Shape of the Geoid as a Function of Mass Displacements Deep within the Earth.

Local and regional changes in the shape of the geoid result from mass displacements at or near the Earth's surface. Thus, geoidal variations may result from fluctuations of the groundwater table, from intrusive and extrusive activity, from tectonic movements caused by uplift, and from land movements associated with gradation and diastrophism.

But what about geoidal variations due to mass movements deep within the Earth? These are the low-order geoidal undulations that do not show general correlation with the surficial distribution of oceans and continents. To explain these lower harmonics, an anomalous density distribution must be assumed to exist deep within the Earth's interior, i. e., at depths beneath the oceans and continents (See, e. g., Vogel, 1968).

Specifically, the lack of correlation between the shape of the geoid and the global pattern of oceanic-continental features suggests that geoidal features are due to: (1) density differences (and therefore mass transport) deep within the mantle, i. e., below the low-viscosity zone by virtue of which isostatic balance is sometimes postulated to be effected; (2) density differences at shallower depths maintained by convection, or by an equivalent mechanism; or (3) by a combination of (1) and (2), i. e., density differences effected along rheid surfaces at more than one depth within the tectonosphere (Chapter6).

618

The Shape of the Geoid as a Function of the Earth's Elongate
Linear Features and of Its Quadrate Features.

Geosynclines, mountains, and related rifting and quasilinear
uplift represent over 100,000 km of elongate linear features
on the Earth's surface. The interrelationships among the various members of this linear family is not completely understood,
but there appears to be an antithetical Phoenix-like association among them (Chapter 6). The nature of this enigmatic association becomes more meaningful when the linear family of surficial features is analyzed in conjunction with the quadrate
features of the Earth's surface, i. e., the continents and
oceans.

From this standpoint, the evolution of the Earth's continents during the past 4.6 b. y. may be described as an accretionary process whereby more-or-less quadrate central bodies
(the cratons) grew outward along their peripheries by the action
of linear elongate features, or orogens (Chapter 7). It is
not surprising that the ocean-continent panorama displays a
quadrate pattern when analyzed globally, nor that the geoid
presents a similar pattern offset spatially therefrom by an
amount equivalent to the temporal lag between the Earth's surficial features and the internal causal mechanism motivating
the evolution of these features.

619

The Shape of the Geoid According to the Tectonospheric Earth Model.

The shape of the geoid, as a measure of the Earth's gravity field, is identifiable, according to the tectonospheric Earth model, with the wedge-belts of activity and with their associated thermal and tectonic activity (Chapter 6).

Using geographical coordinates based on the Earth's pole of rotation, the shape of the geoid may be expressed as a function of geographic colatitude θ_k and of geographical longitude ϕ_k, the value of the geoidal undulation at any point being expressible as

$$g = g\ (\theta_k,\ \phi_k)$$

This expression may be expanded to any degree of resolution or order N:

$$g = \sum_{n=0}^{N} \sum_{m=0}^{m} \big[A_n^m\ \cos m\ \phi_k + B_n^m\ \sin m\ \phi_k \big]\ P_n^m\ (\cos \theta_k)$$

for values of k from 1 to N (See, e. g., Kreyszig, 1962; Merritt, 1962; Abramowitz and Stegun, 1964; Heiskanen and Vening Meinesz, 1958; Caputo, 1967; or any reference discussing associated Legendre functions of the second kind).

The shape of the geoid, according to the dual primeval

620

planet hypothesis, is "harmonic" with respect to the geotectonospheric coordinate system based on the geotectonospheric behavior of the Earth, rather than with respect to the geographical
coordinate system based on the rotational axis of the Earth.
That is, the dual primeval planet hypothesis and the tectonospheric Earth model postulate that the Earth's geoidal undulations are manifestations of its teotectonospheric structure and
behavior during the past 4.6 b. y., rather than of its undefined
association with the Earth's axis of rotation.

Preliminary analyses have verified this hypothesized "harmonic association" by means of transforming the coordinates
from the geographical to the geotectonospheric system described
in Chapter 3 (Tatsch, 1963, 1964, 1969). This transformation,
briefly stated, involves substituting geotectonospheric coordinates for the values of θ_k and ϕ_k in the above expression for
geoidal undulations, prior to performing the spherical harmonic
analysis of the shape of the geoid. Analyses made with the
transformed data generally confirm the predicted correlation
between the shape of the geoid and other geophenomena and features when considered on a global scale.

The Shape of the Geoid as a Function of the Evolutionary Modes of the Tectonosphere Postulated by the Dual Primeval Planet Hypothesis.

Depending upon what basic assumptions are made, the Earth's tectonosphere developed, according to the dual primeval planet hypothesis, through any one of 3 basic modes of accretionary evolution: octantal-fragment, multiple-fragment, and composite-fragment (Chapter 3). The evolution of the Earth's gravity field during the past 4.6 b. y. would have been different in each of these possible accretionary modes of tectonospheric evolution.

In the octantal-fragment mode, the 5 octants of Earth Prime destined for the Earth are not completely fragmented prior to accretion onto the basic 5400-km promordial Earth (Chapter 3). In such case, parts of each of the 5 octants of Earth Prime might still exist today as unmodified portions of the Earth's present tectonosphere, perhaps even as subsurficial portions of presently-existing continental shields.

The multiple-fragment mode differs from the octantal-fragment mode primarily in the degree to which the 5 "terrestrial" octants of Earth Prime were fragmentized prior to accretion onto the basic 5400-km primordial Earth. If the degree of pre-accretion fragmentation was fairly complete, then it is not very likely that any of the actual surficial rocks of present

622

continental shields would now be identifiable with <u>unmodified</u> portions of the original octants of Earth Prime because tectonic, magmatic, and metamorphic activity during the past 4.6 b. y. would have modified these multiply-fragmentized primordial rocks.

The <u>composite-fragment</u> <u>mode</u> may be considered as an intermediate mode lying between the octantal-fragment mode and the multiple-fragment mode. Depending upon the specific degree of fragmentation assumed for the octants of Earth Prime prior to accretion onto the basic 5400-km primordial Earth, this mode occupies any one of a myriad of intermediate positions within the entire spectrum, or envelope, bounded by the octantal-fragment and multiple-fragment modes as extremes. A similar statement may be made about the modifications of these fragments by tectonic, magmatic, and metamorphic activity on the Earth during the past 4.6 b. y. That is, the nature and amount of tectonic, magmatic, and metamorphic modification undergone by these fragments on the Earth during the past 4.6 b. y. would be a function of the specific degree of fragmentation suffered by the octants of Earth Prime prior to accretion onto the basic 5400-km primordial Earth.

An extension of this analysis implies that the octantal-fragment mode would have had the greatest effect upon the present shape of the geoid; the multiple-fragment mode would have had the least effect; and the composite-fragment mode would have had an effect intermediate between those of the other 2 modes. It will be noticed, however, that the present shape of the geoid will be more a function of the Earth's internal behavior during the past 4.6 b. y. than of the specific accretionary mode through which the tectonosphere evolved.

Among the specific elements of the Earth's internal behavior that would have had a significant effect upon the present shape of the geoid are those dependent upon paths of preferential heat flow within the tectonosphere during the past 4.6 b. y.

<u>The</u> <u>Shape</u> <u>of</u> <u>the</u> <u>Geoid</u> <u>as</u> <u>a</u> <u>Function</u> <u>of</u> <u>Paths</u> <u>of</u> <u>Preferential</u>
<u>Heat</u> <u>Flow</u> <u>Predicted</u> <u>by</u> <u>the</u> <u>Tectonospheric</u> <u>Earth</u> <u>Model.</u>

Within the tectonospheric wedge-belts of activity, the
nature and location of geoidal undulations depend largely upon:
(1) the nature of the magmatic activity and (2) the paths of
preferential heat flow. These two elements are not mutually
exclusive when considered on a global basis.

If an energy source at some point along the subtectonospheric
fracture system consisted of heat, and if the tectonosphere
above that point were homogeneous, then it would be relatively
simple to determine the most probable paths that the heat might
follow through the tectonosphere to the surface of the Earth.
Because the tectonosphere (and specifically the wedge-belts
of activity) is not homogeneous, the preferential heat-flow
paths (from the surface of the subtectonosphere to the surface
of the Earth) might be quite circuitous in order to circumvent
heterogeneities of low thermal conductivity. The exact paths
followed by the heat would influence the nature and location
of the resulting magmatic activity, whether intrusive or extrusive. This, according to the model, accounts for the circuitous
routes taken by some conduits to volcanoes and to intrusive-
body emplacements. This, in turn, has an influence on the
gravity anomalies within wedge-belts of activity.

On a global scale, the salient influence on the shape of

the geoid is the subtectonospheric fracture system. This system
divides the subtectonosphere into octants with the result that
the shape of the geoid would be expected to present characteristics inherent therein, i. e., 8 "poles" due to the 8 tectonospheric geoblocks.

The Shape of the Geoid as a Function of Asthenospheric Flow
Patterns Predicted by the Tectonospheric Earth Model.

Most geoidal features are interpretable as functions of
epiasthenospheric behavior motivated by asthenospheric flow
occurring underneath the lithosphere (See, e. g., Navidi et
al., 1970). Specifically, ocean rises, trenches, and island
arcs represent mass excesses, whereas ocean basins, areas of
recent glaciation, and the Asian portion of the Alpide belt
represent mass deficiencies.

These and other positive and negative mass displacements
may result from: (1) the piling up or removal of material at
the Earth's surface; (2) the replacement of less-dense material
by more-dense material or vice versa, i. e., of more-dense
material by less-dense material within the tectonosphere; (3)
thermal contraction or thermal expansion; (4) transition of
tectonospheric material from one density to a higher or lower
density; (5) petrological fractionation in which a component
of a fixed density is separated from the material before it
enters the region; and (6) a combination of two or more of these
types of mass displacement.

Thus, it may be seen that mass anomalies and consequent
changes in the shape of the geoid can be produced only by an
actual transfer of material from one location to another. From
this requirement, it follows that densifications or rarefactions

627

of material, regardless of how great, without actual movement of the material, will not produce mass anomalies and consequent changes in the shape of the geoid.

This requirement that mass actually be moved to produce a mass anomaly is analogous to the requirement that an <u>unbalanced</u> force vector must operate in order to produce motion; or that a hydraulic head must exist before a fluid will flow regardless of how large the pressure; or that a potential <u>difference</u> must exist before an electrical current will flow regardless of how high the voltage. On the other hand, any unbalanced force, regardless of how small, will produce motion, just as any difference in head, however small, will produce fluid flow, or that even the smallest voltage difference will cause electrical current to flow.

Extending this analysis, it follows that whenever asthenospheric motion occurs, an unbalanced force must be operating. The direction of the instantaneous flow within a given area of the asthenosphere is in the direction of the instantaneous unbalanced force. The direction of the asthenospheric flow changes with changes in direction of the unbalanced force.

Most mass transport within the tectonosphere must be related, in the last analysis, to flow within the asthenosphere. According to present concepts, most asthenospheric flow is produced by incipient melting (See, e. g., Lambert and Wyllie,

1970). The same mechanism that causes this melting is responsible for intrusive and extrusive activity (Chapter 10).

Summarizing, asthenospheric flow within the complex network of "horizontal intrusions" between asthenospheric plates (Chapter 6) moves horizontally whenever there are insufficient vertical conduits to convey upward all protomagma and accociated heat (Chapter 10). Magma unable to proceed either sideward or upward will remain in situ. Consequently the shape of the geoid is inseparately related to asthenospheric structure and behavior and to the paths of preferential heat flow and incipient melting associated therewith.

<u>The Shape of the Geoid as a Function of the Driving Mechanism</u>
<u>of the Tectonospheric Earth Model</u>.

On the basis of present evidence, it appears that the Earth's gravity field and the shape of the geoid are surficial manifestations of the internal behavior of the Earth during the past 4.6 b. y. However, it appears that mass redistributions lag behind the forces causing the Earth's internal behavior. The shape of the geoid, therefore, should correlate with the tectonospheric driving mechanism rather than with the surficial manifestations of the Earth's internal behavior.

Briefly, according to the tectonospheric Earth model, the basic behavior of the primordial Earth, i. e., the present subtectonosphere, is assumed to be such that it tends to equilibrate itself to a state of minimum energy. The driving mechanism may be expressed as a function of the disequilibration energy inherent in its initial state. In simplest terms, then, the driving mechanism that has determined the shape of the geoid during the past 4.6 b. y. consists essentially of the resultant of two factors: (1) the potential energy inherent in the geogenetically disequilibrated shape of the basic 5400-km primordial Earth; and (2) the selectively-channelled energy from the preferential flow of heat, and possibly of volatiles, outward from the subtectonospheric fracture system (Chapter 6).

This driving mechanism has been responsible for practically

630

all salient changes in the shape of the geoid that have occurred during the past 4.6 b. y. and is responsible for the present shape of the geoid and will motivate future geoidal changes.

Sea-Floor Spreading, Crustal-Plate Motions, and Continental
Drift as Manifestations of Geoidal Equilibrations Predicted
by the Tectonospheric Earth Model.

When viewed on a global scale and over a long period of
time, sea-floor spreading and other forms of crustal rifting
represent surficial manifestations of the Earth's attempt to
equilibrate its shape (the geoid) to a figure of minimum potential through tension and resulting crustal thinning in certain
areas. Crustal-plate motion, subsurface-plate and block motion,
and continental drift, similarly, represent the Earth's attempt
to equilibrate its shape by moving plates, blocks, and continents with respect to a datum therein.

In short, sea-floor spreading, crustal-plate motion, subsurface-plate and block motion, and continental drift are surficial manifestations of geoidal equilibrations. The shape of the
geoid, therefore, is a temporarily delayed attempt of the Earth's
gravitational field to adjust itself to the ceaseless equilibrating behavior of the Earth's interior. In some cases a block may
move in an attempt to effect equilibration; in others, a plate may
move. In still others, a rift may develop to effect the equilibration. Sometimes, worldwide mountain ranges grow from these
rifts; at other times, a plate sinks into an undersea rift or
trench. At still other times, an earthquake occurs when a block
or plate adjusts itself to a more nearly equilibrated position
in the never-ending attempt of the Earth to assume a surface of
minimum potential.

632

REFERENCES

Abramowitz, M. and Stegun, I. A. (Editors), 1964. Hand-book of Mathematical Functions. National Bureau of Standards, Washington, 1046 pp.

Artemjev, M. E. and Artyushkov, E. V., 1971. Structure and isostasy of the Baikal rift and the mechanism of rifting. J. Geophys. Res., 76: 1197-1211.

Caputo, M., 1967. The Gravity Field of the Earth. Academic, New York, 202 pp.

Challinor, J., 1964. A Dictionary of Geology, 2d. ed. Oxford Univ. Press, New York, 289 pp.

Fotiadi, E. E., Moiseenko, U. I., Sokolova, L. S., and Duchkov, A. D., 1970. Geothermal investigations in some regions of western Siberia. Tectonophysics, 10: 95-101.

Gaposchkin, E. M. and Lambeck, K., 1970. 1969 Smithsonian standard Earth (II). Smithson. Astrophys. Observ. Spec. Rep. 315, 93 pp.

Guier, W. H. and Newton, R. R., 1965. The Earth's gravity field as deduced from the Doppler tracking of five satellites. J. Geophys. Res., 70: 4613-4626.

Heiskanen, W. A. and Moritz, H., 1967. Physical Geodesy. Freeman, San Francisco, 364 pp.

Heiskanen, W. A. and Vening Meinesz, F. A., 1958. The Earth and Its Gravity Field. McGraw Hill, New York, 470 pp.

Hide, R., 1970. On the Earth's core-mantle interface. Quart. J. R. Met. Soc., 96: 579-590.

Hide, R. and Malin, S. R. C., 1970. Novel correlations between global features of the Earth's gravitational and magnetic fields. Nature, 225: 605-609.

Izack, I. G., 1965. A new determination of non-zoned harmonics by satellites. In: G. Veis (editor), The Use of Artificial Satellites for Geodesy, vol. 2. North-Holland, Amsterdam, pp. 223-229.

Kaula, W. M., 1967. Geophysical implications of satellite determinations of the Earth's gravity field. Space Sci. Rev., 7: 769-794.

Kaula, W. M., 1969. A tectonic classification of the main features of the Earth's gravitational field. J. Geophys. Res., 74: 4807-4826.

Kaula, W. M., 1970. Earth's gravity field: Relation to global tectonics. Science, 169: 982-985.

Khan, M. A., 1969. Figure of the Earth and mass anomalies defined by satellite orbital perturbations. In: P. J. Hart (Editor), The Earth's Crust and Upper Mantle. Am. Geophys. Union, Washington, pp. 293-304.

Kreyszig, E., 1962. Advanced Engineering Mathematics. Wiley, New York, 856 pp.

Lambert, I. B. and Wyllie, P. J., 1970. Low-velocity zone

of the Earth's mantle: incipient melting caused by water. Science, 196: 764-766.

Lee, W. H. K. and MacDonald, G. J. F., 1963. The global variation of terrestrial heat flow. J. Geophys. Res., 68: 6481-6492.

Lubimova, E. A., 1969. Thermal history of Earth. In: P. J. Hart (Editor), The Earth's Crust and Upper Mantle. Am. Geophys. Union, Washington, pp. 63-77.

Merritt, F. S., 1962. Mathematics Manual. McGraw Hill, New York, 378 pp.

Navidi, M. H., Brittain, H. G., and Heller, A., 1970. Earth's gravity field: Relation to global tectonics. Science, 169: 982-985.

Orlin, H. (Editor), 1966. Gravity Anomalies: Unsurveyed Areas. Am. Geophys. Union, Washington, 142 pp.

Strange, W. E., 1966. Comparisons with surface gravity data. In: C. A. Lundquist and G. Veis (Editors), Geodetic Parameters for a 1966 Smithsonian Standard Earth, vol. 3. Smithson. Astrophys. Observ. Spec. Rept. 200, pp. 15-20.

Tatsch, J. H., 1963. Certain geodetic implications of applying a dual primeval planet model to the Earth. Submitted for presentation at 3d Western National Meeting, Am. Geophys. Union, Boulder, Colo.

Tatsch, J. H., 1964. Global distribution patterns of gravity anomalies: Preliminary results of a spherical harmonic analysis. J. Geophys. Res., submitted.

Tatsch, J. H., 1967. Global geomagnetic evidence supporting the existence of a geophysical equator predicted as a consequence of applying a dual primeval planet model to the Earth. Trans. Am. Geophys. Union, 48: 59.

Tatsch, J. H., 1969. Sea-floor spreading, continental drift, and plate tectonics unified into a single global concept by the applications of a dual primeval planet hypothesis to the Earth. Trans. Am. Geophys. Union, 50: 672.

Tatsch, J. H., 1970. Global seismicity patterns as interpreted in accordance with a dual primeval planet hypothesis. Geol. Soc. Am. Abstracts with Programs, 2: 153.

Toksöz, M. N., Chinnery, M. A., and Anderson, D. L., 1967. Inhomogeneities in the Earth's mantle. Geophys. J., 13: 31-59.

Toksöz, M. N., Arkani-Hamed, J., and Knight, C. A., 1969. Geophysical data and large-wave heterogeneities of the Earth's mantle. J. Geophys. Res., 74: 3751-3770.

Uotila, U. A., 1962. Harmonic analysis of world-wide gravity material. Helsinki Publ. Isostat. Inst. Int. Assoc. Geod. number 33.

Verma, R. K., Hamsa, V. M., and Panda, P. K., 1970. Further study of correlation of heat flow with age of basement rocks. Tectonophysics, 10: 301-320.

Vogel, A., 1968. The question of secular variations in the Earth's gravity field from mass displacements in the Earth's deep interior. Uppsala Univ. Geodetic Inst. Report 486, 5 pp.

Woollard, G. P., 1969a. Tectonic activity in North America as indicated by earthquakes. In: P. J. Hart (Editor), The Earth's Crust and Upper Mantle. Am. Geophys. Union, Washington, pp. 125–133.

Woollard, G. P., 1969b. Standardization of gravity measurements. Ibid., pp. 283–293.

Woollard, G. P., 1969c. Regional variations in gravity. Ibid., pp. 320–341.

Woollard, G. and Khan, M., 1970. A review of satellite-derived figures of the geoid and their geophysical significance. Pac. Sci., 24: 1–28.

Chapter 14

GEOMAGNETISM AND POLARITY REVERSALS

Geomagnetism and polarity reversals constitute the most-studied and least-understood surficial manifestation of the Earth's internal behavior. These manifestations are closely associated with the evolution of the Earth's tectonosphere and with other geophenomena, including the Earth's convective field and its electrical field. What is this association, when did it start, and what motivates it? How do changes in the geometrical, mechanical, thermal, and chemical characteristics of the tectonosphere correlate with the behavior of the geomagnetic field? Can the many reversals of this field during the past 4.6 b. y. be associated with any of the geometrical, mechanical, thermal, and chemical aspects of tectonospheric evolution during this time? Tentative answers to these and related questions are examined in this chapter.

The Earth's Convective Field.

The existence of the Earth's magnetic field suggests that mantle convection occurs; but the nature of this convection has defied definition. Evidence indicates that thermal, gravitational, and hydromagnetic forces contribute to the Earth's convective field, but little else is known about it.

Various hypotheses of mantle convection attempt to explain how the internal energy of the Earth causes the dynamic processes observed in the crust and upper mantle. The postulated mantle convection is usually supposed to have a large-scale cellular pattern with mean-flow vectors linear over distances not exceeding an Earth radius (6371 km), or about 57.3° on the surface of the Earth, covering temporal intervals comparable with geological periods (roughly 100 m. y.).

Movements and stresses in crustal layers associated with mountain building (Chapter 12) appear to be an order of magnitude greater than those that could result from simple cooling of the Earth since primordial times. Mantle convection seems the most plausible physical process within the Earth's interior to explain these observed crustal movements and stresses (Runcorn, 1969a). But what motivates the postulated mantle convection?

Thermal processes are most usually proposed as the causative mechanism for mantle convection. Certain geophysical

639

measurements show that the mantle has properties that are not radially symmetric, such as would be expected from an Earth that had formed cold as a randomly accreted sphere or from a fluid (Runcorn, 1969b). Thermal convection may embrace any form of instability that results from (1) heating within the core; (2) heating within the mantle; (3) the outward flow of heat through the tectonosphere; or (4) a combination of 2 or more of these.

Chemical or gravitational convection may be postulated on the basis of the release of gravitational energy from the downward segregation of heavier material in the mantle (Runcorn, 1969a). A density variation of 10 ppm will produce convective flow at a rate of 1 cm/yr, equivalent to a temperature difference of $1^{o}C$ (Runcorn, 1969b).

Coode (1966) lent support to the convective hypothesis by showing that both tensional features (oceanic ridges and related features and activity) and compressional features (mountain building and related activity) have a single deep-seated global cause. Continental drift proponents used similar arguments (Runcorn, 1969b). These and other arguments were weakened by the lack of a specific geometrico-mechanical motivation to show how mantle convection was started and sustained. Although Maxwell, Kelvin, and others proved the feasibility of elasto-viscous behavior of solids, other evidence showed

that no simple convective model can be generally applied without
serious contradictions (See, e. g., Jeffreys, 1962).

Other analyses showed that the short-term behavior of the
mantle cannot be described in terms of simple Newtonian viscosity
(Runcorn, 1969a). Over intermediate-term temporal scales, stress
and strain relations and laws governing them appear to become
increasingly complex with increase in time. For long-term temporal scales, the mechanical behavior of the mantle is even
less well-behaved. Classical theory is not adequate, therefore,
to use as a basis for analyzing the origin, evolution, and behavior of the Earth's convective field during the past 4.6 b. y.
(See, e. g., Elsasser, 1963; Orowan, 1965).

The Earth's crust, when analyzed over temporal scales of
millions of years, acts like an elastic solid, fracturing when
the stresses exceed the breaking stress but not flowing under
the influence of lesser stresses, whereas the asthenosphere,
being in isostatic equilibrium at various depths throughout
the Earth, behaves more nearly like a viscous fluid (Runcorn,
1969a). An analysis of convection within the tectonosphere
(Chapter 6) must therefore consider convective behavior at a
minimum of seven tectonospheric levels: (1) the surface of
the Earth; (2) within and among lithospheric plates; (3) at
the lithosphere-asthenosphere interface; (4) within and among
asthenospheric plates; (5) at the interface between the astheno-

641

sphere and subasthenosphere; (6) within and among subasthenospheric plates and blocks; and (7) at the interface between the tectonosphere and subasthenosphere.

A complete analysis of all facets of convection at these seven levels is not essential to a discussion of the evolution of the Earth's tectonosphere. The essential facets are discussed in the last section of this chapter where the tectonospheric Earth model is considered.

Other analyses regarding mantle convection have been summarized by Munk and MacDonald (1960), Runcorn (1964, 1967), Girdler (1967), Orowan (1969), Clark (1969), Blackwell (1970), Kanamori (1970), and Elsasser (1970), to name only a few. Depending upon which model is used for tectonospheric evolution and behavior during the past 4.6 b. y., some evidence fails to support the possibility of a convective field within the Earth (See, e. g., Jeffreys, 1970).

The Earth's Electrical Field.

The Earth's "electrical" field is extremely weak and, because it defies direct measurement, has not been analyzed to any great extent. Indirect measurements are, however, possible by noting changes in some characteristics of the tectonosphere when changes occur in the geoelectrical field.

Changes in mechanical stress within the tectonosphere produce changes in both the geoelectrical and the geomagnetic field. Consequently, the accumulation and release of mechanical stresses within tectonically active regions change the magnetization of crustal rocks (See, e. g., Nagata, 1950).

In each seismic region of the Earth (Chapter 8), either the greatest relative compression or the greatest relative tension is horizontal. In these regions the vectors of greatest relative tension or compression are normal to the trend of regional structure (Balakina, et al., 1969). From this, the development of tectonic structures, the local discharge of seismic stresses in earthquakes, and changes in the geoelectrical field appear to be maintained by the continuous generation of stresses by a single global deep-seated long-leved mechanism.

The observation of unusual electrical displays in the immediate vicinity of most volcanic eruptions and many earthquakes may possibly indicate the occurrence of variations in the potential of the geoelectrical field during tectonomagmatic activity. Other

643

possible causes for lightning displays at these times have been proposed (See, e. g., Finkelstein and Powell, 1970; Goldstein et al., 1970; Chatelain et al., 1970).

Gough and Porath (1970) found two magnetic anomalies associated with transient electrical currents in north-south conductors beneath the southern Rockies in Colorado and beneath the Wasatch fault belt in Utah. These magnetic anomalies correlate closely with the heat-flow pattern of that area (Chapter 9). Gough and Porath postulated a causal relationship between these variations and a long-lived deep-seated highly-conducting structure beneath the western United States. The coincidence of the heat-flow and induction anomalies, in their view, greatly increases the probability that this causal relationship has existed for at least 200 m. y.

Keller (1971) concluded from studies above "subduction" zones in New Zealand, Central America, Indonesia, and California that an anomalously high electrical conductivity in these areas is "produced by" the water and other volatiles associated with subduction zones. In a related but more generalized study, Molyneux (1971) analyzed the relationship between the geomagnetic field and differences in electrical potential for various points on the surface of the Earth.

The electrical conductivity of molten tectonospheric material exceeds that of the same material in unmolten state by 2 to 3

orders of magnitude (See, e. g., Lubimova and Feldman, 1970),
emphasizing the strong association of the geoelectrical field
with the geothermal field (Chapter 9). In analyzing the close
interrelation between the geoelectrical and geothermal fields,
it is necessary to distinguish closely and continually between
average temperatures at a given depth and **actual** temperatures
within relatively narrow paths of preferential heat flow at
that depth (Chapter 6). Average temperatures at a given depth,
considered globally or regionally, are approximately the same,
but actual temperatures along the relatively-narrow paths of
preferential heat flow may be several orders of magnitude
greater than those of other parts of the tectonosphere at the
same depth.

Deep sounding in various parts of the Earth confirm a close
relationship between high electrical conductivity and the pillow-
shaped low-viscosity zone beneath grabens, emphasizing the association between the geoelectrical field and the global belts
of intrusive and extrusive activity (See, e. g., Illes and
Mueller, 1970).

Other studies regarding the Earth's electrical field have
been made by Bullard (1970), Caner (1970), Madden and Swift
(1969), Madden (1970), Pecova et al. (1970), Rikitake (1969),
Untiedt (1970), and Uyeda and Rikitake (1970).

From these and other observations and analyses, the elec-
trical-conductivity characteristics related to tectonospheric

evolution may be summarized: (1) electrical conduction in the lower mantle may result from the semiconducting properties of silicates at elevated temperatures; (2) the crust generally is less conductive than are sedimentary rocks; (3) conductivity is increased by the quantity and salinity of interstitial waters; (4) the mantle exhibits a rapid increase in conductivity with depth; (5) upper-mantle conductivities may vary by several orders of magnitude within small horizontal distances (e. g., the upper-mantle conductivity under Arizona is 50 to 100 times greater than that under adjacent New Mexico); and (6) conductivities are highest in regions of high heat-flow rates and low seismic velocities.

The Earth's Magnetic Field.

The Earth's magnetic field has many aspects of a field produced by a large permanent magnet, with the magnetic axis offset from the Earth's axis of rotation. Closer scrutiny of the geometrical, mechanical, thermal, and chemical parameters associated with the Earth's magnetic field shows that this simple concept is hopelessly inadequate to explain the observed complexity of this field. These complexities are so intricately integrated with other parameters of the Earth's tectonosphere that a closed solution explaining the observed behavior of the Earth's magnetic field has not been found (See, e. g., Doell and Cox, 1971a, 1971b).

Because an analysis of the evolution of the Earth's tectonosphere must include an analysis of the evolution of the geomagnetic field, it is well to summarize what is known about the Earth's magnetic field during the past 4.6 b. y. This is done in the following sections.

The Relationship between the Geomagnetic Field and the Earth's Axis of Rotation.

Symmetry and other geometrical considerations seem to enter into the origin, evolution, and present characteristics of the geomagnetic field. For example, the observation that the geomagnetic field is roughly symmetrical about the Earth's axis of rotation poses important constraints upon hypotheses regarding the origin and perpetuation of the Earth's magnetic field during the past 4.6 b. y.

Within the geological time scale, the angular variation (inclination) between the geomagnetic axis and the Earth's rotational axis appears as a relatively small transient. Consequently, there is little reason to search for asymmetric elements in the Earth or to hypothesize geomagnetic mechanisms involving asymmetric characteristics. Although the principles of symmetry are normally applied more-or-less intuitively in analyses involving physical cause-and-effect situations, a more explicit approach should be used in analyzing the geomagnetic field where cause-and-effect relationships are closely intermingled and even, on occasion, interchanged (Elsasser, 1966).

The basic causes for the origin and perpetuation of the geomagnetic field appear to be simply (1) the rotation of the Earth which, for all practical purposes, is axisymmetrical and (2) various motions, temperature gradients, and related effects in the core, all of which, apart from the effect of rotation, appear to be spherisymmetric. If the geomagnetic field were to deviate in a consistent asymmetric manner from the Earth's axis of rotation, then the situation would be that of a field with lower symmetry than that attributable to a combination of known contributing causes. This would require the existence of some other (currently unknown) contributing cause of lower symmetry. Because there is no consistent inclination of the dipole field, such cause of lower symmetry does not exist. It is reasonable to assume, therefore, that the geomagnetic axis is constrained to remain within relatively small angles of the Earth's rotational axis as an axis of symmetry. The cause of this constraint is not known, but observational evidence supports its existence, and successful hypotheses must embody its origin and consequences.

Summarizing, hypotheses for the origin, evolution, and present behavior of the geomagnetic field should include (1) an explanation for the long-lived close association between the Earth's axis of rotation and the axis of the geomagnetic field, (2) the cause for the transient variations between the two axes, and (3) the geometrico-mechanical nature of a suitable driving mechanism to produce and sustain these closely-constrained relationships over a long period of time, i. e., billions of years.

The Dipole Nature of the Earth's Magnetic Field.

The Earth's magnetic field is resolvable into dipole and non-dipole components. The non-dipole components reflect the characteristics of deep-seated contributions, the exact nature of which is not understood. Spherical harmonic analyses of the geomagnetic field do not produce individual harmonic terms identifiable with the non-dipole characteristics of the field. Independent analyses suggest that the field may be composed of a multiplicity of deep-seated dipoles (See, e. g., Bullard, 1956).

It is difficult to quantify the dipolar multiplicity. Most analyses suggest that 8 non-central radial dipoles reasonably represent the Earth's geomagnetic field on a global scale, with the dipoles at about one-quarter Earth radius, or possibly farther, from the center of the Earth (See, e. g., Lowes and Runcorn, 1951; Alldredge and Hurwitz, 1964).

Hide and Malin found a close but temporally-offset correlation between the non-dipole component of the geomagnetic field and the shape of the geoid (Hide and Malin, 1970; Hide, 1970), suggesting that these two tectonospheric parameters are motivated by the same global deep-seated long-lived mechanism.

During the past 2 m. y., the Earth's dipole field has closely approximated a geocentric axial dipole. The present boreal trace of this magnetic dipole is at $89°N$ $211°E$ (See, e. g., Cox and Opdyke, 1971), or roughly 1200 km from the boreal

trace of the geocentric axial dipole. The non-antipodal position
of the austral trace of the magnetic dipole is analyzed in the
last section of this chapter where the origin, evolution, and
present behavior of the geomagnetic field is interpreted in
terms of the tectonospheric Earth model.

Paleomagnetically Smooth Areas of the Geomagnetic Field.

On both sides of the Atlantic Ocean (i. e., in Europe-Africa and in the Americas) as well as in other parts of the world, paleomagnetically "smooth" areas are found with peak-to-trough amplitudes of less than 50 gammas (See, e. g., Pitman, 1971; Heirtzler and Vogt, 1971).

Various causes have been postulated for these "smooth" zones: (1) the lack of magnetic polarity reversals during extended periods of time; (2) low-latitude positions of the zones at the time of the paleomagnetic imprinting; or (3) the absence of highly-magnetized pillow basalts or other material at the time of the hypothesized reversal episodes. The latter origin is supported by the observation that the smooth zones are not necessarily coeval in all parts of the globe. Other origins are supported by other data, suggesting that the actual origin of magnetically smooth zones is a combination of several contributing causes.

The Age of the Earth's Magnetic Field.

Direct evidence for a quantitative determination of the age of the geomagnetic field does not exist. The available evidence neither confirms nor denies that the Earth's magnetic field is as old as the Earth.

Blackwell (1971) has postulated that the Earth's accretionary process was completed no later than approximately 4 b. y. ago in order to satisfy an assumption that the geomagnetic field existed before the emplacement of the oldest rocks exhibiting remanent magnetism. But the assumption that this temporal constrain actually existed is not necessary, because the Earth's magnetic field could have existed before the accretionary process was actually completed. In the tectonospheric Earth model (Chapter 3), the accretion of the 5 terrestrial octants of Earth Prime onto the primordial Earth was not necessarily completed prior to the existence of the Earth's magnetic field. The existence of remanent magnetism on the present Moon (Chapter 2) supports the hypothesis that both the primordial Earth and its prime had magnetic fields approximately 4.6 b.y. ago, or approximately 0.6 b. y. earlier than that postualted by Blackwell.

Variations in the Intensity of the Earth's Magnetic Field.

The intensity of the geomagnetic field is highly variable and may change by 100% within the relatively short geologic span of a few thousand years (See, e. g., Kitazawa, 1970). These rapid secular variations are cyclical but aperiodic, suggesting that they are motivated by a composite of several causes. These causes may be associated with a myriad of geometrical, mechanical, thermal, and chemical parameters, e. g., the heterogeneity of the mantle, the nature of the core-mantle interface, the possible heterogeneity of the core, the tectonomagmatic behavior of the Earth's interior, and the nature of the Earth's ultimate driving mechanism.

Geomagnetic Analyses Related to the Evolution of the Earth's Tectonosphere.

In an analysis of the evolution of the Earth's tectonosphere, many geomagnetic studies provide valuable constraints within which the ultimate model must operate. Not all of these geomagnetic studies need be considered in detail, but it is well to summarize the more salient of these by Adams (1970), Ball et al. (1968), Beck (1970), Cox and Opdyke (1971), Creer et al. (1970), Dickson et al. (1968), Doell (1970), Doell and Cox (1971a, 1971b), Dunn et al. (1971), Ewing and Ewing (1967), Hanus and Krs (1970), Heirtzler et al. (1968), Heirtzler (1970), Holcomb (1970), Jacobs (1970), Johnson and Stacey (1969), Jones (1971), Kahle et al. (1969), Kellogg et al. (1970), Kennett et al. (1971), Le Pichon (1968), Le Pichon and Heirtzler (1968), McElhinny and Luck (1970), Morgan (1968), Nair et al. (1970), Niblett and Whitman (1970), Phillips et al. (1969), Pitman and Hayes (1968), Rex and Randall (1970), Strangway (1970), Thomsen (1970), Van Andel and Moore (1970), Vine (1966), Vine and Matthews (1963).

From these and other analyses and observations, the salient characteristics of the geomagnetic field bearing on the evolution of the Earth's tectonosphere may be summarized: (1) belts of "sea-floor spreading" and of other tectonomagmatic activity display patterns of magnetic anomalies; (2) generally Precambrian blocks and other old inactive parts of the Earth show

655

few large magnetic anomalies except those associated with a minimum of tectonomagmatic activity found in old stable regions; (3) the Precambrian blocks frequently are bounded by narrow zones of intense positive anomalies suggesting "crush zones"; (4) in the epi-Hercynian platforms and folded areas, the anomaly zones are consistently parallel to the trend of exposed and buried fold complexes; (5) some anomalies define previously undetected tectonic structure; (6) in some areas, magnetic anomalies do not correlate with surface geology, suggesting that surface geology is not simply an image of subsurface tectonics; and (7) the magnetic field considered on a global scale over a long period of time is very complex.

The Geomagnetic Dynamo.

The geomagnetic field, on first analysis, appears to be caused by a global-sized permanent magnet. But this mechanism is not feasible because a permanent magnet loses its magnetic properties at $580^{\circ}C$, a temperature correspnding to a depth of less than 20 km beneath the Earth's surface.

An alternate possibility is that the Earth behaves as an electromagnet. But a current of 10^9 amps would be required unless the postulated geoelectromagnet involves the mechanical motion of a conductor such as is provided by a geomagnetic dynamo in which the "conductor" may be a conducting fluid within the core or mantle.

Most hypothesized geomagnetic dynamos fall into one of two categories: (1) convective geomagnetic dynamos which derive their driving force from convective processes deep within the Earth and (2) precessional dynamos which derive their driving force from the Earth's precession.

Theoretically, a convective dynamo could have evolved from an adiabatic temperature gradient in the outer core, resulting from the latent heat of a progressively solidifying inner core (See, e. g., Verhoogen, 1961, 1965). Such a mechanism would not, however, have been able to sustain a geomagnetic dynamo over a period of 4.6 b. y. For this, and other, reasons, convective dynamos, whether thermally or gravitationally induced,

657

do not appear entirely satisfactory.

Consequently, a _precessional_ _dynamo_ might be preferred to a convective dynamo (Malkus, 1963, 1968). There is the possibility that the Earth's dynamo is neither convective nor precessional but a combination of both.

In order to determine the true nature of the geomagnetic dynamo, it is necessary to examine the history of the Earth's internal behavior during the past 4.6 b. y., with particular attention to the thermal behavior. If, for example, the Earth's core is in a state of thermal convection, then it is cooling at a rate of about $100^{o}C/b$. y. In the case of precessional dynamos, nothing need be said about deep cooling, although lunar precession implies a decrease in the internal heat generated by the precessional tourques. Thus, the thermal flux from the Earth's core need not be appreciably different in the two basic types of dynamos because a "stirred" outer core implies an adiabatic temperature gradient against thermal conduction in both types.

Geomagnetic polarity reversals appear to occur with greater frequency during periods of rapid polar wander and/or continental drift (See, e. g., Irving, 1966). This intriguing observation suggests that the mantle configuration has a direct influence upon core motions which are considered to be responsible for the geomagnetic field. An alternate suggestion is that

658

both geomagnetic polarity reversals and continental drift (and/or polar wander) are driven by the same geometrico-mechanical system deep within the Earth.

In attempting to resolve this paradox, several tentative conclusions may be drawn. First, it is difficult to explain the correlation between frequency of polarity reversals and rate of polar wander and/or continental drift on the basis of ordinary convective-dynamo processes, but this correlation appears to be somewhat relatable to a precessional dynamo, because the process of polar wander would appear to cause the axis of symmetry of the core to depart somewhat from the Earth's axis of rotation and, thereby, alter the differential precessional torques acting on the core and mantle.

How could the geomagnetic dynamo, with a computed time constant of only about 10^3 years, be responsive to axial changes of a time scale of 10^6 years, or 3 orders of magnitude greater?

The lack of answers to this and other questions suggests that a satisfactory geomagnetic dynamo has not yet been defined.

The Driving Mechanism for the Earth's Geomagnetic Field.

The present dipole moment of the Earth is changing by approximately 1/20 of 1% per year, indicating a time constant of only about 2×10^3 years, from which it follows that the geomagnetic field cannot be simply a decaying vestige of the primordial Earth but must be continuously motivated by some means. What is the exact nature of this motivation, and how has it been sustained during the past 4.6 b. y.?

Bullard (1949) concluded that this motivation poses no problem if the Earth's rotation contributes energy to the driving mechanism for the geomagnetic dynamo. He postulated that, because the core is probably ellipsoidal, rather than purely spherical, it may not follow mantle precession exactly, with the result that consequent internal motions may be adequate to drive the geomagnetic dynamo. However, the issue is in doubt because the geomagnetic currents appear to be of thermo-electric origin rather than purely thermal. Furthermore, the dominance of the dipole field and its axial character over periods of tens of thousands of years suggest that the Earth's rotation (or some other axisymmetric characteristic) exercises a controlling influence. This leads to the conclusion that the motive force for the geomagnetic dynamo may be a linear combination of several contributing factors, of the form $aA + bB + - - - + jJ$, where $a + b + - - - + j = 1$, and A, B, - - -, J are contributions

660

from several factors combining to motivate the geomagnetic field.

Electrical conduction in the lower mantle, such as would be required in a thermo-electrical dynamo, may result from the semi-conducting properties of silicates at elevated temperatures, such as would be expected within shells, planes, and surfaces of preferential heat flow (Chapter 6).

Other analyses regarding possible driving mechanisms for hypothesized geomagnetic dynamos have been made by Creer and Ispir (1970), Suffolk (1970), and Mansinha et al. (1969).

An analysis of the above indicates that the following tentative conclusions may be drawn: (1) the geomagnetic dynamo has evolved, and is presently driven, by the same global causal mechanism that has produced the Earth's internal behavior during the past 4.6 b. y.; (2) the geomagnetic field correlates now, as it did in the past, with global characteristics of the Earth's internal driving mechanism; and (3) the cyclical but aperiodic nature of polarity reversals is definable in terms of the geometry and mechanics of the Earth's internal driving mechanism. On this basis, it appears that viewed on a global scale geomagnetism and polarity reversals are surficial manifestations of the Earth's internal behavior and that such has been the case for possibly all of the 4.6 b. y. of the Earth's existence.

Geomagnetic Polarity Reversals.

Geomagnetic polarity reversals are known to have occurred for at least the past 4.5 million years. There is no reason to doubt that they have been occurring since the Earth began. A geometrico-mechanical explanation in "closed" form for one reversal would serve to explain any number of reversals that might have occurred during the 4.6 billion years that the Earth has been in existence. But a tractable explanation for even one reversal has not yet been found.

The complexity of the reversal process may be seen when the applicable observational evidence is analyzed in terms of the most probable spatial and temporal characteristics of the reversal process. When such analyses are made, the salient spatial characteristics of the reversal process appear to be (1) the total number of magnetized rocks is about equally divided between normal and reversed polarities; (2) each reversal is a single complete 180° change rather than a succession of increments totaling 180°; (3) all basic contributors to the geomagnetic field appear to be axisymmetric without polar symmetry, i. e., there is no distinction between the opposite axial directions, north and south polarities being equally probable; (4) a marked correlation exists between polarity reversals observed on different continents as well as within different rocks on a single continent; (5) sedimentary rocks, baked by contact with subsequently-emplaced igneous rocks, have acquired, in

662

Probabilistic Models for Geomagnetic Polarity Reversals.

Because the problem of the planetary hydromagnetic dynamo has not been solved in closed form, numerous probabilistic models have been postulated for the complex and seemingly intractable problem of geomagnetic polarity reversals.

Most of these models fall into one of two categories: (1) deterministic two-disc dynamos (Rikitake, 1958; Allan, 1962; Mathews and Gardner, 1963); and (2) statistical models (Cox, 1968, 1969, 1970; Parker, 1969; Nagata, 1969).

Other analyses regarding the behavior of the Earth's tectonosphere during cycles of geomagnetic polarity reversals have been made by Cox and Opkyke (1971), Creer and Ispir (1970), Heirtzler and Vogt (1971), Ito and Fuller (1970), McElhinny (1971), and Watkins (1968).

<u>Geomagnetism</u> <u>and</u> <u>Polarity</u> Reversals <u>as</u> Interpreted <u>by</u> <u>the</u> Tec-
<u>tonospheric</u> <u>Earth</u> <u>Model</u>.

Geomagnetism and polarity reversals are identifiable, according to the tectonospheric Earth model, with the subtectonospheric fracture system and with the thermal and tectonic activity associated therewith. The origin, evolution, and perpetuation of the Earth's magnetic field depend upon the flow
of heat, and possibly of volatiles, outward from the subtectonospheric fracture system through the tectonosphere along paths
of preferential heat flow. The migration of the field and its
polarity reversals are motivated by the migration of tectonospheric plates and blocks with respect to the "fixed" subtectonospheric fracture system.

Thus the tectonospheric Earth model provides a basic motive
power and a stable geometrico-mechanical framework for initia-
ting and sustaining a geomagnetic dynamo with a long time constant during the past 4.6 b. y. The origin, evolution, and
stabilization of the geomagnetic field are, therefore, closely
associated with the evolution of the Earth's tectonosphere; and
a study of tectonospheric behavior provides a means of analyzing
the Earth's magnetic behavior including polarity reversals.

The Earth's Magnetic Field and Plate-Block Motion as Interpreted by the Tectonospheric Earth Model.

The plate-tectonics concept postulates that the lithosphere moves in large continental-sized plates. Movement of any plate with respect to another is specified by a rotation pole and a magnitude, the instantaneous values of which may be calculated at any time from magnetic-anomaly data refined by topographic and seismic data (See, e. g., Morgan, 1968; Le Pichon, 1968).

Over an extended period of time, these instantaneous values are not fixed parameters. The number of plates is almost un-limited and continually changing depending upon the minimum plate size used as a lower limit. Consequently, the number of "poles and magnitudes" is similarly unlimited and produces a system having a multiplicity of shifting datums if the surface of the Earth is viewed over an extended period of time, as it must be in an analysis of the evolution of the tectonosphere during the past 4.6 b. y.

The undesirable situation of multiple mobile datums does not occur in the tectonospheric Earth model, because all plate and block positions and motions are stated with respect to a single long-lived fixed datum, i. e., the framework established by the three mutually-orthogonal central planes through the 6 fixed points of the subtectonospheric fracture system (Chapter 6).

667

The globally "fixed" nature of a datum for plate and block motions within the tectonosphere is substantiated by the observation that the present "rotational poles" for most large surficial plates lie near the Earth's rotational poles (See, e. g., Heirtzler et al., 1968; Heirtzler and Vogt, 1971).

Because the motion of tectonospheric plates and blocks, including that of the surficial lithospheric plates identified by the poles and rotations of the plate-tectonics concept, is described in the tectonospheric Earth model with respect to the poles of the Earth's tectonosphere, the behavior of the Earth's magnetic field and the motion of tectonospheric blocks are referred to the same datum. This common datum, the axis of the Earth's subtectonospheric fracture system, is a geogenetic characteristic of the Earth that has been geometrically immutable during the past 4.6 b. y., according to the model (Chapter 6).

The geometrical immutability of the framework of the subtectonospheric fracture system during the evolution of the Earth's tectonosphere therefore provides a long-lived fixed framework within which to analyze the behavior of the Earth's electric field and to study its association with geomagnetism and polarity reversals on a global scale.

The Earth's Electrical Field as Interpreted by the Tectonospheric Earth Model.

Because the electrical properties of the Earth's deep interior are highly sensitive to temperature variations (See, e. $g.$, Lubimova and Feldman, 1970), it may be assumed that the electrical conductivity pattern of the Earth correlates generally with its thermal pattern on a global scale. Thus, the Earth's thermal field may be assumed to represent a first approximation to the Earth's electrical field on a global scale.

The tectonospheric Earth model uses this approximation in selecting the subtectonospheric axis as a "working-hypothesis" axis for the Earth's electrical field. Spatio-temporal separations exist between the actual geoelectrical axis and the subtectonospheric axis, but these separations are negligible transients over any extended period of time such as that involved in the evolution of the Earth's tectonosphere, i. e., billions of years.

Geomagnetic Polarity Reversals as Interpreted by the Tectonospheric Earth Model.

In any analysis of geomagnetic polarity reversals, observations regarding the geomagnetic and geoelectric fields must
be supplemented by considerations concerning the applicable
frames of reference as well as the geometrical and mechanical
environments and constraints in which these fields exist. An
analysis of the origin, evolution, and present behavior of the
geomagnetic field must, therefore, consider not only its geomagnetic and geoelectric attributes but also its geometric
characteristics.

This analysis may be facilitated by assuming that each of
these three attributes has both normal and reversed polarity.
A reversal of the polarity of the geomagnetic attribute of the
system will be defined as a 180° change in the direction of
the dipole vector of the geomagnetic field; a reversal of the
polarity of the geoelectric attribute, as a 180° change in the
direction of the geoelectric field; and a reversal of the polarity of the geometric attribute, as a change in the sequential
direction of the traces of the 3 boreal vectors of the system
on the surface of the Earth. These three vectors are (1) the
Earth's boreal (north) rotational pole, (2) the associated boreal
geomagnetic pole (not necessarily of "north" magnetic polarity
but geographically near the boreal rotational pole directed

670

roughly toward the star Polaris), and (3) the boreal geoelectric
pole.

Fig. 14-10-1 shows the present "positive" configuration
of the geometric attribute of the Earth's magnetic field. In
this configuration the surficial traces of the geoelectric,
geomagnetic, and georotational vectors form a counterclockwise
(positive) sequence $P_E P_M P_R$ on the surface of the Earth in the
vicinity of the north georotational pole. O is in the vicinity of the Earth's center.

Fig. 14-10-2 shows 4 possible configurations in which the
surficial traces of these vectors form a clockwise sequence
and thereby produce a "negative" configuration of the geometric
attribute of the geomagnetic field, it being understood that
in any configuration, whether positive or negative, the sequence
of the subscripts E, M, and R is always read in alphabetical
order. The configuration $\underline{d}$ is "almost positive" because a relatively small amount of northward migration of either P_E or P_M
would convert the triangular configuration $P_E P_M P_R$ to a positive
(counterclockwise) sequence.

Fig. 14-10-3 shows 4 other possible configurations in which
the surficial traces of the geoelectric, geomagnetic, and georotational vectors in the vicinity of the north georotational
pole might have been arranged in the past to form a counterclockwise sequence and thereby to have produced other positive

671

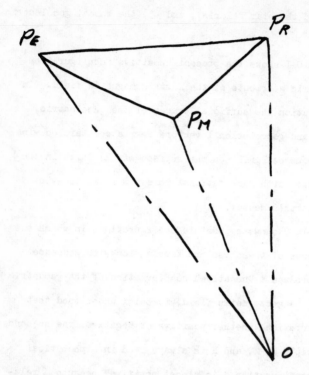

Figure 14-10-1. Schematic representation of the present positive configuration of the geometric attribute of the Earth's magnetic field in the vicinity of the boreal georotational pole. P_E, P_M, and P_R are on the surface of the Earth; O, near the Earth's center. See text for details.

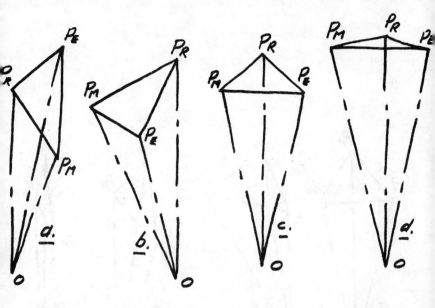

Figure 14-10-2. Schematic representation of a few of the many
possible "negative" configurations of the geometric attribute
of the Earth's magnetic field. In each case, the triangle
$P_E P_M P_R$ forms a negative (clockwise) configuration on the surface of the Earth in the vicinity of the north georotational
pole. The polarity of P_M, according to the tectonospheric Earth
model, is "south" in each of these cases, i. e., the geomagnetic
field is of "reversed" polarity. See text for details.

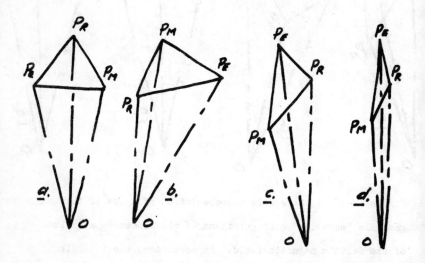

Figure 14-10-3. Schematic representation of a few of the many
possible positive configurations of the geometric attribute of
the Earth's magnetic field. In each case, the triangle $P_E P_M P_R$
forms a "positive" (counterclockwise) configuration on the surface of the Earth in the vicinity of the north georotational
pole. The polarity of P_M, according to the tectonospheric Earth
model, is "north" in each of these cases, i. e., the geomagnetic
field is of "normal" polarity. See text for details.

configurations of the geometric attribute of the geomagnetic field. The configuration $\underline{d}$ in this figure is "almost negative" because a relatively small amount of eastward migration of either P_E or P_M would convert the triangular configuration $P_E P_M P_R$ to a negative (clockwise) sequence.

It is a premise of the tectonospheric Earth model that the reversal of the polarity of any one of the three attributes of the Earth's geomagnetic field will simultaneously reverse the polarity of one of the other two attributes. Thus, the polarity of the geomagnetic attribute of the system may be reversed (1) by reversing the georotational attribute; (2) by reversing the geoelectric attribute; or (3) by reversing the geometric attribute of the system.

It is not likely that the spin of the Earth has been reversed more than once or twice, if at all, during the past 4.6 b. y. Reversal of the georotational vector is not adequate, therefore, to account for the hundreds of times that the Earth's magnetic field has changed polarity during even the past few million years.

The reversal of the Earth's electrical field would require the application of forces beyond the scope of those predicted by the driving mechanism of the tectonospheric Earth model. This approach as a means of reversing the Earth's magnetic field need not, therefore, be discussed.

675

From the above, it follows that a reversal of the <u>geometric</u> attribute of the system provides the best means for reversing the Earth's magnetic field.

Within the framework of the model, the blocks and plates of the tectonosphere are "free" to move with respect to each other and with respect to the subtectonospheric fracture system (Chapter 6). Consequently, the geoelectric and geomagnetic vectors would be expected to migrate, in time, with respect to each other and with respect to the Earth's axis of rotation. This migration, in turn, would cause the geometric attribute of the geomagnetic field to change. Whenever such a change involves a change in polarity of the geometric attribute, the polarity of the geomagnetic attribute also reverses, i. e., the magnetic field reverses its polarity. This has occurred, according to the model, in the cyclical but aperiodic manner expected from the described migrations of the geoelectric and geomagnetic vectors with respect to the Earth's axis of rotation during the past 4.6 b. y. Between reversals, the geomagnetic field is continuously "drifting" in response to the ceaseless re-configuration of the geometric attribute of the Earth's magnetic field.

REFERENCES

Adams, A., 1970. Some results of the magnetotelluric survey in the Carpathian Basin and its complex interpretation. _J. Geomagnet. Geoelect._, 22: 223-233.

Allan, D. W., 1962. On the behavior of systems of coupled dynamos. _Proc. Cambridge Phil. Soc._, 58: 671-693.

Alldredge, L. R. and Hurwitz, L., 1964. Radial dipoles as the source of the Earth's main magnetic field. _J. Geophys. Res._, 69: 2631-2640.

Balakina, L. M., Misharina, L. I., Shirokova, E. I., and Vvedenskaya, A. V., 1969. The field of electric stresses associated with earthquakes. In: P. J. Hart (Editor), _The Earth's Crust and Upper Mantle_. Am. Geophys. Union, Washington, pp. 166-171.

Ball, R. H., Kahle, A. B., and Vestine, E. H., 1968. Variations in the geomagnetic field and in the rate of the Earth's rotation. _Rand Corp. Rept._ F44620-67-C-0045, 33 pp.

Beck, M. E., 1970. Paleomagnetism and Keweenawan intrusive rocks, Minnesota. _J. Geophys. Res._, 75: 4985-4996.

Blackwell, D. D., 1970. Heat flow near Marysville, Montana. _Trans. Am. Geophys. Union_, 51: 824.

Blackwell, D. D., 1971. Heat flow. _Trans. Am. Geophys. Union_, 52: I135-I139.

677

Bullard, E. C., 1949. The magnetic field within the Earth. Proc. Roy. Soc. A, 197: 433.

Bullard, E. C., 1956. Edmund Halley (1656-1742). Endeavour, 15: 189.

Bullard, E. C., 1970. Geophysical consequences of induction anomalies. J. Geomagnet. Geoelect. 22: 73-74.

Caner, B., 1970. Electrical conductivity structure in western Canada and petrological interpretation. J. Geomagnet. Geoelect. 22: 113-129

Chatelain, A., Kolopus, J. L., and Weeks, R. A., 1970. Radiation effects and oxygen vacancies in silicates. Science, 168: 570-571.

Clark, S. P., 1969. Heat conductivity in the mantle. In: P. J. Hart (Editor), The Earth's Crust and Upper Mantle. Am. Geophys. Union, Washington, pp. 622-626.

Coode, A. M., 1966. Analysis of major tectonic features. Geophys. J., 12: 55-66.

Cox, A., 1968. The lengths of geomagnetic polarity intervals. J. Geophys. Res., 73: 3247-3260.

Cox, A., 1969. Geomagnetic reversals. Science, 163: 237-245.

Cox, A., 1970. Reconciliation of statistical models for reversals. J. Geophys. Res., 75: 7501-7503.

Cox, A. and Opdyke, N., 1971. Geomagnetic reversals and

long-period secular variations. <u>Trans</u>. <u>Am</u>. <u>Geophys</u>. <u>Union</u>, <u>52</u>: I210-I213.

Creer, K. M. and Ispir, Y., 1970. An interpretation of the behavior of the geomagnetic field during polarity transitions. <u>Phys</u>. <u>Earth</u> <u>Planetary</u> <u>Interiors</u>, <u>2</u>: 283-293.

Creer, K. M., Collison, D. W., and Runcorn, S. K. (Editors), 1967. <u>Methods</u> <u>in</u> <u>Paleomagnetism</u>. Elsevier, Amsterdam, 609 pp.

Dickson, G. O., Pitman, W. C., and Heirtzler, J. R., 1968. Magnetic anomalies in the South Atlantic and ocean floor spreading. <u>J</u>. <u>Geophys</u>. <u>Res</u>., <u>73</u>: 2087-2100.

Doell, R. R., 1970. Preliminary paleomagnetic results, leg 5. In: D. A. McManus et al. (Editors), <u>Report</u> <u>of</u> <u>Deep</u> <u>Sea</u> <u>Drilling</u> <u>Project</u>, vol. 5. Govt. Printing Off., Washington, pp. 523-524.

Doell, R. R. and Cox, A., 1971a. Pacific geomagnetic secular variation. <u>Science</u>, <u>171</u>: 248-254.

Doell, R. R. and Cox, A., 1971b. The Pacific geomagnetic secular variation anomaly and the question of lateral uniformity in the lower mantle. In: E. Robertson (Editor), <u>The</u> <u>Nature</u> <u>of</u> <u>the</u> <u>Solid</u> <u>Earth</u>. McGraw-Hill, New York, in press.

Dunn, J. R., Fuller, M. D., Ito, H., and Schmidt, V. A., 1971. On a paleomagnetic study of a reversal of the Earth's magnetic field. <u>Trans</u>. <u>Am</u>. <u>Geophys</u>. <u>Union</u>, <u>52</u>: 187.

Elsasser, W. M., 1963. Early history of the Earth. In:

G. Veis and A. D. Goldberg (Editors), Earth Science and Meteoritics. North Holland, Amsterdam, pp. 1-30.

Elsasser, W. M., 1966. Thermal structure of the upper mantle. In: P. M. Hurley (Editor), Advances in Earth Sciences. MIT Press, Cambridge, pp. 461-471.

Elsasser, W. M., 1970. Non-uniformity of crustal growth. Trans. Am. Geophys. Union, 51: 823.

Ewing, J. and Ewing, M., 1967. Sediment distribution on the mid-ocean ridges with respect to the spreading of the sea floor. Science, 156: 1590-1592.

Finkelstein, D. and Powell, J., 1970. Earthquake lightning. Nature, 228: 759-760.

Girdler, R. W., 1967. A review of terrestrial heat flow. In: S. K. Runcorn (Editor), Mantles of the Earth and Terrestrial Planets. Wiley, New York and London, pp. 549-566.

Goldstein, Y., Cohen, M., and Abeles, B., 1970. Strain-induced electric fields in superconducting aluminum. Phys. Rev. Letters, 25: 1571-1575.

Gough, D. L. and Porath, H., 1970. Long-lived thermal structure under the southern Rocky Mountains. Nature, 226: 837-839.

Hanus, V. and Krs, M., 1970. Proposal for paleomagnetic research of endogenous processes in eastern and south-eastern Africa - in the broader adjacent region of the rift valley of eastern Africa. <u>J</u>. <u>Geomagnet</u>. <u>Geoelect</u>., <u>22</u>: 165–167.

Heirtzler, J. R., 1970. The paleomagnetic field as inferred from marine magnetic studies. <u>J</u>. <u>Geomagnet</u>. <u>Geoelect</u>., <u>22</u>: 197–211.

Heirtzler, J. R. and Vogt, P. R., 1971. Marine magnetic anomalies and their bearing on polar wander and continental drift. <u>Trans</u>. <u>Am</u>. <u>Geophys</u>. <u>Union</u>, <u>52</u>: I220–I223.

Heirtzler, J. R., Dickson, G. O., Herron, E. M., Pitman, W. C., and Le Pichon, X., 1968. Marine magnetic anomalies, geomagnetic field reversals, and motions of the ocean floor and continents. <u>J</u>. <u>Geophys</u>. <u>Res</u>., <u>73</u>: 2119–2136.

Hide, R., 1970. On the Earth's core–mantle interface. <u>Quart</u>. <u>J</u>. <u>Roy</u>. <u>Met</u>. <u>Soc</u>., <u>96</u>: 579–590.

Hide, R. and Malin, S. R. C., 1970. Novel correlations between global features of the Earth's gravitational and magnetic fields. <u>Nature</u>, <u>225</u>: 605–609.

Holcomb, R. W., 1970. Astrophysics: Model proposed for galactic magnetic field. <u>Science</u>, <u>168</u>: 811.

Jacobs, J. A., 1970. The evolution of the Earth's core and its magnetic field. <u>Phys</u>. <u>Earth</u> <u>Planetary</u> <u>Interiors</u>, <u>3</u>: 513–518.

Illes, J. H. and Mueller, S. (Editors), 1970. <u>Graben</u> <u>Problems</u>. E. Schweizerbartsche Verlagsbuchhandlung, Stuttgart, 316 pp.

Irving, E., 1966. Paleomagnetism of some carboniferous rocks from New South Wales and its relation to geological events. J. Geophys. Res., 71: 6025-6051.

Ito, H. and Fuller, M., 1970. A paleomagnetic study of the reversal process of the geomagnetic field. In: S. K. Runcorn (Editor), Paleophysics. Academic, New York, pp. 133-138.

Jeffreys, H., 1962. The Earth: Its Origin, History, and Physical Constitution, 4th. ed. Cambridge Univ. Press, Cambridge, 438 pp.

Jeffreys, H., 1970. Imperfections of elasticity and continental drift. Nature, 225: 1007-1008.

Johnson, M. J. S. and Stacey, F. D., 1969. Transient magnetic anomalies accompanying volcanic eruptions in New Zealand. Nature, 224: 1289-1290.

Jones, J. G., 1969. Aleutian enigma: a clue to transformation in time. Nature, 229: 400-403.

Kahle, A. B., Ball, R. H., and Cain, J. C., 1969. Confirmation of prediction of geomagnetic secular change. Rand Corp. Rept. AD-684-116, 6 pp.

Kanamori, H., 1970. Seismological evidence for heterogeneity of the mantle. J. Geomagnet. Geoelect., 22: 53-58.

Keller, G. V., 1971. Electrical methods for the study of subduction tectonics. Trans. Am. Geophys. Union, 52: 354-355.

682

Kellogg, E., Larson, E. E., and Watson, W. E., 1970. Thermochemical remanent magnetization and thermal remanent magnetization: comparison in basalt. Science, 170: 628–630.

Kennett, J. P., Watkins, N. D., and Vella, P., 1971. Paleomagnetic chronology of Pliocene–Early Pleistocene climates and the Plio–Pleistocene boundary in New Zealand. Science, 171: 276–279.

Kitazawa, K., 1970. Intensity of the geomagnetic field in Japan for the past 10,000 years. J. Geophys. Res., 75: 7403–7411.

Le Pichon, X., 1968. Sea floor spreading and continental drift. J. Geophys. Res., 73: 3661–3697.

Le Pichon, X. and Heirtzler, J. R., 1968. Magnetic anomalies in the Indian Ocean and sea-floor spreading. J. Geophys. Res., 73: 2101–2117.

Lowes, F. J. and Runcorn, S. K., 1951. The analysis of geomagnetic secular variation. Phil. Trans. Roy. Soc. A, 243: 525–535.

Lubimova, E. A. and Feldman, I. S., 1970. Heat flow, temperature, and electrical conductivity of the crust and upper mantle in the U. S. S. R. Tectonophysics, 10: 245–281.

Madden, T., 1970. Geoelectric upper mantle anomalies in the U. S. J. Geomagnet. Geoelect., 22: 91–95.

Madden, T. R. and Swift, C. M., 1969. Magnetotelluric

studies of the electrical conductivity structure of the crust
and upper mantle. In: P. J. Hart (Editor), The Earth's Crust
and Upper Mantle. Am. Geophys. Union, Washington, pp. 469-479.

Malkus, W. V. R., 1963. Precessional torques as the cause
of geomagnetism. J. Geophys. Res., 68: 2871-2886.

Malkus, W. V. R., 1968. Precession of the Earth as the
cause of geomagnetism. Science, 160: 259-260.

Mansinha, L., Smylie, D. E., and Beck, A. E. (Editors),
1969. Earthquake Displacement Fields and the Rotation of the
Earth. Reidel, Dordrecht, 310 pp.

Mathews, J. H. and Gardner, W. K., 1963. Field reversals
of "paleomagnetic" type in coupled disc dynamos. U. S. Naval
Res. Lab. Rept. 5886.

McElhinny, M. W., 1971. Geomagnetic reversals during the
Phanerozoic. Science, 172: 157-159.

McElhinny, M. W. and Luck, G. R., 1971. Paleomagnetism
and Gondwanaland. Science, 168: 830-832.

Molyneux, L., Richards, M., Runcorn, S. K., and Strens,
R., 1971. Earth currents and the geomagnetic field. Trans.
Am. Geophys. Union, 52: 193.

Morgan, W. J., 1968. Rises, trenches, great faults, and
crustal blocks. J. Geophys. Res., 73: 1959-1982.

Munk, W. H. and MacDonald, G. J. F., 1960. The Rotation
of the Earth. Cambridge Univ. Press, Cambridge, 323 pp.

Nagata, T., 1950. Summary of the geophysical investigations on the great earthquakes in southwestern Japan on December 21, 1946. Trans. Am. Geophys. Union, 31: 1-6.

Nagata, T., 1969. Length of geomagnetic polarity intervals. J. Geomagnet. Geoelect., 21: 701-704.

Nair, K. N., Rastogi, R. G., and Sarabhai, V., 1970. Daily variations of the geomagnetic field at the dip equator. Nature, 226: 740-741.

Niblett, E. R. and Whitman, K., 1970. Multi-disciplinary studies of geomagnetic variation anomalies in the Canadien arctic. J. Geomagnet. Geoelect., 22: 99-111.

Orowan, E., 1965. Convection in a non-Newtonian mantle, continental drift, and mountain building. Phil. Trans. Roy. Soc. London A, 258: 284-313.

Orowan, E., 1969. The origin of oceanic ridges. Sci. Am., 221: 103-119.

Parker, E. N., 1969. The occasional reversal of the geomagnetic field. Astrophysical J., 158: 815-827.

Pecova, J., Petr, V., and Praus, O., 1970. Depth distribution of the electric conductivity on the Czechoslovak territory. J. Geomagnet. Geoelect., 22: 235-240.

Phillips, J. D., Woodside, J., and Bowin, C. O., 1969. Magnetic and gravity anomalies in the central Red Sea. WHOI Contribution 2182, 22 pp.

Pitman, W. C., 1971. Sea-floor spreading and plate tectonics. Trans. Am. Geophys. Union, 52: I130-I135.

Pitman, W. C. and Hayes, D. E., 1968. Sea-floor spreading in the Gulf of Alaska. J. Geophys. Res., 73: 6571-6580.

Rex, R. W. and Randall, W., 1970. Possible sea-floor spreading of the Imperial Valley of California, V: Present centers. Trans. Am. Geophys. Union, 51: 422.

Rikitake, T., 1958. Oscillations of a system of disk dynamos. Proc. Cambridge Phil. Soc., 54: 89-105.

Rikitake, T., 1969. Conductivity anomaly of the upper mantle. In: P. J. Hart (Editor), The Earth's Crust and Upper Mantle. Am. Geophys. Union, Washington, pp. 463-469.

Runcorn, S., K., 1962a. Convection currents in the Earth's mantle. Nature, 195: 1248-1249.

Runcorn, S. K., 1962b. Towards a theory of continental drift. Nature, 193: 311-314.

Runcorn, S. K., 1964. Satellite gravity measurements and a laminar viscous flow model of the Earth's mantle. J. Geophys. Res., 69: 4389-4394.

Runcorn, S. K., 1967. Flow in the mantle inferred from the low degree harmonics of the geopotential. Geophys. J., 14: 375-384.

Strangway, D. W., 1970. History of the Earth's Magnetic Field. McGraw-Hill, New York and London, 168 pp.

Suffolk, G. C. J., 1970. Precession in a disk dynamo madel of the Earth's dipole field. Nature, 226: 628-629.

Thomsen, D. E., 1970. Searching for monopoles. Sci. News, 98: 183-184.

Untiedt, J., 1970. Conductivity anomalies in central and southern Europe. J. Geomagnet. Geoelect., 22: 131-149.

Uyeda, S. and Rikitake, T., 1970. Electrical conductivity anomaly and terrestrial heat flow. J. Geomagnet. Geoelect., 22: 75-90.

Van Andel, T. H. and Moore, T. C., 1970. Magnetic anomalies and seafloor spreading rates in the northern South Atlantic. Nature, 226: 328-330.

Verhoogen, J., 1961. Heat balance of the Earth's core. Geophys. J., Roy. Astr. Soc., 4: 276.

Verhoogen, J., 1965. Phase changes in convection in the Earth's mantle. Phil. Trans. Roy. Soc. London A, 258: 276-283.

Vine, F. J., 1966. Spreading of the ocean floor: New evidence. Science, 154: 1405-1415.

Vine, F. J. and Matthews, D. H., 1963. Magnetic anomalies over ocean ridges. Nature, 199: 947-949.

Watkins, N. D., 1968. Short period geomagnetic polarity events in deep-sea sedimentary cores. Earth. Planet. Sci. Letters, 4: 341-349.

Chapter 15

CONTINENTS AND OCEANS

An analysis of the evolution of the Earth's tectonosphere must consider the origin, evolution, and present behavior of the continents and oceans, their similarities and differences, and the nature of the driving mechanism that has caused large horizontal and vertical movements of continents and oceans at various times during the past 4.6 b. y.

Most of the Earth's surficial area, including parts of Eurasia and Antarctica and the oceanic areas south of 40°S latitude (1/3 of the Earth's surface), have been inadequately studied to permit a truly comprehensive analysis of the ocean–continent panorama during the past 4.6 b. y. But tentative conclusions may be drawn on the basis of evidence for the less–than–half of the Earth's surface that has been studied.

The evolution of the individual continents during the past 4.6 b. y. was considered in Chapter 7. The relatively short-term behavior of individual ocean basins will be considered in Chapter 16; the horizontal movements of individual continents, in Chapter 17. This chapter takes a global look at the ever-changing behavior of the ocean–continent panorama during the past 4.6 b. y.

<u>Large Ocean Basins</u>.

In analyzing the ocean—continent interrelationships during the past 4.6 b. y. two elements require special attention: (1) the order—of—magnitude difference in age of oceans and continents; and (2) the inherent surficial heterogeneity within each of the ocean basins. The age—difference of oceans and continents is considered in another section. The intra—ocean surficial heterogeneity is discussed below.

Each ocean floor may be described as a young basaltic surface interspersed with granitic remnants such as islands and related features. Such is the case for the Atlantic sea floor, which corresponds essentially with the top of the basaltic layer, interrupted by patches of granitic islands and ridges. The western half of the Indian Ocean, which resembles the Atlantic floor in physical characteristics, also is essentially a basaltic surface containing granitic remnants ranging in size from Madagascar to small coral islands resting upon rocky foundations plus some granite submarine ridges. Similarly, the western Pacific is largely a basaltic floor interspersed with granitic islands and other features between the Marianas arc and the Asiatic mainland as well as in the New Zealand and Polynesian areas.

The most notable examples of these granitic or partly granitic islands are Kermadec, Kerguelen, Solomon, New Hebrides,

Ascension, Bouvet, Reunion, Fiji, Madagascar, New Zealand, Seychelles, New Caledonia, and many of the Arctic islands.

The Pacific Ocean Basin.

About 93% of the oceanic areas and about 65% of the Earth's surface is covered by 4 large ocean basins: the Pacific, Atlantic, Indian, and Arctic. The largest of these, the Pacific, is best defined in terms of the Andesite Line that traces a path of approximately 50,000 km roughly peripheral to the entire Pacific ocean complex and separates truly oceanic areas (deep and basaltic) from quasi-continental oceanic areas (shallower and andesitic). This line circumscribes roughly 3/8 of the Earth's surface, with sharp 90° turns in Indonesia and the Tonga-Kermadecs area and with poorly-defined boundaries in the southern Pacific.

The Andesite Line, as indicated on Fig. 12-10-3, runs parallel and close to the west coast of the Americas; then oceanward of the Aleutians, Kamchatka, the Kuriles, Japan, the Marianas, New Guinea, the Fijis; then landward of Ellice Island (a basaltic island inside the deep Pacific); south between Tonga and Cook islands; then oceanward of New Zelanad; and finally southward to meet the South American branch at an undetermined point somewhere within the austral oceanic area.

A satisfactory explanation for the origin, evolution, and geographical configuration of the Andesite Line does not exist.

Some possibilities are discussed in the last section of this chapter where the evolution of ocean basins is interpreted in terms of the tectonospheric Earth model.

If the Pacific Ocean has evolved during the past 200 m. y., its average rate of growth was 10 cm/yr at the equator. Very little is known about the geometrico-mechanical processes that created the Pacific Ocean. In some relevant hypotheses, the Pacific is a remnant of a larger basin, Panthalassa (See, e. g., Carey, 1958; Rodolfo, 1971); in others, it evolved from a smaller ancestral basin (Chapter 16). Most hypotheses do not discuss the evolution of the Pacific during the first 95% of the Earth's existence (See, e. g., Kamen-Kaye, 1970).

Some hypotheses for the evolution of the Pacific include a post-Mesozoic counterclockwise rotation of about 80° at a peripheral rate of 5 cm/yr. Kawai et al. (1969) found that the island of Honshu has suffered a bend of approximately this amount since about 120 m. y. ago. Prior to that, Japan was unbent, suggesting that only the northern part of Honshu has participated in the hypothesized Pacific rotation. Many other Pacific island chains appear to have undergone rotations of approximately this same amount since the Mesozoic.

The Atlantic Ocean Basin.

The Atlantic Ocean probably evolved as two distinct entities separated along the Galapagos–Gibraltar arc of the extended Tethys. The North Atlantic probably subsided from a shallow-sea environment to abyssal depths and has been subjected to violent undersea erosion. Labrador, Greenland, Iceland, Rockall Plateau, the British Isles, and Spain delineate large almost-isolated portions of the North Atlantic; and the Mid-Atlantic Ridge bisects the Labrador Sea and the Bay of Biscay.

The eastern coast of the Atlantic is significantly younger than is the western coast, suggesting that the present Atlantic evolved from a proto-Atlantic basin. The oldest sediments sampled from the abyssal margins of South America (late Jurassic) now lie 3 or 4 km deeper than when deposited in the hypothesized proto-Atlantic (See, e. g., Fox et al., 1970).

Temporal gaps of 60 m. y. in eastern-Atlantic sediments suggest that the Atlantic basin evolved in a manner more complex than that of simple sea-floor spreading (Chapter 16), particularly since the "missing" sediments correspond to the age of mountain building along the Tethys zone to the east, i. e., the Alps of Europe and the Atlases of Africa.

Ziegler (1970) concluded from Silurian analyses that the Atlantic Ocean was opened during the Lower Paleozoic, closed, and then reopened along a slightly different line during the

Mesozoic. Since the Atlantic opening involved both north-south and east-west movement, this rifting and drifting probably involved a wrenching along the Tethys zone as well as surficial motion along two lines at right angles to the Tethys to suggest a "double" triple-junction such as that shown schematically in Fig. 15-1-1. The point G in the figure is roughly 1000 km above the Gibraltar point of the Earth's subtectonosphere (Chapter 6). Similar double triple-junctions exist roughly 1000 km above the Galapagos, Kermadecs, and Bengal point os the subtectonosphere.

The complexity of the evolution of the Atlantic basin is substantiated by other analyses including those by Maxwell et al. (1970), Anon. (1971), Macdougall (1971), and Gibson and Towe (1971).

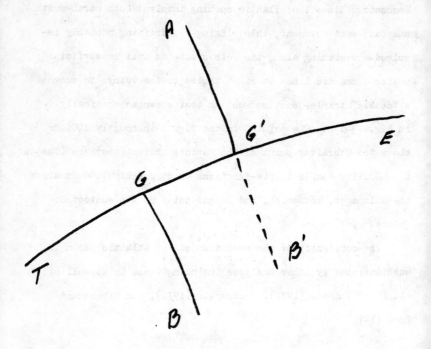

Figure 15-1-1. Hypothesized evolution of double triple-junction
(G and G') as result of wrenching along the extended Tethys
zone TE in the Atlantic Ocean. GA and GB are segments of great-
circular arcs Gibraltar-Aleutians and Gibraltar-Bouvet, respectivel
See text for details.

The Indian Ocean Basin.

The Indian Ocean basin apparently formed in a complex sequence of events extending over more than 100 m. y. (See, e. g., LePichon, 1968; LePichon and Heirtzler, 1968; Vinogradov et al., 1969; and McElhinny, 1970).

One example of this complex evolution is the fact that the intrusive and extrusive activity within peri-Indian continents, such as the copious Deccan flows of India, is temporally relatable with similar activity associated with the opening of the Indian Ocean. Other complexities of Indian Ocean evolution will be discussed in Chapter 17. There it will be shown, for example, that the riftogenic belts of the Indian and Atlantic oceans apparently originated by rifting within "continental" crust, whereas the Pacific Ocean apparently developed by rifting within "oceanic" crust (See, e. g., Vinogradov et al., 1969).

The Arctic Ocean Basin.

The Arctic Ocean, like other large-ocean basins, evolved in a highly complex manner. Demenitskaya and Karasik (1969) found, for example, that the worldwide system of rifts and ridges includes the Gakkel Ridge of the Arctic basin. The Lomonosov Ridge, on the other hand, is not a part of that system. The Mendeleyev Ridge probably is an ancient, or fossil, mid-ocean ridge.

The Gakkel Ridge and other linear features of the quasi-quadrate Arctic basin apparently continue into continental areas surrounding the Arctic basin. Thus, the Nomsk Trough of East Siberia appears to be a continuation of the Gakkel Ridge. In this and other respects, the Arctic Ocean resembles the other 3 large-ocean basins. Like them, it presents a dissymmetry of modern and ancient structures (See, e. g., Churkin, 1969), suggesting a long-lived complex evolution motivated by a deep-seated global driving mechanism.

There are two alternate possibilities for the evolution of the Arctic basin: (1) the basin is a permanent or old feature of the Earth's crust, dating from early Paleozoic or late Pro-terozoic; or (2) it is a relatively young Meso-Cenozoic feature formed by subsidence of continental crust or by rifting and drifting of continental masses. Neither hypothesis explains all the complexities observable in the Arctic basin: (1) its

separation into several basins by submarine mountain ranges;
(2) the Nansen Cordillera (in the Eurasian Arctic) as an extension of the Atlantic ridge; (3) the Lomonosov Ridge as a fragment
from the Eurasian landmass; (4) the Alpha Cordillera (in the
Canadian Arctic) as a fossil ridge; (5) discontinuously exposed
Paleozoic fold belts in Canada, trending into both ends of the
Lomonosov Ridge; (6) the close similarity of Alaskan and Siberian
geology; (7) remnants of a proto-Arctic basin, structurally
separate from the Atlantic and Pacific.

These and other complexities of the Arctic basin have been
analyzed by Churkin (1969), Dietz and Shumway (1961), Johnson
and Heezen (1967a, 1967b), Sykes (1965), Vogt and Ostenso (1970),
Vogt et al. (1969), and Whitman (1964), to name only a few.

Small Ocean Basins.

Most of the small ocean basins are even more complex than are the large ocean basins (See, e. g., Drake, 1970). Many are composites of several anomalous entities of obscure origin (Menard, 1967; Vogt et al., 1971). The youngest of the small ocean basins are only a few million years old (Ryan, 1970). Many of these basins lie near the interface between large crustal plates (Menard, 1967; Vogt et al., 1971). Some appear to have basements transitional between continental and oceanic crust or to be anomalous structures of undetermined evolution. Salt diapirs, found in the Gulf of Mexico and the Mediterranean, are associated with many of these basins (Ryan, 1970).

698

The Origin and Evolution of Small Ocean Basins.

The myriad complexities of small ocean basins have produced many hypotheses attempting to explain their origin, evolution, and present behavior (See, e. g., Van Bemmelen, 1969; Beloussov, 1969).

These hypotheses suggest 5 basic models for the formation of small basins: (1) as remnants of an originally larger ocean encroached by surrounding continents in an environment that produced sialic crust; (2) as a disruption of continental crust and a drifting apart or rotation of continental blocks relative to each other; (3) as surface erosion and tectonic denudation of a continental crust, followed by subsidence to an oceanic level; (4) as an upheaval of an originally oceanic floor above sealevel, followed by later subsidence; and (5) as subcrustal erosion and tectonic transport, chemical assimilation, and/or basification of continental crust, followed by isostatic subsidence.

Application of these hypotheses may be examined by considering the origin and evolution of a few representative ocean basins.

The Caribbean Basin.

The Caribbean basin fails to fit into any one hypothesis for the origin and evolution of ocean basins. Geologically, the Caribbean structure is partly oceanic and partly continental, e. g., the crust beneath the Caribbean is thicker than normal oceanic crust, being over twice as thick as normal oceanic oceanic crust beneath the Beata and Nicaragua ridges. Granitic material has been recovered from Aves Ridge. These and other observations suggest that the Caribbean may be a downdropped continental mass. But reconstructed "pre-drift fits" of the Americas and Eurasia make no provision for a pre-drift continental mass between North and South America (Chapter 17).

Dr. Saunders, co-chief of Leg 15 of the Deep Sea Drilling Project summarized the problem: "The Caribbean was always a problem area. No matter how you join the other land masses, there's always a question of what to do with the Caribbean". Most hypotheses simply ignore the Caribbean area (See, e. g., Bullard, 1965, and other hypotheses of "pre-drift" configurations discussed in Chapter 17).

Molnar and Sykes (1969) defined the Caribbean in terms of two relatively small, nearly aseismic plates. One of these plates, the Cocos, is bordered by the East Pacific Rise, the Galapagos Rift Zone, the north-trending Panama fracture zone near 82°W, and the Middle America arc. The other plate, the

Caribbean, is bounded by the Middle America arc, the Cayman trough, the West Indies arc, and the seismic zone through northern South America. Molnar and Sykes found that (1) the Cocos plate is underthrusting Mexico; (2) the Caribbean plate is moving easterly with respect to the Americas plate; (3) left-lateral strike-slip motion is occurring along steeply dipping fault planes in the Cayman trough; and (4) the Americas plate is under-thrusting the Caribbean in a westerly direction at the Lesser Antilles near Puerto Rico. Molnar and Sykes did not give the geometrico-mechanical details of a driving mechanism that would motivate these four elements of the tectonic behavior related to the Caribbean basin.

Other analyses of the problematical behavior of the Carib-bean include those by Anon. (1971a, 1971b), Deuser (1970), Edgar (1968), Epp et al. (1970), Ewing et al. (1962, 1967), Fink (1971), Fox et al. (1970a, 1970b), Johnson and Headington (1971), Nol-timier (1970), Purrett (1971), Ryan (1970), Sykes and Ewing (1965), Young (1970).

Close scrutiny of these and other analyses show that the Caribbean is a geological "misfit" and that it is easier to prove that the Caribbean does not exist than it is to prove that it does. This suggests (1) a defect in current continental-drift hypotheses (Chapter 17); (2) a defect in current plate-tectonic hypotheses (Chapter 18); or (3) a combination of both.

Small Basins within the Tethys Zone.

In an analysis of the evolution of the Earth's tectonosphere during the past 4.6 b. y., it is important that consideration be given to the geometrico-mechanical behavior of the Tethys zone during that time. The origin, evolution, and present behavior of small basins in that zone provide a key to an understanding of that part of the tectonosphere.

The salient basins of the Tethys zone are the Madeira Basin, the Mediterranean, the Black Sea, the Red Sea, the Caspian Sea, the Persian Gulf, the Gulf of Oman, the Bay of Bengal, the Gulf of Siam, and the Indonesian seas. It is not essential that this analysis consider all these basins in detail, only that enough of the salient features of some of them be interpreted in terms of the Tethys zone as a long-lived global feature. The largest entity of the Eurasian Tethys zone is the Mediterranean, considered in the following section.

The Mediterranean Sea.

During the early Pliocene about 12 m. y. ago, the Mediterranean apparently began acting like a giant rain pond, alter-
nately filling with water and evaporating repeatedly as its
access to the Atlantic opened and closed repeatedly. This process, according to scientists on Leg 13 of the Joides program,
ceased approximately 5.5 m. y. ago. The Mediterranean has remained geologically unstable since then.

Other analyses of the Mediterranean show that (1) the sea-
floor is underthrusting the Eurasian continent in the vicinity
of the Hellenic Trench; (2) during the opening of the Atlantic
130 to 200 m. y. ago, the Iberian Peninsula, severed from the
North American land mass prior to the Eurasian-American separation, was subsequently sutured at the Pyrennes; (3) during
the cyclical but aperiodic drying episodes, the ancestral Mediterranean assumed many aspects similar to those of the present-
day Dead Sea and Death Valley; (4) the multilayered cores from
the Mediterranean suggest repeated rapid evaporations; (5) the
Straits of Gibraltar might have closed and opened through cyclical but aperiodic episodes of mountain-building and erosion;
and (6) the present Mediterranean floor suggests an embryonic
stage of mountain-building especially along the trenches off
Crete and Greece and other quasi-geosynclinal areas (Chapter 12).

These and other features of the Mediterranean have been

analyzed by Anon. (1970), Antoine and Pyle (1970), Caputo et al. (1970), Fahlquist (1963), Harrison (1955), Menard (1967), Muraour (1970), Rabinowitz and Ryan (1970), Ritsema (1970), and Vogt et al. (1971).

The Black and Caspian Seas.

Lying to the east of the Mediterranean and closely related to its evolutionary sequences are the Black and Caspian Seas.

The Black Sea was isolated from the Mediterranean for a period from about 20,000 to 10,000 years ago as a result of the most recent worldwide lowering of sea-level. During the latter part of the interval, the melting of the ice increased and enough fresh water entered the Black Sea to make it a freshwater lake. The re-connection of the Black Sea with the Mediterranean through the Bosporus, approximately 10,000 years ago, re-converted the Black Sea to a saline ocean basin (See, e. g., Ross et al., 1970).

The Caspian is similar to the Black Sea except that intensive sinking and accumulation of sediments began earlier in the Caspian before extending westward to the Black and Mediterranean seas. The Caspian has begun the next stage of tectonic evolution characterized by the formation of folds, mud volcanoes, and diapiric structures (Neprochnov et al., 1970).

Other analyses of the Black and Caspian seas have been made by Garkalenko (1970), Garkalenko et al. (1969), and Sollogub (1970)

704

The Red Sea.

The Red Sea is a nascent ocean basin, associated with the Tethys zone, that has been widening 1 to 1.5 cm/yr during the past several million years. The details of the evolution of this basin have been studied by Phillips et al. (1969), Girdler (1970), and others. In Girdler's model, the Red Sea and the Gulf of Aden were formed when Arabia, moving northward to form the seismically active mountains of Turkey and Iran, left these two basins in its wake.

The Persian Gulf and Other Basins of the Asian Tethys Zone.

The salient basins of the Asian Tethys zone are the Persian Gulf, the Gulf of Oman, the Bay of Bengal, the Gulf of Siam, and the Indonesian seas.

The evolution of these basins, like those of the European Tethys, is closely related to the development of the entire Tethys zone, which is analyzed in the last section of this chapter.

A Generalized Model for the Evolution of Small Ocean Basins.

Most hypotheses for the formation of small ocean basins may be expressed in the form $aA + bB + C$, where A represents vertical motion such as subsidence, or downdropping, motivated by downward-operating unbalanced force vectors; B represents horizontal motion embodying translation and rotation, either one of which may approach zero; C represents the constraints of the local environment; and a and b are constants, either one of which may be zero provided $a + b = 1$.

Hypotheses embodying largely horizontally-operating unbalanced force vectors (i. e., $a \doteq 0$ in the above equation) have been reviewed by Zijderveld et al. (1971) and Vogt et al. (1971). Hypotheses embodying largely vertically-operating unbalanced force vectors (i. e., $b \doteq 0$ in the above equation) have been reviewed by Van Bemmelen (1969).

As expected, deep drilling in various parts of the world supports values of a and b ranging from 0 to 1. Values of $a \doteq 1$ (and $b \doteq 0$) apply to the area immediately west of Sardinia; values of $b \doteq 1$ (and $a \doteq 0$) apply to the Alboran Sea; values of $a \doteq \frac{1}{2}$ (and $b \doteq \frac{1}{2}$) apply in other parts of the Mediterranean. The results of drilling in the Caribbean and other composite small basins show similar variations of a and b from 0 to 1.

The results of drilling in some small-basin areas leaves

the issue in doubt, and more data are required before values for $\underline{a}$ and $\underline{b}$ can be assigned for each small basin on the surface of the Earth. Values of $\underline{a}$ and $\underline{b}$ for fossil basins are even more obscure.

Other aspects of the formation of inner seas and the subsidence of oceanic peripheries have been discussed by Beloussov (1969) and Van Bemmelen (1969).

Similarities and Differences in Large and Small Ocean Basins.

It is not known whether large and small ocean basins are motivated by the same mechanism. Some hypotheses favor separate mechanisms, but Wilson (1969) favors a single mechanism for both large and small basins. In his hypothesis, small basins are embryonic basins that grow progressively larger in an evolutionary sequence: (1) embryonic (East African Rift Valley); (2) young (Red Sea, Gulf of Aden, Norwegian Sea, and Baffin Bay); (3) mature (Atlantic and Indian oceans); (4) declining (Western Pacific); (5) closing (Mediterranean, Black, and Caspian seas); (6) scar or geosuture marking location of former ocean (Indus line in Himalayas, edge of Caledonian thrust in Scandinavia, and Ural Mountains). The spectrum or sequence of motions corresponding with the above genetic stages is as follows: (1) uplift; (2) uplift and spreading; (3) spreading; (4) compression; (5) compression and uplift; and (6) compression

707

and uplift. Unlike most other models, Wilson's does not include any stages or sequences involving subsidence, although 4 sequences in his model embody uplift.

The igneous rocks associated with Wilson's stages of the life-cycle of ocean basins run the complete gamut from tholeiitic flood basalts with alkalic basalt centers, through tholeiitic sea-floor with basaltic islands, to andesitic volcanics with granodiorite-gneiss plutonics. The most basaltic rocks in this model are associated with spreading, whereas the most acidic intrusives and extrusives are associated with compression. This correlation between pressure and type of effusives was discussed in Chapter 10.

<u>The</u> <u>Ocean-Continent</u> <u>Panorama</u> <u>as</u> <u>Interpreted</u> <u>by</u> <u>the</u> <u>Tectonospheric</u>
<u>Earth</u> <u>Model</u>.

An analysis of the evolution of the Earth's tectonosphere
must consider the development of the ocean-continent panorama
during the past 4.6 b. y.; and any hypothesis for the evolution
of the Earth's tectonosphere must account for features such as
continental margins, islands, and the global distribution pattern of oceans and continents on the surface of the Earth.

The Earth's internal behavior is controlled, according
to the tectonospheric Earth model, by a single global long-lived
deep-seated mechanism completely independent of continental
margins, island arcs, and other surficial features motivating
the ocean-continent panorama, except insofar as these features
may be surficial manifestations of the same internal driving
mechanism (Chapter 6).

It is meaningless, therefore, to analyze oceans independently of continents, or to study their individual features
as separate entities except insofar as they have contributed
to the evolution and behavior of the entire global ocean-continent
panorama over a extended period of time, i. e., billions of
years.

One of the features of the ocean-continent complex that
apparently contributed to the evolution of the present ocean-
continent pattern is the antipodal arrangement of the oceans

with respect to the continents. This unique feature of the ocean-continent panorama, although known for hundreds of years, has never been explained.

The Antipodal Arrangement of the Oceans with Respect to the Continents as Interpreted by the Tectonopsheric Earth Model.

Over 90% of the continents lie antipodal to oceanic areas. Most of the Eurasian land-mass lies opposite the South Pacific Ocean; Antarctica, opposite the Arctic; North America, opposite the Indian; and Africa, opposite the North Pacific.

This antipodal arrangement of oceans and continents suggests several speculations: (1) that oceans and continents are isostatically compensated on a global scale, the "heavier" oceanic areas having forced the "lighter" antipodean continents upward by piston-like action, with the "fluid" core serving as the reservoir in this global hydrostatic system (See, e. g., Tatsch, 1964, 1966a, 1966b, 1969); (2) that the Earth is gradually evolving into a tetrahedral shape, with the edges as continents and the flat surfaces as oceans (See Clayton, 1967, p. 59, for a schematic sketch); or (3) that the present ocean-continent pattern represents a combination of (1) and (2).

The concept of the tetrahedral Earth is relatable to the fact that a tetrahedron represents the terminal shape assumed by a shrinking planet (Chapter 1; Tatsch, 1963; Tombaugh, 1963).

Types of Ocean-Continent Interfaces as Interpreted by the Tectonospheric Earth Model.

An explanation for the similarities and differences observed in ocean-continent interfaces, or margins, presents an enigma. Because oceans and continents are not fixed features (Chapters 16 and 17), the interfaces between them must be changing continually. Consequently, there should be many types of ocean-continent interfaces reflecting the ceaseless changes in the ocean-continent panorama. But all interfaces fall actually into two distinct contrasting types, suggesting that a single global mechanism is operating to create this clear-cut separation of all margins into just two basic types, i. e., one marginal type where the hypothesized mechanism is effective, another where it is not.

What is this long-lived deep-seated mechanism, and how has it managed to remain so uniquely effective along some ocean-continent interfaces but not along others?

The sharp contrast between the two basic types of ocean-continent interfaces has been known for many years and is usually expressed by the terms "Atlantic" and "Pacific" margins (See, e. g., Beloussov, 1969; Conolly, 1969, 1970; Emery et al., 1970; Karig, 1971; Kosminskaya and Zerev, 1968, 1970; Mitchell and Reading, 1969).

The **Atlantic** type margin has stable unmodified blocks on the landward side and no evidence of post-Paleozoic activity. Ocean-to-continent crustal dimensions are rather abrupt and may, in places, be characterized by deep sedimentations. Atlantic type margins do not normally contain deep trenches, although a trench-like feature may be found beneath the sediments of the associated continental rise. Plaeozoic and earlier features, in many cases, are truncated by the continental margin.

The _Pacific_ type margin, on the other hand, has young tectonic belts paralleling the ocean-continent interface. Shallow-focus earthquakes are very common, and intermediate and deep-focus earthquakes are abundant, particularly along certain segments of the Pacific type margin. Many trenches lie along these margins. The ocean-continent transition is much more varied than it is in the Atlantic type, and sedimentary accumulations do not appear to be as great.

The basic difference between Atlantic and Pacific margins, analyzed on a global scale, appear to be a function of position with respect to an active geosyncline (Chapter 12). Closer scrutiny reveals that ocean-continent margin types are a function of subsurface tectonospheric behavior.

According to the tectonospheric Earth model, Pacific type margins are associated with current wedge-belts of activity (Chapter 6); all other margins are Atlantic type. Thus the west coasts of North and South America, the east coast of Eurasia, the north coast of Australia, being near a current wedge-belt of activity, are examples of Pacific type margins.

An extension of this concept reveals that there may be a complete spectrum of margin types ranging from purely Pacific to purely Atlantic, definable by the relationship aA + bB, where A represents Atlantic type, B represents Pacific type, and $\underline{a}$ and $\underline{b}$ are constants either of which may be zero provided a + b = 1. Then, purely Pacific type margins are representable by conditions in which a = 0; purely Atlantic margins, by those in which b = 0.

Cases of sutured margins (e. g., Eurasia along the Urals; Antarctica along the Transarctic Mountains; and the suture along the Tethys interface) fall within the general expression aA + bB. However, recent orogenies have obliterated much of the evidence regarding the marginal distinctions along sutures. The understructure beneath these sutured interfaces is considered in Chapter 18 where the plate-block structure and behavior of the tectonosphere is considered.

713

<u>The Origin and Evolution of the Hawaiian Islands as Interpreted by the Tectonospheric Earth Model</u>.

Any analysis of the evolution of the ocean-continent panorama during the past 4.6 b. y. must account for the occurrence of "paradoxes" such as the Hawaiian Islands. The complexity of the origin and evolution of these islands makes them "misfits" in any modern concepts such as sea-floor spreading (Chapter 16) and plate tectonics (Chapter 18). This situation may best be summarized by saying that it is easier to "prove" that the Hawaiian Islands do not exist than it is to prove that they do.

The Hawaiian Islands lie along the Hawaiian Ridge that extends approximately 2500 km across the central Pacific as a chain of shield volcanoes that are progressively younger toward the southeast and that have been built up by successive eruptions of tholeiitic basalt followed by smaller eruptions of alkalic basalts, ankaramite, hawaiite, trachyte, and nephelinite. Coarsegrained xenoliths of layered, feldspathic cumulates in these eruptives indicates that floored chambers existed beneath the central areas of the volcanoes during much of their eruptive history.

The textures, structures, mineral assemblages, and physical properties of these xenoliths suggests that they are fragments of subcrustal rocks and that the upper mantle beneath the Hawaiian Ridge is heterogeneous vertically, horizontally,

714

or both, and that it has other complexities associatable with upper mantle material (See, e. g., Jackson, 1968; Kawai et al., 1969; Knopoff et al., 1968; Macdonald and Katsura, 1962; Macdonald and Kuno, 1962; Morgan, 1971; Normark and Shor, 1968; Porter, 1971; Spies et al., 1969).

The Hawaiian Islands, as interpreted by the tectonospheric Earth model, are identifiable with the extrusive and intrusive activity (Chapter 10) associated with the Kermadecs-Aleutians wedge-belt of activity. This association is shown schematically in Fig. 15-10-3-1. The 78° rotation represents a peripheral rotation of about 5 cm/yr for the Pacific during the past 200 m. y. Other islands of the Pacific, according to the model, have undergone similar rotations.

As may be seen from the figure, the northern Pacific, according to the model, consists of two basic surficial plates, temporarily sutured along the line KR-AL, except for sufficient flow to sustain the volcanic flow of the islands. Francheteau et al. (1969) have discussed evidence for the composite nature of the Pacific as well as for (1) post-Cretaceous motion along the present suture between the northwestern and northeastern Pacific plates, and (2) volcanic aseismic ridges as fossils remaining from earlier extrusions along sutured boundaries. Chapter 10 discusses the age relationships among extrusions of different geological periods as a function of the current position of

715

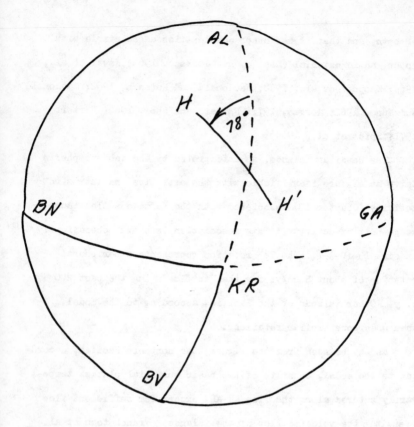

Figure 15-10-3-1. The origin and evolution of the Hawaiian
Islands as interpreted by the tectonospheric Earth model. HH',
Hawaiian Islands. AL, BN, BV, GA, and KR are the surficial
traces of the Aleutians, Bengal, Bouvet, Galapagos, and Kermadecs points of the subtectonosphere. See text for details.

716

wedge-belts of activity of the model.

The behavior of the individual oceans and continents as separate entities is discussed in Chapter 16 and in Chapter 17, respectively; their understructures as a cause of that behavior, in Chapter 18.

The Tethyan Zone and Its Global Extension as Interpreted by the Tectonospheric Earth Model.

The basic features of the Tethyan Zone were described in an earlier section of this chapter. These features as well as their origin and evolution must be accounted for in any analysis of the behavior of the global ocean–continent panorama during the past 4.6 b. y.

The Tethyan Zone and its physiography were almost completely obliterated and recast by the Alpine orogeny, but a pre–Cenozoic reconstruction suggests that this Paleo–Mesozoic Eurasian belt formed a great–circular arc of approximately 13,000 km, including an ancestral proto–Caribbean as well as the basic Eurasian Tethys extending from the Straits of Gibraltar through the East Indies (See, e. g., Challinor, 1964; Fuster, 1968; Fuster et al., 1968; Bosshard and MacFarlane, 1970; Anon., 1971a, 1971b).

According to the tectonospheric Earth model, an "extended" Tethys is identifiable along a 40,000–km belt formed by extending the basic Eurasian Tethys in both directions to form a globe–encircling great–circular belt lying approximately 1000 km radially above the "equator" of the subtectonospheric fracture system (Chapter 6).

The extended Tethys, like the basic Eurasian Tethys, embodies more–or–less enclosed seas; these include the Gulf of Mexico, the Caribbean, the Eurasian Tethys basins, the East Indies basins,

and the Galapagos basins. This "Mesogeic" zone, or extended
Tethys, separates the northern lands from the southern lands,
i. e., Europe from Africa, Asia from Australia, the northern
Pacific islands from the southern Pacific islands, and North
America from South America.

It is along the extended Tethys that the northern lands
appear to be displaced (See, e. g., Suess, 1909; Zijderveld
et al., 1970). The tectonospheric Earth model provides a
geometrico-mechanical framework within which this torsion might
have occurred and thereby produced the 4 sets of double triple-
junctions shown earlier in Fig. 15-1-1.

The most probable causal mechanism for producing the hypothesized Tethyn torsion is provided by the basic driving mechanism of the tectonospheric Earth model (Chapter 4); i. e., the
Earth, in its ceaseless attempt to equilibrate its geogenetically
disequilibrated shape, twisted its northern and southern hemi-
tectonospheres with respect to each other, along a central plane
through the globally-extended 40,000-km great-circular Tethys.

The behavior of the underlying tectonospheric plates, as well
as the geometrico-mechanical framework for that behavior, is discussed in Chapter 18, after considering the basic behavior of
the individual oceans and continents as separate entities in
Chapters 16 and 17.

REFERENCES

Anon., 1970. Probing the Mediterranean's hidden geological past. Sci. News, 98: 20–21.

Anon., 1971a. On to the Caribbean. Geotimes, 16(1): 27–30.

Anon., 1971b. Deep sea drilling project Leg 14. Geotimes, 16(2): 14–17.

Antoine, J. W. and Pyle, T. E., 1970. Crustal studies in the Gulf of Mexico. Tectonophysics, 10: 477–494.

Beloussov, V. V., 1969. Interrelations between the Earth's crust and upper mantle. In: P. J. Hart (Editor), The Earth's Crust and Upper Mantle. Am. Geophys. Union, Washington, pp. 698–712.

Bosshard, E. and MacFarlane, D. J., 1970. Crustal structure of the eastern Canary Islands from seismic refraction and gravity data. J. Geophys. Res., 75: 4901–4918.

Bullard, E. C., Everett, J. E., and Smith, A. S., 1965. The fit of the continents around the Atlantic. Phil. Trans. Roy. Soc. Lond. A, 258: 41–51.

Caputo, M., Panza, G. F., and Postpischl, D., 1970. Deep structure of the Mediterranean basin. J. Geophys. Res., 75: 4919–4923.

Carey, W. S., 1958. A tectonic approach to continental drift. In: W. S. Carey (Editor), Continental Drift: A Symposium. Univ. of Tasmania, Hobart, pp. 177–355.

Challinor, J., 1964. A Dictionary of Geology, 2nd ed. Oxford Univ. Press, New York, 289 pp.

Churkin, M., 1969. Paleozoic tectonic history of the Arctic basin north of Alaska. Science, 165: 549-555.

Clayton, K., 1967. The Crust of the Earth. Natural History Press, Garden City, N. Y., 155 pp.

Conolly, J. R., 1969. Western Tasman Sea floor. New Zealand J. Geol. Geophys., 12: 310-343.

Conolly, J. R., 1970. Sedimentary history of the continental margin of Australia. N. Y. Acad. Sci. Trans. ser. 2, 32: 364-380.

Demenitskaya, R. M. and Karasik, A. M., 1969. The worldwide pattern of midoceanic ridges and its northern part. Noveyshiye dvizheniya, vulcanizm, i zemletryaseniya materikov i dna okeanov, Izdatelstvo Nauka, pp. 249-257.

Deuser, W. G., 1970. Hypothesis of the formation of the Scotia and Caribbean seas. Tectonophysics, 10: 391-401.

Dietz, R. S. and Shumway, G., 1961. Arctic basin geomorphology. Bull. Geol. Soc. Am., 72: 1319-1329.

Drake, C. L., 1970. A long-range program of solid Earth studies. Trans. Am. Geophys. Union, 51: 152-159.

Edgar, N. T., 1968. Seismic refraction and reflection in the Caribbean. PhD thesis, Columbia Univ.

Emery, K. O., Uchupi, E., Phillips, J. D., Bowin, C. O.,

and Bunce, E. T., 1970. Continental rise off eastern North America. Am. Assoc. Geologists Bull., 54: 44-108.

Epp, D., Grim, P. J., and Langseth, M. G., 1970. Heat flow in the Caribbean and Gulf of Mexico. J. Geophys. Res., 75: 5655-5669.

Ewing, J. I., Worzel, J. L., and Ewing, M., 1962. Sediments and ocean structural history of the Gulf of Mexico. J. Geophys. Res., 67: 2509-2527.

Ewing, J., Talwani, M., Ewing, M., and Edgar, N. T., 1967. Sediments of the Caribbean. Studies in Tropical Oceanography, Univ. Miami, 5: 88-102.

Fahlquist, D. A., 1963. Seismic refraction measurements in the western Mediterranean Sea. PhD thesis, Mass. Inst. Tech.

Fink, K. L., 1971. Evidence in the eastern Caribbean for Mid-Cenozoic cessation of sea-floor spreading. Trans. Am. Geophys. Union, 52: 251.

Fox, P. J., Heezen, B. C., and Johnson, G. L., 1970a. Jurassic sandstone from the tropical Atlantic. Science, 170: 1402-1404.

Fox, P. J., Ruddiman, W. F., Ryan, W. B. F., and Heezen, B. C., 1970b. The gology of the Caribbean crust, I: Beata Ridge. Tectonophysics, 10: 495-513.

Francheteau, J., Harrison, C. G. A., Sclater, J. S., and Richards, M. L., 1969. Magnetization of Pacific seamounts. Trans. Am. Geophys. Union, 50: 634.

Fuster, J. M., 1968. Gran Canaria. In: Geology and Volcanology of the Canary Islands. Lucas Mallada, Madrid, 243 pp.

Fuster, J. M., Cendrero, A., Gastesi, P., Ibarrola, E., and Lopez-Ruiz, J., 1968. Fuerteventura. In: Geology and Volcanology of the Canary Islands. Lucas Mallada, Madrid, 243 pp.

Garkalenko, I. A., 1970. The deep-seated crustal structure in the western part of the Black Sea and adjacent areas: seismic reflection measurements. Tectonophysics, 10: 539-547.

Garkalenko, I. A., Nikiforuk, B. S., Mikhaylov, V. M., and Chekunov, A. V., 1969. Deep structure and main features of the evolution of the northwestern sector of the Black Sea and adjacent regions. Sov. Geol., 8: 37-47.

Gibson, T. G. and Towe, K. M., 1971. Eocene volcanism and the origin of Horizon A. Science, 172: 152-153.

Girdler, R. W., 1970. The structure and evolution of the Red Sea and the nature of the Red Sea, Gulf of Aden, and Ethiopian Rift Junction. Tectonophysics, 10: 579-582.

Harrison, J. C., 1955. An interpretation of gravity anomalies in the eastern Mediterranean. Phil. Trans. Res. Soc. London, 248: 283-324.

Jackson, E. D., 1968. The character of the lower crust and upper mantle beneath the Hawaiian Islands. In: K. Benes (Editor), Upper Mantle (Geological Processes). Academia, Prague, pp. 135-150.

Johnson, M. S. and Headington, E., 1971. Panama: exploration history and petroleum potential. Oil and Gas J., 69(15): 96-100.

723

Johnson, G. L. and Heezen, B. C., 1967a. Morphology and evolution of the Norwegian-Greenland Sea. Deep-Sea Res., 14: 755-771.

Johnson, G. L. and Heezen, B. C., 1967b. The Arctic mid-oceanic ridge. Nature, 215: 724-725.

Johnson, H. and Smith, B. L. (Editors), 1970. The Mega-tectonics of Continents and Oceans. Rutgers Univ. Press, New Brunswick, N. J., 284 pp.

Kamen-Kaye, M., 1970. Age of Basins. Geotimes, 15(7): 6-8.

Karig, D. E., 1971. Origin and development of marginal basins in the western Pacific. J. Geophys. Res., 76(11): 2542-2561.

Kawai, N., Hirooka, K., and Nakajima, T., 1969. Paleomag-netic and potassium-argon age information supporting Cretaceous-Tertiary hypothetic bend of the main island of Japan. Paleogeog. Paleoclim. Paleoecol., 6: 277-282.

Knopoff, L., Drake, C. L., and Hart, P. J. (Editors), 1968. The Crust and Upper Mantle of the Pacific Area. Am. Geophys. Union, Washington, 522 pp.

Kosminskaya, I. P. and Zvrev, S. M., 1968. Deep seismic sounding in the transition zones from continents to oceans. In: L. Knopoff et al. (Editors), The Crust and Upper Mantle of the Pacific Areas. Am. Geophys. Union, Washington, pp. 122-130.

Kosminskaya, I. P. and Zverev, S. M., 1970. Structure of continental margins. _J. Geomagnet. Geoelect._, _22_: 179-195.

Le Pichon, X., 1968. Sea floor spreading and continental drift. _J. Geophys. Res._, _73_: 3661-3697.

Le Pichon, X. and Heirtzler, J. R., 1968. Magnetic anomalies in the Indian Ocean and sea-floor spreading. _J. Geophys. Res._, _73_: 2101-2117.

Macdonald, G. A. and Katsura, T., 1962. Relationships of petrographic suites in Hawaii. In: G. A. Macdonald and H. Kuno (Editors), The Crust of the Pacific Basin. Am. Geophys. Union, Washington, pp. 187-195.

Macdonald, G. A. and Kuno, H. (Editors), 1962. The Crust of the Pacific Basin. Am. Geophys. Union, Washington, 195 pp.

Macdougall, D., 1971. Deep sea drilling: age and composition of an Atlantic basaltic intrusion. _Science_, _171_: 1244-1245.

Maxwell, A. E., Von Herzen, R. P., Hsu, K. J., Andrews, J. E., Saito, J., Percival, S. F., Milow, E. D., and Boyce, R. E., 1970. Deep sea drilling in the South Atlantic. _Science_, _168_: 1047-1059.

McElhinny, M. W., 1970. Formation of the Indian Ocean. _Nature_, _228_: 977-979.

Menard, H. W., 1967. Transitional types of crust under small ocean basins. _J. Geophys. Res._, _72_: 3061-3073.

725

Mitchell, A. H. and Reading, H. G., 1969. Continental margins, geosynclines, and ocean floor spreading. J. Geol., 77: 629-646.

Molnar, P. and Sykes, L. R., 1969. Tectonics of the Caribbean and middle America regions from focal mechanisms and seismicity. Geol. Soc. Am. Bull., 80: 1639-1684.

Morgan, W. J., 1971. Parallel seamount chains in the northeast Pacific. Trans. Am. Geophys. Union, 52: 236.

Muraour, P., 1970. Considerations sur genese de la Mediterranae occidentale et du Golfe de Gascogne (Atlantique). Tectonophysics, 10: 663-677.

Neprochnov, Y. P., Kosminskaya, I. P., and Malovitsky, Y. P., 1970. Structure of the crust and upper mantle of the Black and Caspian seas. Tectonophysics, 10: 517-538.

Noltimier, H. C., 1970. A proposed structural evolution of the Caribbean and Central America since the Jurassic. Trans. Am. Geophys. Union, 51: 825.

Normark, W. R. and Shor, G. C., 1968. Seismic reflection study of the shallow structure of the Hawaiian arch. J. Geophys. Res., 73: 6991-6998.

Phillips, J. D., Woodside, J., and Bowin, C. O., 1969. Magnetic and gravity anomalies in the central Read Sea. WHOI Contribution 2182, 22 pp.

Porter, S. C., 1971. Holocene eruptions of Mauna Kea Volcano, Hawaii. Science, 172: 375-377.

Purrett, L., 1971. The rock bottom. Science News, 99(1): 31.

Rabinowitz, P. D. and Ryan, W. F. B., 1970. Gravity anomalies and crustal shortening in the eastern Mediterranean. Tectonophysics, 10: 585-608.

Ritsema, A. R., 1970. On the origin of the western Mediterranean sea basin. Tectonophysics, 10: 609-623.

Rodolfo, K. S., 1971. Contrasting geometric adjustment styles of drifting continents and spreading sea floors. J. Geophys. Res., 76: 3272-3281.

Ross, D. A., Degens, E. T., and MacIlvaine, A. B., 1970. Black Sea: recent sedimentary history. Science, 170: 163-165.

Ryan, W. F. B., 1970. Deep Sea Drilling Project: Leg 13. Geotimes, 15(10): 12-15.

Sollogub, V. B., 1970. On certain regularities of crustal structure associated with the major geologic features of southeastern Europe. Tectonophysics, 10: 549-559.

Spiess, F. N., Luyendyk, B. P., Larson, R. L., Normark, W. R., and Mudie, J. D., 1969. Detailed geophysical studies on the northern Hawaiian arch using a deeply towed instrument package. Marine Geology, 7: 501-527.

Suess, E., 1909. The Face of the Earth. Clarendon, Oxford, 2 vols., 400 and 673 pp.

Sykes, L. R., 1965. The seismicity of the Arctic. Bull. Seismol. Soc. Am., 55: 501-518.

Sykes, L. R. and Ewing, J., 1965. The seismicity of the Caribbean region. J. Geophys. Res., 70: 5065-5074.

Tatsch, J. H., 1963. Certain selenophysical implications of applying a dual primeval planet model to the Earth. Trans. Am. Geophys. Union, 44: 877.

Tatsch, J. H., 1964. Certain oceanographic implications of applying a dual primeval planet model to the Earth. Trans. Am. Geophys. Union, 45: 74.

Tatsch, J. H., 1966a. Certain correlations between oceanographic observations along the South Atlantic Ridge and deductions arrived at from applying a dual primeval planet model to that region of the Earth. Trans. Am. Geophys. Union, 47: 123.

Tatsch, J. H., 1966b. Certain correlations between oceanographic observations along the North Atlantic Ridge and deductions made from applying a dual primeval planet model to that region of the Earth. Trans. Am. Geophys. Union, 47: 477.

Tatsch, J. H., 1969. Sea-floor spreading, continental drift, and plate tectonics unified into a single global concept by the application of a dual primeval planet hypothesis to the Earth. Trans. Am. Geophys. Union, 50: 672.

Tombaugh, C. W., 1963. Private discussions.

Vinogradov, A. P., 1969. The structure of the mid-oceanic rift zone of the Indian Ocean and its place in the world rift system. Tectonophysics, 8: 377-401.

Van Bemmelen, R. W., 1969. Origin of western Mediterranean
728

Sea. Verh. Kong. Mynbowk. Gen., 26: 13-52.

Vogt, P. R., and Ostenso, N. A., 1970. Magnetic and gravity
profiles across the Alpha cordillera and their relation to Arctic
sea-floor spreading. J. Geophys. Res., 75: 4925-4937.

Vogt, P. R., Schneider, E. D., and Johnson, G. L., 1969.
The crust and upper mantle beneath the sea. In: P. J. Hart
(Editor), The Earth's Crust and Upper Mantle. Am. Geophys.
Union, Washington, pp. 556-617.

Vogt, P. R., Higgs, R. H., and Johnson, G. L., 1971. Hypotheses on the origin of the Miditerranean basin: magnetic data.
J. Geophys. Res., 76: 3207-3228.

Waters, A. C., 1962. Basalt magma types and their tectonic
associations: Pacific Northwest of U. S. G. A. Macdonald and
H. Kuno (Editors), The Crust of the Pacific Basin. Am. Geophys.
Union, Washington, pp. 158-170.

Wilson, J. T., 1969. Aspects of the different mechanisms
of ocean floors and continents. Tectonophysics, 8: 281-284.

Young, J. C., 1970. Caribbean Island Arc. Geotimes, 15(9):
17-19.

Ziegler, A. M., 1970. Geosynclinal development of the
British Isles during the Silurian Period. J. Geol., 78: 445-
479.

Zijderveld, J. D. A., Hazeu, G. J. A., Nardin, M., and
Van der Voo, R., 1970. Shear in the Tethys and the Permian
paleomagnetism in the southern Alps, including new results.
Tectonophysics, 10: 639-661.

Chapter 16

SEA-FLOOR SPREADING

The surface of the Earth, according to modern concepts, is composed of a number of rigid crustal plates each in motion with respect to all others. The interfaces between these surficial plates constitute regions of the Earth's strongest seismic, orogenic, and tectonic activity. Along those interfaces that are separating from each other, extrusions occur and are termed "sea-floor spreading". These extrusions form ridges of new oceanic crust. Where transcurrent motion occurs between plates, the interfaces are referred to as "transform faults". According to the sea-floor spreading hypothesis, trenches, island arcs, and/or mountain chains form along those interfaces that are in compression (See, e. g., Anon, 1971a, 1971b, 1971c, 1971d; Bell, 1971; Bird and Dewey, 1970; Blackwell and Roy, 1971; Chase et al., 1970; Cook, 1970; Dewey and Bird, 1970a, 1970b; Dietz and Holden, 1970; Dietz and Sproll, 1970; Elders et al., 1970, 1971; Emery et al., 1970; Grim, 1970; Hall, 1970; Hamblin and Petersen, 1971; Hamilton, 1970; Hayes and Pitman, 1970; Heirtzler and Vogt, 1971; Hess, 1968; Karig, 1970, 1971; Larson and Chase, 1970; Martin, 1971; Maxwell et al., 1970; Moores, 1970; Oriel, 1971; Phillips and Luyendyk, 1970; Pitman, 1971; Price and Mountjoy, 1970, 1971; Robinson, 1971; Rona et al., 1970; Talwani et al., 1971; Vine and Hess, 1971; Vogt and Ostenso, 1970; Vogt et al., 1970, 1971; Zietz, 1971; and the many earlier references cited in these).

Because the concept of sea-floor spreading involves only a relatively small part of the upper tectonosphere during the past 200 m. y., an analysis of the evolution of the Earth's tectonosphere need not include a detailed analysis of this hypothesis. It is well to consider, however, some of the problematical areas of sea-floor spreading by examining the basic concept in terms of the tectonospheric Earth model.

Sea-Floor Spreading as Interpreted by the Tectonospheric Earth
Model.

When viewed on a global scale and over a long period of time,
sea-floor spreading and other forms of crustal rifting apparently
represent the surficial behavior of the Earth in those areas
where subsurface activity causes crustal tension. Crustal plate
motion, subsurface plate and block motion, and continental drift,
in a similar manner, represent the surficial behavior of the
Earth in those areas where the resultant of the Earth's internal
behavior is representable as a tangential unbalanced force vector
acting on a plate, on a block, or on a continent. This produces
motion of that block, that plate, or that continent with respect
to the bulk of the Earth, i. e., to a fixed datum within the
Earth.

Sea-floor spreading, crustal plate motion, subsurface plate
and block motion, and continental drift are closely related
phenomena, but the causal relationships among these phenomena
is not known. Each of these phenomena, according to the tectonospheric Earth model, represents a manifestation of the delayed
attempt by the Earth's tectonosphere to adjust itself to the
ceaseless equilibrating behavior of the interior as it attempts
to place the geoid into a condition of minimum energy.

732

<u>Mantle</u> <u>Convection</u> <u>as</u> <u>Interpreted</u> <u>by</u> <u>the</u> <u>Tectonospheric</u> <u>Earth</u>
<u>Model</u>.

The exact association between mantle convection, continental
drift, and sea-floor spreading is not known. Hypotheses that
link these phenomena in a causal manner have not proved satisfactory.

If it is assumed that continental drift has not been a
continuing process since the Earth began, then convection must
be considered to have started from a "stand-still" state at time
"t minus k", where <u>t</u> is the time when continental drift started
and <u>k</u> is the number of years required (1) to put the convective
system into operation and (2) to permit it to start moving continents. This assumption poses an almost intractable problem
of supplying a large motivating force at time "t minus k".

If, on the other hand, it is assumed that convection started
4.6 b. y. ago, i. e., when the Earth began, then any segment
of the convective streamline could have traveled 46,000 km, since
then, at a rate of 1 cm/yr. In this case, there is no problem
of supplying an impossibly large motivating force at time "t
minus k" to start the system.

Nearly all hypotheses for mantle convection are based on
the Benard-Rayleigh concept (Benard, 1900; Rayleigh, 1916; Foster,
1969; Takeuchi and Sakate, 1970; Ichiye, 1971) and assume that
such convection caused sea-floor spreading to start about 200
m. y. ago. These hypotheses make one of two assumptions: (1)

733

that the value of the constant $\underline{k}$ is about $4.6 - 0.2 = 4.4$ b. y.; or (2) that a large unbalanced force became available at some time between 4.6 and 0.2 b. y. ago.

In the tectonospheric Earth model, it is assumed that the Earth's behavior has been more-or-less the same during at least the past 3.6 b. y. and that phenomena analogous to sea-floor spreading and continental drift have been occurring in a cyclical but aperiodic manner since that time. Mantle convection, in this concept, may be motivated by the preferential flow of heat, and possibly of volatiles, outward from the subtectonospheric fracture system.

Summarizing, the tectonospheric Earth model, unlike most hypotheses, does not invoke mantle convection as a mechanism for causing phenomena such as sea-floor spreading and continental drift. Rather, in this concept, mantle convection is motivated by the same basic driving mechanism that motivates sea-floor spreading and continental drift.

<u>Oceanic</u> <u>Ridges</u> <u>as</u> <u>Interpreted</u> <u>by</u> the Tectonospheric Earth Model.

Oceanic ridges, as interpreted by the tectonospheric Earth model, are associated with the wedge-belts of activity of the model (Chapter 6). Active extrusive ridges are associated with current wedge-belts; inactive ridges, with fossil wedge-belts. Thus, the active ridges of the Atlantic and Indian oceans, which are associated with current wedge-belts, differ from the aseismic ridges of the southern and eastern Pacific, which are associated with fossil (Mesozoic) wedge-belts.

The "active" (post-Mesozoic) ridge of the central Pacific is associated with the Kermadecs-Aleutians wedge-belt. The Hawaiian and other post-Mesozoic chains of the central Pacific have been rotated about 78° since the Mesozoic (Chapter 15).

Relationship between Sea-Floor Spreading and Extrusive-Intrusive
Activity as Interpreted by the Tectonospheric Earth Model.

Sea-floor spreading, according to the tectonospheric Earth
model, is closely related to the magmato-tectonic activity related to the wedge-belts of activity of the model (Chapter 6).
One of the salient types of this magmato-tectonic activity involves extrusive and intrusive activity (Chapter 10).

Many ridges associated with the current wedge-belts of
activity of the tectonospheric Earth model mark linear belts
of intrusion and extrusion of basaltic magma similar to that
of mid-ocean rises. One of these, according to the model, is
the 300-km ridge through the center of the Tonga-Lau basin
(Sclater et al., 1971). Other ridges of this type are (1) the
submerged east-west Méso-Cenozoic ridge associated with the
Cape Verde Islands (See, e. g., Dash et al., 1971); (2) many
of the seamount chains in the Pacific (See, e. g., Hey and Morgan,
1971); (3) a succession of NE-SW linear ridges associated with
the Burma Rise (Holcombe, 1971); (4) "slab" interfaces in the
Japanese area (Carr et al., 1971); (5) linear belts beneath
the Basin and Range Province (Scholz et al., 1971); (6) numerous
island chains, including the Hawaiian, Austral, and Laccadive-
Maldive islands; (7) aseismic ridges such as the Walvis, Iceland-
Faeroe, and Carnegie; and (8) linear intrusions such as the
Deccan traps, parts of the Afar system, and the Snake River

736

basalts.

Differences in the characteristics of the extrusive-intrusive features of linear attributes of the Earth's surface are discussed in Chapters 10 and 12.

Because the driving mechanism of the tectonospheric Earth model lies far beneath the surface of the oceans and continents, the model predicts that there is no clear-cut distinction between the tectonomagmatic activity of the oceans and that of the continents.

What observational evidence is there to support this prediction of the tectonospheric Earth model?

Relationship between Magmato-Orogenic Processes on the Continents and Sea-Floor Spreading as Interpreted by the Tectonospheric Earth Model.

Any attempt to interpret continental magmato-orogenic processes in terms of basic "sea-floor spreading" concepts encounters a fundamental discordance between the steady state nature of the inferred mechanism of sea-floor spreading and the episodic characteristics of igneous and tectonic processes observed on the continents (See, e. g., McBirney, 1969). This suggests either (1) that a separate mechanism is driving oceanic and continental magmato-orogenic processes or (2) that sea-floor spreading is occurring in an episodic manner similar to that observed on the continents.

According to the tectonospheric Earth model, (1) a single global long-lived deep-seated mechanism is driving magmato-orogenic processes in both oceanic and continental areas; (2) this single mechanism is the basic driving mechanism of the tectonospheric Earth model (Chapter 6); (3) magmato-orogenic activity in both oceans and continents is occurring in an episodic manner, i. e., cyclically but with aperiodicity; and (4) this activity in oceans as well as in continents is occurring along current wedge-belts of activity (Chapter 6) in close association with active seismicity (Chapter 8) and with geosynclinal development (Chapter 12).

The Migration of Spreading Axes as Interpreted by the Tectonospheric Earth Model.

Within certain constraints, tectonospheric plates and blocks
are free to move with respect to the subtectonospheric fracture
system (Chapter 3). The surficial plates of the tectonosphere,
between which sea-floor spreading extrusions occur, may become
displaced with respect to lines and points radially above associated segments of the subtectonospheric fracture system.

These "migrations" of the spreading axes may consist of
translations, rotations, or a combination of both. These may
be expressed generally as aA + bB, where A represents translation,
B represents rotation, and a and b are constants either of which
may be zero provided a + b = 1. Migrations of pure translation
occur when b = 0; of pure rotation, when a = 0; and of combinations of translation and rotation, when neither a nor b is
zero.

Fig. 16-10-5-1 shows a migration of the spreading axis
involving only a translation, i. e., b = 0. The spreading axis
AB, on the surface of the Earth, has undergone a translation a
with respect to the surficial radial projection A'B' of the
associated segment of the subtectonospheric fracture system
lying 1000 km radially beneath A'B' (Chapter 6).

Fig. 16-10-5-2 shows migration of the spreading axis involving only rotation, i. e., a = 0. The spreading axis AB,

739

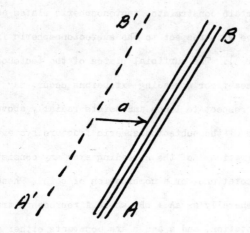

Figure 16-10-5-1. Migration of spreading axis involving only
translation. The spreading axis AB has undergone a translation
_a_ with respect to the radial surficial projection A'B' of the
associated segment of the subtectonospheric fracture system
lying roughly 1000 km beneath A'B'.

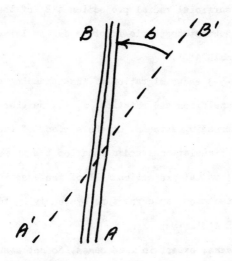

Figure 16-10-5-2. Migration of spreading axis involving only
rotation. The spreading axis AB has underfone a rotation <u>b</u>
with respect to the radial surficial projection A'B' of the
associated segment of the subtectonospheric fracture system
lying roughly 1000 km radially beneath A'B'.

749

on the surface of the Earth, has undergone a rotation $\underline{b}$ with respect to the surficial radial projection A'B' of the associated segment of the subtectonospheric fracture system lying 1000 km radially beneath A'B'.

Fig. 16-10-5-3 shows migration of the spreading axis involving both translation and rotation, i. e., neither $\underline{a}$ nor $\underline{b}$ is zero. The spreading axis AB, on the surface of the Earth, has undergone a translation $\underline{a}$ and a rotation $\underline{b}$ with respect to the surficial radial projection A'B' of the associated segment of the subtectonospheric fracture system lying 1000 km radially beneath A'B'.

Spreading axes, except in rare cases, do not assume the simple idealized configurations shown schematically in the above 3 sketches. Rather, depending upon the effect of the "local" unbalanced force vector on the individual tectonospheric plates and blocks (Chapter 6), the actual spreading axis, on the surface of the Earth, will appear more like that shown in Fig. 16-10-5-4. Each segment of the spreading axis has undergone a separate translation and rotation with respect to the surficial radial projection A'B' of the associated segment of the subtectonospheric fracture system lying 1000 km radially beneath A'B'.

When the spreading axis is associated with one of the 6 basic points of the subtectonosphere (Chapter 6), the basic pattern shown in Fig. 16-10-5-4 may become modified. Two cases

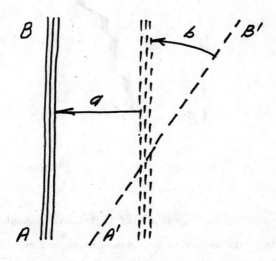

Figure 16-10-5-3. Migration of spreading axis involving both translation and rotation. The spreading axis AB has undergone a translation a and a rotation b with respect to the radial surficial projection A'B' of the associated segment of the subtectonospheric fracture system lying roughly 1000 km radially beneath A'B'.

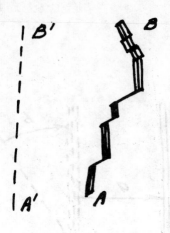

Figure 16-10-5-4. Migrations of individual segments of the
spreading axis involving both translation and rotation. Each
segment of the spreading axis AB has undergone a separate translation and rotation with respect to the surficial radial projection A'B' of the associated segments of the subtectonospheric
fracture system lying roughly 1000 km radially beneath A'B'.

744

of such modification may be considered: (1) that in which no hemispheric torsion (Chapter 3) is associated with the subtectonospheric point; and (2) that in which there is hemispheric torsion associated with the subtectonospheric point beneath the region being studied.

Fig. 16-10-5-5 shows the first of these two cases, i. e., the one in which no hemispheric torsion is associated with the subtectonospheric point P. The segments of the spreading axes, CB, CA, BD, and AD, on the surface of the Earth, have migrated with respect to the surficial projections, A'B' and C'D', of the associated segments of the tectonospheric fracture system lying 1000 km radially beneath A'B' and C'D' and passing through the subtectonospheric point P.

A further modification of the spreading pattern may occur if hemispheric torsion is associated with one of the fractures through the underlying subtectonospheric point, as shown in Fig. 16-10-5-6. In this case, a double triple-junction (Chapter 15, Fig. 15-1-1) is produced on the surface of the Earth. The hemispheric torsion produces, in addition to the double triple-junction, a distortion in the associated spreading axis. Several variants of the myriad possible configurations of the spreading axis are shown schematically in the figure. For example, as shown in the figure, the segments CA and AD of the spreading axis have assumed the configuration CAD because there are no

745

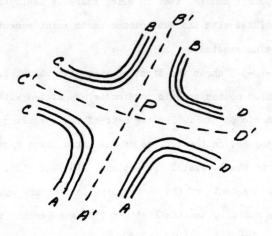

Figure 16-10-5-5. Migration of spreading axes when spreading
is associated with one of the 6 basic points of the subtectonosphere, P. The segments of the spreading axes, CB, CA, BD,
and AD, on the surface of the Earth, have migrated with respect
to the radial surficial projections A'B' and C'D' of the associated segments of the subtectonospheric fracture system lying
roughly 1000 km radially beneath A'B' and C'D' and passing through
P.

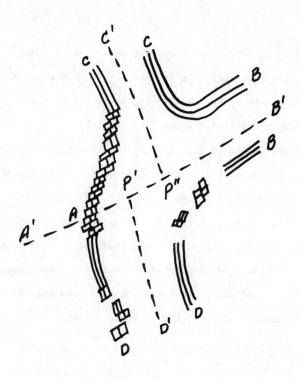

Figure 16-10-5-6. Migration of the spreading axis in the vicinity of an epicenter of one of the 6 basic points of the subtectono-sphere through which hemispheric torsion has occurred. See text for details.

sea-floor-spreading extrusions along the segment A'P'. The segment CB of the spreading axis is "normal". The segment BD of the spreading axis is broken near the point P'' because the extrusive conduits are closed there (Chapter 10).

In a varient of the idealized representative configuration shown in Fig. 16-10-5-6, the individual segments of the spreading axis CD may undergo rotation (See Fig. 16-10-5-2) in addition to the translations shown, thereby approximating the Mid-Atlantic Ridge. These and other real-world cases observed on the surface of the Earth are discussed in Chapter 18 where the global geometry and mechanics of sea-floor spreading are discussed in conjunction with continental drift and plate tectonics.

Descending Lithospheric Plates as Interpreted by the Tectonospheric Earth Model.

One of the basic assumptions of the sea-floor spreading
hypothesis concerns the descent of lithospheric plates into the
mantle. These descending plates, according to that hypothesis,
create the trenches into which they descend. These and other
complexities involved in the descent of these plates have been
analyzed by Elsasser (1971), Griggs (1970), Hanks and Whitcomb
(1971), Hasebe et al. (1970), Ichiye (1971), Jacob (1970), Julian
(1970), Luyendyk (1971), Minear and Toksoz (1970a, 1971b), Magnitsky (1971), McKenzie (1971), Sykes et al. (1970), Taylor et
al. (1971), Toksoz et al. (1971), Torrance and Turcotte (1971),
Watanabe et al. (1970), and many other earlier researchers cited
by these.

The details of these and other studies regarding the behavior
of the descending plates need not be discussed in an analysis
of the evolution of the Earth's tectonosphere. Suffice it here
merely to mention one salient difference in the assumptions of
the sea-floor-spreading hypothesis and the predictions of the
tectonospheric Earth model regarding the creation of the Earth's
deep trenches associated with these descending lithospheric
plates.

These trenches, according to the tectonospheric Earth model,
are produced not by the descending plates but by the deep-seated

749

tectonospheric behavior resulting from the basic driving mechanism of the model (Chapter 6). In this concept, the plates descend because there is an <u>unbalanced</u> force acting on them in a downward direction, i. e., the force of gravity plus an already existing open trench <u>cause</u> the plate to descend, according to the tectonospheric Earth model. In short, according to the model, the lithosphere descends <u>because</u> a trench is <u>already</u> there. This differs from the sea-floor-spreading hypothesis in which the trench is there <u>because</u> the lithosphere "digs" a trench into which to descend. These differences in the two concepts are considered in greater detail in Chapter 18 where plate tectonics is discussed.

REFERENCES

Anon., 1971a. On to the Caribbean. *Geotimes*, 16(1): 27-30.

Anon., 1971b. Deep sea drilling project: Leg 14. *Geotimes*, 16(2): 14-17.

Anon, 1971c. Geodynamics Project: Development of a U. S. program. *Trans. Am. Geophys. Union*, 52: 396-405.

Anon., 1971d. Deep sea drilling project: Leg 16. *Geotimes*, 16(6); 12-14.

Bell, J. S., 1971. Sea-floor spreading through gravity sliding. *Geol. Soc. Am. Abs. Programs*, 3: 368.

Benard, H., 1900. Tourbillions cellulaires dans une nappe liquide. *Rev. Gen. Sci.*, 11:1261-1271, 1328.

Bird, J. M. and Dewey, J. F., 1970. Lithosperic plate: Continental margin tectonics and the evolution of the Appalachian orogen. *Bull. Geol. Soc. Am.*, 81: 1031-1060.

Blackwell, D. D. and Roy, R. F., 1971. Geotectonics and Cenozoic history of the western United States. *Geol. Soc. Am. Abs. Programs*, 3: 84-85.

Bullard, E. C., Everett, J. E., and Smith, A. S., 1965. The fit of the continents around the Atlantic. *Phil. Trans. Roy. Soc. Lond., Ser. A*, 258: 41-51.

Carey, S. W., 1958. A tectonic approach to continental drift. In: W. S. Carey (Editor), *Continental Drift: A Symposium*.

751

Univ. of Tasmania, Hobart, pp. 177-355.

Carr, M. J., Stoiber, R. E., and Drake, C. L., 1971. A model for the upper mantle below Japan. Trans. Am. Geophys. Union, 52: 279.

Chase, C. G., Menard, H. W., Larson, R. L., Sherman, G. F., and Smith, S. M., 1970. History of sea-floor spreading west of Baja California. Geol. Soc. Am. Bull., 81: 491-498.

Conolly, J. R., 1969. Western Tasman Sea floor. New Zealand J. Geol. Geophys., 12: 310-343.

Cook, K. L., 1970. Active rift system in the Basin and Range province. Tectonophysics, 8: 469-511.

Dash, B. P., Ball, M. M., King, G. A., Butler, L. W., and Rona, P. A., 1971. Geophysical measurements in the Cape Verde archipelago. Trans. Am. Geophys. Res., 52: 235.

Dewey, J. F. and Bird, J. M., 1970a. Mountain belts and the new global tectonics. J. Geophys. Res., 75: 2625-2647.

Dewey, J. F. and Bird, J. M., 1970b. Plate tectonics and geosynclines. Tectonophysics, 10: 625-638.

Dietz, R. S. and Holden, J. C., 1970. East Indian basin (Wharton basin) as pre-Mesozoic ocean crust. Geol. Soc. Am. Abs. Programs, 2: 537.

Dietz, R. S. and Sproll, W. P., 1970. Fit between Africa and Antarctica: A continental drift reconstruction. Science, 167: 1612-1614.

Elsasser, W. M., 1971. Sea-floor spreading as thermal convection. J. Geophys. Res., 76: 1101-1112.

Elders, W. A., Rex, R. W., Meidav, T., Robinson, P. T., and Biehler, S., 1971. A plate tectonic model for the Salton Trough. Geol. Soc. Am. Abs. Programs, 3: 116-117.

Elders, W. A., Meidav, T., Rex, R. W., and Robinson, P. W., 1970. The Imperial Valley of California: The product of oceanic spreading centers acting on a continent. Geol. Soc. Am. Abs. Programs, 2: 545.

Emery, K. O., Uchipi, E., Phillips, J. D., Bowin, C. O., Bunce, E. T., 1970. Continental rise off eastern North America. Am. Assoc. Geologists Bull., 54: 44-108.

Fink, K. L., 1971. Evidence in the eastern Caribbean for Mid-Cenozoic cessation of sea-floor spreading. Trans. Am. Geophys. Union, 52: 251.

Foster, T. D., 1969. Convection in variable fluid heated from within. J. Geophys. Res., 74: 685-693.

Francheteau, J., Harrison, C. G., Sclater, J. S., and Richards, M. L., 1969. Magnetization to Pacific seamounts. Trans. Am. Geophys. Union, 50: 634.

Griggs, D. T., 1970. The sinking lithosphere and the focal mechanism of deep earthquakes. In: Symposium on the Nature of the Solid Earth, in honor of Professor F. Birch. McGraw-Hill, New York, in press.

753

Grim, P. J., 1970. Bathymetric and magnetic anomaly profiles from a survey south of Panama and Costa Rica. ESSA Tech. Memo ERLTM-AOML, 9.

Hall, J. K., 1970. Arctic Ocean geophysical studies: The Alpha cordillera and Mendeleyev ridge. Columbia Univ. CU-2-70, 1.

Hamblin, W. P. and Petersen, M. S., 1971. A possible subduction zone in western U. S. Geol. Soc. Am. Abs. Programs, 3: 385.

Hamilton, W., 1970. The Uralides and the montion of the Russian and Siberian platforms. Bull. Geol. Soc. Am., 81: 2553-2576.

Hanks, T. C. and Whitcomb, J. H., 1971. Comments on paper by J. W. Minear and M. N. Toksoz, "Thermal regime of a downgoing slab and new global tectonics". J. Geophys. Res., 76: 613-616.

Hasebe, R., Fujii, N., and Uyeda, S., 1970. Thermal processes under island arcs. Tectonophysics, 10: 335-355.

Hayes, D. E. and Pitman, W. C., 1970. Magnetic lineations in the North Pacific. Geology of the North Pacific Basin, Geol. Soc. Am. Mem. 126.

Heirtzler, J. R. and Vogt, P. R., 1971. Marine magnetic anomalies and their bearing on polar wander and continental drift. Trans. Am. Geophys. Union, 52: I 220 - I 223.

Hess, H. H., 1968. Reply (to paper by A. A. Meyerhoff,

"Arthur Holmes: Originator of spreading ocean floor hypothesis").
J. Geophys. Res., 73: 6569.

Hey, R. N. and Morgan, W. J., 1971. Parallel seamounts
in the northeast Pacific. Trans. Am. Geophys. Union., 52: 236.

Holcombe, T. L., 1971. Evolution of the sea-floor relief
of the Burma Rise and adjoining provinces. Trans. Am. Geophys.
Union, 52: 250.

Ichiye, T., 1971. Continental breakup by nonstationary
mantle convection generated with differential heating of the
crust. J. Geophys. Res., 76: 1139-1153.

Isacks, B. and Molnar, P., 1969. Mantle earthquake mechanisms and the sinking of the lithosphere. Nature, 223: 1121-1124.

Jacob, K. H., 1970. P-residuals and global tectonic structures investigated by three-dimensional seismic ray tracings
with emphasis on Longshot data. Trans. Am. Geophys. Union,
51: 359.

Julian, B. R., 1970. Regional variations in upper mantle
structure in North America. Trans. Am. Geophys. Union, 51: 359.

Karig, D. E., 1970. Ridges and basins of the Tonga-Kermadec
island arc systems. J. Geophys. Res., 75: 239-254.

Karig, D. E., 1971. Origin and development of marginal
basins in the western Pacific. J. Geophys. Res., 76: 2542-2561.

Knopoff, L., Drake, C. L., and Hart, P. J. (Editors), 1968.
The Crust and Upper Mantle of the Pacific Area. Am. Geophys.

755

Union, Washington, 522 pp.

Knopoff, L., Heezen, B. C., and MacDonald, G. J. F. (Editors), 1969. The World Rift System, Internat. Upper Mantle Project Sci. Rept. 19. Tectonophysics, 8, 586 pp.

Larson, R. L. and Chase, C. G., 1970. Relative velocities of the Pacific, N. America, and Cocos plates in the Middle America region. Earth and Planetary Sci. Letters, 7: 425-428.

Le Pichon, X., 1968. Sea floor spreading and continental drift. J. Geophys. Res., 73: 3661-3697.

Le Pichon, X. and Heirtzler, J. R., 1968. Magnetic anomalies in the Indian Ocean and sea-floor spreading. J. Geophys. Res., 73: 2101-2117.

Luyendyk, B. P., 1971. Comments on paper by J. W. Minear and M. N. Toksoz, "Thermal regime of a downgoing slab and new global tectonics". J. Geophys. Res., 76: 605-606.

Magnitsky, V. A., 1971. Geothermal gradients and temperatures in the mantle and the problem of fusion. J. Geophys. Res., 76: 1391-1396.

Martin, R., 1971. Cretaceous-Lower Tertiary rift basin of Baffin Bay: Continental drift without sea-floor spreading. Geol. Soc. Am. Abs. Programs, 3: 393.

Maxwell, A. E., Von Herzen, B. P., Hsu, K. J., Andrews, J. E., Saito, T., Percival, S. F., Milow, E. D., and Boyce, R. E., 1970. Deep sea drilling in the South Atlantic. Science, 168: 1047-1059.

756

McBirney, A. R., 1970. Cenozoic igneous events of the Circum-Pacific. Geol. Soc. Am. Abs. Programs, 2: 749-751.

McElhinny, M. W., 1970. Formation of the Indian Ocean. Nature, 228: 977-979.

McKenzie, D. P., 1971. Comments on paper by J. W. Minear and M. N. Toksoz, "Thermal regime of a downgoing slab and new global tectonics". J. Geophys. Res., 76: 607-609.

Menard, H. W., 1967. Transitional types of crust under small ocean basins. J. Geophys. Res., 72: 3061-3073.

Minear, J. W. and Toksoz, M. N., 1970a. Thermal regime of a downgoing slab and new global tectonics. J. Geophys. Res., 75: 1397-1419.

Minear, J. W. and Toksoz, M. N., 1970b. Thermal regime of a downgoing slab. Tectonophysics, 10: 367-390.

Mitchell, A. H. and Reading, H. G., 1969. Continental margins, geosynclines, and ocean floor spreading. J. Geol., 77: 629-646.

Moores, E. M., 1970. Patterns of continental fragmentation and reassembly: some implications. Geol. Soc. Am. Abs. Programs, 2: 629.

Morgan, W. J., 1971. Island chains, aseismic ridges, and hot-spots. Trans. Am. Geophys. Union, 52: 371.

Noltimier, H. C., 1970. A proposed structural evolution of the Caribbean and Central America since the Jurassic. Trans.

Am. Geophys. Union, 51: 825.

Oriel, S. S., 1971. Southern Idaho thrust belt. Geol. Soc. Am. Abs. Programs, 3: 400-401.

Oliver, J. and Isacks, B., 1967. Deep earthquake zones, anomalous structures in the upper mantle, and the lithosphere. J. Geophys. Res., 72: 4259-4275.

Phillips, J. D. and Luyendyk, B. P., 1970. Central North Atlantic plate motions over the last 40 million years. Science, 170: 727-729.

Pitman, W. C., 1971. Sea-floor spreading and plate tectonics. Trans. Am. Geophys. Union, 52: I130-I135.

Price, R. A. and Mountjoy, E. W., 1970. Geologic structure of the Canadian Rocky Mountains between Bow and Athabasca rivers. Geol. Assoc. Canada Spl. Paper 6: 7-25.

Price, R. A. and Mountjoy, E. W., 1971. The Cordilleran foreland thrust and fold belt in the southern Canadian Rockies. Geol. Soc. Am. Abs. Programs, 3: 404-405.

Rayleigh, Lord, 1916. On convective currents in a horizontal layer of fluid when the higher temperature is on the under side. Phil. Mag., 32: 529-546.

Robinson, G. D., 1971. Montana disturbed belt segment of Cordilleran orogenic belt. Geol. Soc. Am. Abs. Programs, 3: 408-409.

Rodolfo, K. S., 1971. Contrasting geometric adjustment

styles of drifting continents and spreading sea floors. J. Geophys. Res., 76: 3272-3281.

Rona, P. A., Brakl, J., and Heirtzler, J. R., 1970. Magnetic anomalies in the northeast Atlantic between the Canary and Cape Verde islands. J. Geophys.Res., 75: 7421-7425.

Scholz, C. H., Barazangi, M., and Sbar, M., 1971. Cenozoic evolution of the Basin and Range province. Trans. Am. Geophys. Union, 52: 350.

Sclater, J. G., Hawkins, J. W., and Chase, C. G., 1971. Crustal extension between the Tonga and Lau ridges: Petrologic and geophysical evidence. Trans. Am. Geophys. Union, 52: 194.

Sykes, L. R., Kay, R., and Anderson, D. L., 1970. Mechanical properties and processes in the mantle. Trans. Am. Geophys. Union, 51: 874-879.

Takeuchi, H. and Sakata, S., 1970. Convection in a mantle of variable viscosity. J. Geophys. Res., 75: 921-927.

Talwani, M., Windisch, C. C., and Langseth, M. G., 1971. Reykjanes ridge crest: A detailed geophysical study. J. Geophys. Res., 76: 473-517.

Tatsch, J. H., 1966a. Certain correlations between oceanographic observations along the South Atlantic Ridge and deductions arrived at from applying a dual primeval planet model to that region of the Earth. Trans. Am. Geophys. Union, 47: 123.

Tatsch, J. H., 1966b. Certain correlations between oceanographic observations along the North Atlantic Ridge and deductions made from applying a dual primeval planet model to that region of the Earth. _Trans. Am. Geophys. Union_, 47: 477.

Tatsch, J. H., 1969. Sea-floor spreading, continental drift, and plate tectonics unified into a single global concept by the application of a dual primeval planet hypothesis to the Earth. _Trans. Am. Geophys. Union_, 50: 672.

Taylor, S. R., White, A. J. R., Ewart, A., and Duncan, A. R., 1971. Nickel in high-alumina basalts: A reply. _Geochim. Cosmochim. Acta_, 35: 525-528.

Toksoz, M. N., Minear, J. W., and Julian, B. R., 1971. Temperature field and geophysical effects in a downgoing slab. _J. Geophys. Res._, 76: 1113-1138.

Torrance, K. E. and Turcotte, D. L., 1971. Structure of convection cells in the mantle. _J. Geophys. Res._, 76: 1154-1161.

Utsu, T., 1967. Anomalies in seismic wave velocity and attenuation associated with a deep earthquake zone. _J. Fac. Sci. Hokkaido Univ. Ser._ 7, 3: 1-25.

Utsu, T. and Okada, H., 1968. Anomalies in seismic wave velocity and attenuation associated with a deep earthquake zone, 2. _J. Fac. Sci. Hokkaido Univ. Ser._ 7, 3: 65-84.

Uyeda, S. and Horai, K., 1964. Terrestrial heat flow in Japan. _J. Geophys. Res._, 69: 2121-2141.

Uyeda, S. and Vacquier, V., 1968. Geothermal and geomagnetic data in and around the island arc of Japan. In: L. Knopoff, C. Drake, and P. Hart (Editors), The Crust and Upper Mantle of the Pacific Area. Am. Geophys. Union, Washington, pp. 349-366.

Vine, F. J. and Hess, H. H., 1971. Sea-floor spreading. In: A. E. Maxwell (Editor), The Sea, vol. 4. Wiley-Interscience, New York, 1455 pp., in 2 parts.

Vinogradov, A. P., et al., 1969. The structure of the mid-oceanic rift zone of the Indian Ocean and its place in the world rift system. Tectonophysics, 8: 377-401.

Vogt, P. R. and Ostenso, N. A., 1970. Magnetic and gravity profiles across the Alpha cordillera and their relation to Arctic sea floor spreading. J. Geophys. Res., 75: 4925-4937.

Vogt, P. R., Schneider, E. D., and Johnson, G. L., 1969. The crust and upper mantle beneath the sea. In: P. J. Hart (Editor), The Earth's Crust and Upper Mantle. Am. Geophys. Union, Washington, pp. 556-617.

Vogt, P. R., Anderson, C. N., Bracey, D. R., and Schneider, E. D., 1970. North Atlantic magnetic smooth zone. J. Geophys. Res., 75: 3955-3966.

Vogt, P. R., Higgs, R. H., and Johnson, G. L., 1971. Hypotheses on the origin of the Mediterranean basin: magnetic data. J. Geophys. Res., 76: 3207-3228.

Watanabe, T., Epp, D., Uyeda, S., Langseth, M., and Yasui,

761

M., 1970. Heat flow in the Philippine Sea. _Tectonophysics_,
_10_: 205-224.

Young, J. C., 1970. Caribbean Island Arc. _Geotimes_, _15(9)_:
17-19.

Zietz, I., 1971. Magnetic anomalies over the continents.
Trans. _Am_. _Geophys_. _Union_, _52_: I 204 - I 209.

Chapter 17

CONTINENTAL DRIFT AND POLAR WANDERING

Any analysis of the evolution of the Earth's tectonosphere must consider continental drift, polar wandering, and related phenomena that have changed the surface of the Earth. It should clarify the relationship between these surficial manifestations of the Earth's internal behavior, and, if possible, it should identify the nature of the driving mechanism that motivates continental drift, polar wandering, and the related phenomena.

Ever since Wegener (1915) discussed some of the details of continental drift and of related phenomena, scientists have tried to formulate a hypothesis to explain these surficial manifestations of the Earth's internal behavior. No one hypothesis has succeeded in answering all the critical questions. There is no agreement (1) on pre-drift configurations and reconstructions, (2) on whether there was only one drifting episode or many, and (3) on the exact sequence of the breakup that preceded the drifting.

A closer scrutiny of evidence from deep drilling within the oceans and of evidence already available from within the continents should serve to identify the nature of the discrepancy suggested by this lack of agreement among the various hypotheses for continental drift, polar wandering, and related phenomena.

763

The Evolution of the Concept of Continental Drift.

The hypothesis of continental drift postulates that continent-sized plates of the Earth's crust have retained their form but have moved, or drifted, relative to one another during at least the past 200 m. y. This concept is fairly well established on the basis of global paleomagnetic analyses, supplemented by regional and global studies of other types.

Most earlier concepts of continental drift are associated with Du Toit, Taylor, and Wegener (See, e. g., Wegener, 1915). The modern version is usually associated with Runcorn (1962), Bullard (1964), Blackett et al. (1965),Isacks et al. (1968), and more-recent researchers cited in later parts of this chapter.

General concepts and observations supporting continental drift and/or polar wandering fall into three basic categories: (1) pre-drift super-continent configurations; (2) polar-wander curves; and (3) the cyclical association of rocks according to whether they were produced during tectonically active or quiescent periods.

In some pre-drift configurations, the southern continents, Africa, Australia, South America, India, and Antarctica, are grouped around the latter during the Jurassic (about 150 m. y. ago) to form Gondwanaland, while the remaining continents are grouped to form Laurasia. In other configurations, all conti-nents form a single pre-drift super-continent, Pangaea. Some concepts have suggested that a single pre-drift configuration and a single break-up of a super-continent might be too simple and temporally restrictive to the extent that the basic concept must be supplanted by one in which a succession of super-continents has existed and that these were repeatedly broken up and re-joined in different global surficial configurations.

Polar-wander curves for Africa, South America, and Australia follow paths that are almost identical for the 3 continents. These curves have the unique shape of a large Z in which the diagonal is about $90°$, or a linear trace of roughly one-quarter of the Earth's circumference, i. e., 10,000 km. These curves, analyzed temporally, suggest continental-drift rates of about 2 cm/yr along this unique Z-path during the past 500 m. y. Because other evidence suggests that large-scale tectonic processes occur at a few cm/yr, polar wandering of this magnitude may be globally associated with at least some of these large-scale tectonic processes. All polar movement relative to Australia during the 300-m. y.-period from the Upper Silurian to the Mid-Cretaceous apparently occurred within a single 20-m. y. episode during the Carboniferous (Irving, 1966).

An analysis of the ages of rocks suggests a more-or-less cyclical sequence of tectonically active and quiescent periods with a periodicity of about 250 to 500 m. y. Polar-wander and continental-drift episodes apparently correlate with the tectonically active periods of the cyclical sequence.

Present astronomical determinations of longitude are too imprecise to indicate relative movements of a few cm/yr, but latitude measurements suggest a currently steady movement of the pole toward approximately 70°W at a rate of roughly 10 cm/yr (Markowitz et al., 1964; Markowitz and Guinot, 1968).

Other recent analyses regarding continental drift and polar wandering have been made by Francheteau et al. (1969), Hurley (1969), Lyustikh (1969), Crawford and Wilson (1969), Dietz et al. (1969), Goldreich and Toomre (1969), Knopoff (1969), Knopoff and Shapiro (1969), Maxwell (1969), Meservey (1969), Oliver et al. (1969), Salop and Scheinmann (1969), Vine (1969a, 1969b), and Gough and Porath (1970).

From these and other analyses, it may be concluded (1) that continental drift and polar wandering are surficial manifestations of the Earth's internal behavior; (2) that these phenomena have been occurring for at least the past 200 m. y. and possibly during the entire life of the Earth; and (3) that the geometry and mechanics of the interior of the Earth are not sufficiently understood to permit a complete definition of the nature of the driving mechanism for continental drift and polar wandering.

<u>Patterns of Continental Fragmentation and Reassembly.</u>

In considering the behavior of the crust during the past 3.6 b. y., one may assume either (1) that there has been a single breakup of one supercontinent (Pangaea); or (2) that there has been a multiplicity of breakup episodes.

In the simplest multiple-breakup hypothesis, there were two breakup episodes: (1) Pangaea broke into Gondwanaland and Laurasia and (2) Gondwanaland and Laurasia then separated into the present continents.

Both the single and the two-step breakups assume that continental rifting was more or less non-existent in pre-Mesozoic times.

Moores (1970) has suggested that some deformed belts represent the suturing of continents that were formerly separated by an ocean basin. He reconstructed a hypothetical history of continental assembly and fragmentation:

    a. <u>During the Precambrian</u>.

        The suturing of the Trans-African-Baikalian system assembled the dispersed continents into one supercontient, Pangaea I.

    b. <u>During the Cambrian</u>.

        The fragmentation of Pangaea I into 4 "continents: (a) North America plus Europe northwest of the Caledonides; (b) Europe between the Caledonides and Urals; (c) Asia east of the Urals; and (d) Gondwanaland.

c. <u>During the Siluro-Devonian</u>.

The suturing of the Caledonides.

d. <u>During the Pennsylvanian-Permian</u>.

The suturing of the Appalachians-Hercynides.

e. <u>During the Permo-Triassic</u>.

The suturing of the Urals, to produce the super-
continent Pangaea II.

f. <u>During the Jurasso-Cretaceous</u>.

The fragmentation of Pangaea II into Laurasia
and Gondwanaland along the Tethys.

g. <u>During the Cretaceous-Tertiary</u>.

The suturing along the Tethys and the fragmen-
tation of Laurasia and Gondwanaland into the present continents.

Moores suggested that his hypothesis of continental assembl
and breakup permits at least a partial explanation of worldwide
Phanerozoic sea-level fluctuations and (correlated with paleomagnetic data) it permits a provisional reconstruction of Paleozoic paleogeography.

Continental Breakup and Drifting Sequences.

Most hypotheses assume that all present continents have undergone some migrations, or drifting, since the Mesozoic, and there is general agreement on the sequence in which the rifting, breakup, and drifting occurred during the past 200 m. y. (See, e. g., Zietz, 1971; Larson and La Fountain, 1970; Smith and Hallam, 1970).

On the assumption that a single supercontinent, Pangaea, existed during the early Mesozoic, the actual breakup apparently began during the Triassic and consisted of a fairly well-ordered sequence of episodes:

    a. During the Triassic.

        (1). The southwest Indian Ocean rift split South America and Africa away from the rest of Gondwanaland.

        (2). A Y-junction separated India from Antarctica.

        (3). An independent North Atlantic – Caribbean rift split Laurasia (North America and Eurasia) from South America and Africa.

    b. During the early Jurassic.

        (1). Northward and westward extensions of the Triassic rifts increased the sizes of the North Atlantic and Indian Ocean.

        (2). A new rift split South America from Africa.

    c. During the Cretaceous.

(1). The North Atlantic rift increased northward, delineating the Grand Banks and the western margin of
Greenland.

(2). Spain rotated sinistrally, forming the
Bay of Biscay.

(3). An offshoot rift split Madagascar from
Africa, which continued a northward migration.

(4). India continued a northward migration.

(5). Australia split away from Antarctica.

d. During the Cenozoic.

(1). Antarctica rotated farther westward.

(2). Australia migrated rapidly northward.

(3). New Zealand was split from Australia.

(4). The North and South Atlantic continued
to open.

(5). The erstwhile North America – Greenland
rifting shifted eastward and split Greenland from Europe.

(6). This rifting progressed through the Arctic
and split North America from northern Eurasia.

(7). Africa moved slightly northward and rotated
sinistrally.

(8). The Tethyan megashear became dextral.

(9). India collided with and under-ran Asia.

These breakup sequences cover the past five percent of

the life of the Earth. But what about continental drift during the first 95% ? Close scrutiny of the evidence suggests that earlier drifting episodes occurred and that the earliest might have been even in Precambrian times (See, e. g., Briden, 1969; Moores, 1970).

These and other speculations cannot be considered the final answer regarding breakup and drifting sequences. Only actual drilling can determine how many drift episodes occurred and when they occurred.

<u>Continental</u> <u>Drift</u> <u>and</u> <u>Other</u> <u>Geophenomena</u>.

Continental drift always occurs in conjunction with other geophenomena, such as oceanic and continental rifting, magmatic
activity, and other phenomena associated with tectonomagmatic
activity. Rapid polar wandering occurs during intensive tectonomagmatic activity. At such times, more geomagnetic polarity
reversals occur.

What is the relationship between continental drift and
the associated geophenomena such as oceanic and continental
rifting, magmatic activity, rapid polar wandering, and geomagnetic polarity reversals? Why does continental drift not <u>always</u>
occur when there is activity in these other geophenomena? What
are the geometrical, mechanical, thermal, and chemical characteristics of a global driving mechanism that can be so discri-
minating and do it over so long a period of time?

Questions of this nature have been analyzed by Martin (1971),
McElhinny and Luck (1970), Beck (1970), Rodolfo (1971), Yungul
(1971), Hatherton (1969), Schubert et al. (1970), Dietz and
Holden (1970), McElhinny (1970), Fitch (1970), Fairbridge (1969),
Jeffreys (1970), Beloussov (1970), Van Andel and Moore (1970).

The results of these analyses fall into two categories: (1)
magmatic activity and global thermal patterns; and (2) "drifting
oceans" and "spreading continents". These are discussed below.

772

Continental Drift, Magmatic Activity, and Global Geothermal Patterns.

Because continental drift almost always occurs in association with magmatic activity and because magmatic activity, in turn, occurs approximately along global belts of maximum heat flow, this suggests a causal relationship between these three phenomena.

Sclater and Francheteau (1970) concluded, from a global analysis, that the loss of heat used to create "oceanic" lithosphere at the interface between spreading plates may aggregate as much as 45% of the total average heat lost from the Earth. Some of this loss also occurs in connection with downgoing slabs.

Julian (1970) found that partial melting in the low viscosity zone outside the downgoing slab in the vicinity of the Kermadecs produces a rather large velocity increase of at least 4 sec.

Other problems related to the association between continental drift and global goethermal patterns have been analyzed by Utsu (1967), Isacks et al. (1968), McKenzie and Sclater (1968), Sykes et al. (1970), Griggs (1970), Hasebe et al. (1970), Jacob (1970), Minear and Toksoz (1970a, 1970b), Hanks and Whitcomb (1971), Luyendyk (1971), McKenzie (1971), Toksoz et al. (1971), McBirney (1970).

None of these analyses has produced a completely satisfactory explanation for exactly how the Earth's geothermal behavior is related to continental drift and magmatic activity at the interfaces between spreading and subducting plates.

773

Drifting Oceans and Spreading Continents.

Heirtzler and Vogt (1971) have analyzed continental drift and polar wandering on the basis of magnetic anomalies parallel to mid-ocean crests. They concluded, among other things, that the large number of variables involved in such an analysis makes conclusions difficult. But they did generalize their results by defining continental drift as "the special case of movement between plates that have continents embedded in them".

A simple extension of this concept permits "oceanic drift" to be defined as the special case of movement between plates that have oceans embedded in them. At least part of the Pacific Ocean appears to have participated in such a movement. Most evidence shows that at least part of the central Pacific has migrated, perhaps "embedded" in a plate, northward about 20°, or over 2000 km, since the Mesozoic.

This Heirtzler-Vogt concept may be further extended by noting that both oceanic and continental areas are known to undergo rifting and spreading, apparently from deep-seated tensional forces. We may speak, therefore, of "oceanic drift" and "continental spreading" as phenomena antithetical to continental drift and sea-floor spreading.

Evidence gathered by Leg 14 of the Deep Sea Drilling Project suggests that a protocontinent once existed in a small ancestral basin then bounded by the Galapagos, Gibraltar, the

774

eastern coast of the United States, and Africa. No evidence exists, however, to suggest what happened to this protocontinent. Fig. 17-6-1 shows schematically how this protocontinent and ancestral ocean might have appeared during the late Mesozoic.

Evidence of an analogous nature exists in the African rift valleys. Volcanism in these valleys differs in both composition and volume from that in the median valleys of oceanic ridges, suggesting a much lower geothermal gradient under spreading continents than under spreading sea floors, at least in some cases. Murray (1970) interpreted this as an indication of fundamentally different upper-mantle conditions in these two antithetical areas.

Macdougall (1971) analyzed the occurrence of basaltic flows in areas that are not genetically associated with ridge crests. Dewey and Bird (1971) analyzed the role of obducted ophiolites at the plate margins such as that in Newfoundland. Bosshard and MacFarlane (1970) concluded that the Canary Islands do not form a part of the African continent but are independent volcanic edifices that erupted along northeast-southwest fracture zones, i. e., along the extended Tethys. The crustal composition in this area forms a complete spectrum of rock types, ranging from oceanic to continental material, suggesting a complex magma-tectonic history for that area.

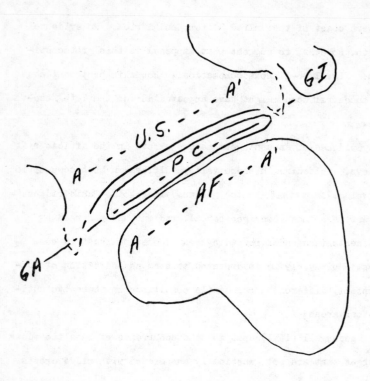

Figure 17-6-1. Schematic representation of a small protocontinent, PC, in the small hypothetical ancestral ocean basin bounded by the Galapagos (GA), Gibraltar (GI), the eastern coast of the United States (U. S.), and Africa (AF). The Appalachian and Atlas mountain ranges are indicated approximately at AA'.

776

Summarizing, the distinction between purely "continental" behavior and purely "oceanic" behavior apparently is not as sharp as once was thought. Much continental behavior is duplicated by analogous behavior in the oceans, and much oceanic behavior is duplicated by analogous behavior in the continents. The causal mechanism for all surficial behavior, whether oceanic or continental, must lie, therefore, deep below the oceans and continents in a globally unified system older than the oldest continent.

<u>Causal</u> <u>Mechanism</u> <u>for</u> <u>Continental</u> <u>Drift</u> <u>and</u> <u>Polar</u> <u>Wandering</u>.

If portions of the Earth's surface are undergoing a certain amount of drifting and if, as a result of this or in addition to it, the Earth's pole of rotation is wandering, then these effects could represent surficial manifestations of the Earth's internal behavior. Unanswered are questions regarding the exact mechanism within the Earth that causes continental drift and polar wandering (See, e. g., Irving and Robertson, 1969).

No answers have been found for the following: (1) Is continental drift something relatively new, or has it been going on for the past 4.6 b. y.? (2) If the latter, <u>what</u> has managed to move continents over the surface of the Earth during all these years? (3) If continental drift started relatively recently, say 200 m. y. ago, what caused it to start then, rather than 300 m. y. ago, or 900 m. y. ago, or 2.9 b. y. ago? (4) Is continental drift proceeding in a strictly random manner, or is it following some definable, quasi-organized pattern? (5) Is pre-drift continental splitting (e. g., Africa from South America) induced by random forces, or are these forces quasi-linear as would be expected if they were induced by an orginized, non-random global driving mechanism effective over a long period of time, say billions of years? (6) Are continental drift and polar wandering independent phenomena? (7) If not, how are they related to each other and to other geophenomena?

Answers to these and other questions regarding continental drift and polar wandering must await the identification of a suitable mechanism for motivating these phenomena. Because many hypotheses for continental drift and polar wandering use mantle convection as the basic driving mechanism, Lee and Mac-Donald (1963) attempted to obtain evidence of a convective pattern by comparing spherical harmonic analyses of gravitational potential with surface heat flow on the assumption that a correlation between geoidal "lows" and thermal "highs" would represent rising limbs in the hypothesized convection pattern. Their results were inconclusive, primarily because the amount of evidence available was too small for the particular type of correlation attempted. Subsequent similar analyses have not clarified the nature of the suspected relationship between continental drift and mantle convection, other than to suggest that both the drift and the hypothesized mantle convection might be manifestations of a single global deep-seated long-lived mechanism. One aspect of this association between continental drift and mantle convection is that the Earth's crust apparently develops rifts, spreads, and drifts whenever the central driving mechanism produces tensile conditions in various regions of the Earth's surface. But the details remain to be defined.

Some researchers circumvent the undesirable features of a poorly understood convective system by using other sources of motive power for causing continental drift. Kane (1971) postulated the differential angular momentum from changes in the Earth's rotation as a motive power for moving continents. Ichiye (1971) used non-stationary mantle convection from differential heating caused by different radioactive heat-production rates in protocontinental material to move older parts of continents with respect to newer parts.

Bell (1971) concluded that the lithospheric plates may be effectively decoupled from the lower part of the mantle and that associated mass transfer patterns at that depth may not be directly detectable as surficial manifestations.

Pan (1971) concluded that the wandering of the pole and the tectonic movements of the Earth's upper layers are closely linked in their driving forces, i. e., "the energy perturbation initiates polar wander, polar wander maintains tectonic movements, and tectonic movements adjust polar wandering". Pan's hypothesis presents a plausible sequence of events, but it does not give the geometrico-mechanical details of what initiated, motivated, and now sustains the "energy perturbations" that are required in his system.

Summarizing these and other analyses, it may be said that a satisfactory driving mechanism for continental drift and polar wandering has not yet been identified and that none can be identified until answers have been found to certain critical questions regarding the associated phenomena.

780

<u>Continental</u> <u>Drift</u> <u>and</u> <u>Polar</u> <u>Wandering</u> <u>as</u> <u>Interpreted</u> <u>by</u> <u>the</u>
<u>Tectonospheric</u> <u>Earth</u> <u>Model</u>.

Continental drift and polar wandering, according to the
tectonospheric Earth model, are closely related to crustal plate
motion and to subcrustal plate and block behavior. These phenomena, according to the model, represent surficial manifestations
of the Earth's internal behavior in those areas where the resultant of such behavior is representable as a tangential unbalanced force vector acting on a plate, on a block, or on a continent, to produce motion of that block, that plate, or that
continent realtive to the bulk of the Earth or to a datum therein.
These phenomena, in short, are surficial adjustments to compensate for the disequilibrated and ever-changing behavior of the
Earth's interior. Sometimes a block moves to effect this compensating equilibration. At other times, a rift may develop,
or an earthquake may occur when a block or plate adusts itself
to a more nearly equilibrated position in the never-ending attempt of the Earth to assume a condition of minimum energy.

Whenever a continent moves to effect such an equilibration,
continental drift occurs, according to the tectonospheric Earth
model.

The most elementary type of continental drift, according
to the model, occurs when a simple translation severs one continent from another. This type of translation occurs outward

781

from a linear rift identifiable with a wedge-belt of activity
of the model described in Chapter 6.

More complex types of continental drift occur whenever the
combination of   both a translation and a rotation severs one
continent from another. This, according to the model, is the
usual form of continental drift, expressible as aA + bB, where
A represents translation, B represents rotation, and $\underline{a}$ and $\underline{b}$
are constants, either of which may be zero provided a + b = 1.

From the discussion in Chapter 6, it follows that the rotational element of the rifting and drifting is closely associated with and controlled by the planes, surfaces, and shells
of preferential rheidity within the Earth's tectonosphere described in that chapter.

<u>Continental</u> <u>Drift</u> <u>and</u> <u>the</u> <u>World</u> <u>Rift</u> <u>System</u> <u>as</u> <u>Interpreted</u> <u>by</u>
<u>the</u> <u>Tectonospheric</u> <u>Earth</u> <u>Model</u>.

Continental drift apparently always originates along a
segment of the world-rift system. The exact causal relationship between this rifting and drifting is not known, but the
association appears to be global, long-lived, and deep-seated.
A close scrutiny of the mechanics and geometry of the world-
rift system should indicate, therefore, the nature of the association between these two phenomena.

Fig. 17-10-2-1 shows the basic elements of the detectable
segments of the world-rift system. On the assumption that only
about half of the world-rift system is detectable, speculations
may be used to predict, or "reconstruct", the undetectable portions. Such speculations are not shown on this figure, but
may be found in most standard texts and on maps of the world-
rift system (See, e. g., Hart, 1969; Stacey, 1969).

Artemjev and Artyushkov (1971) concluded from a global
analysis that all segments of the world-rift system are produced by a single deep-seated mechanism, but they did not specify the geometrico-mechanical details of that mechanism. Mohr's
(1971) analysis correlates rift type with magma type: the faster
rifting rates (i. e., larger tensional gradients) are associated
with "oceanic" tholeiites, even in continental areas; slower
rifting rates (i. e., smaller tensional gradients), with a wider

783

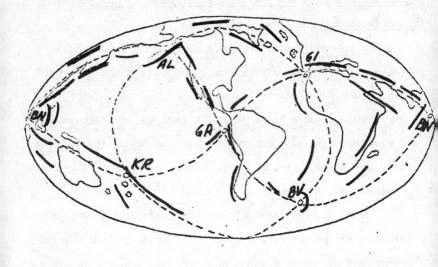

Figure 17-10-2-1. Basic elements of the detectable segments
of the world-rift system superimposed on a Mollweide projection.
AL, BN, BV, GA, GI, and KR are the radial surficial projections
of the 6 basic points of the subtectonosphere (Chapter 6). See
text for details.

spectrum of more–alkalic magmas (cf., e. g., Chapter 10).

Erickson et al. (1970) concluded, from an analysis of the Emperor fracture zone, that an explanation of the "oblique trend" of this zone with respect to the world–rift system requires some rather complex geometry and mechanics. Similar conclusions were drawn by Shor et al. (1971) for the segment of the world–rift system between Australia and New Zealand, and by Grim (1970) for the segment between the Galapagos and Panama.

Rifts in the vicinities of the epicenters of the 6 basic points of the subtectonosphere (AL, BV, BN, GA, GI, and KR) are curved in some cases. There are two contributing causes for these curved surficial rifts: (1) a point of activity on the subtectonosphere may project as a circle on the surface of the Earth; and (2) a straight line on the subtectonosphere may project as a curved line on the surface of the Earth whenever a surficial plate is rotating with respect to that line on the subtectonosphere.

The mechanics and geometry of contributions of type (1) are shown schematically in Fig. 17–10–2–2. Contributions of type (2) may be shown in a similar manner, provided that the projection is made as a temporal sequence covering the duration of the rotation of the surficial plate with respect to a fixed line on the subtectonosphere (Fig. 17–10–2–3). The mechanics and geometry of this type of contribution are similar to those

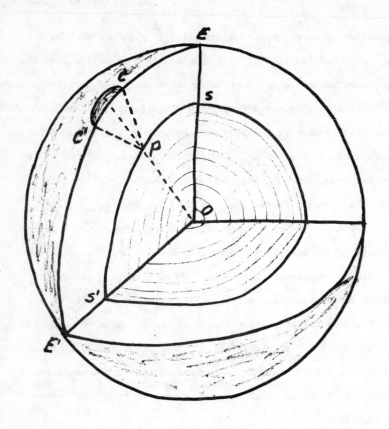

Figure 17-10-2-2. Schematic sketch showing how a point of activity, P, on the surface of the subtectonosphere, SS', may project as a circle, CC', on the surface of the Earth, EE'. O = center of the Earth. See text for details.

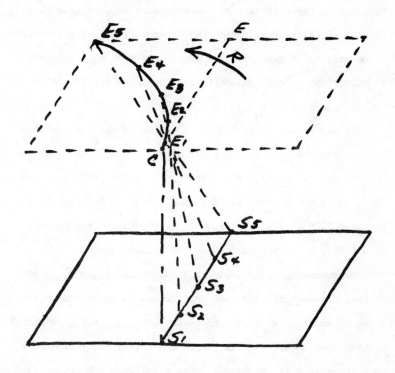

Figure 17-10-2-3. Schematic sketch showing how a sequence of 5 activity points, lying in a straight line, S_1 to S_5, on the surface of the subtectonosphere, may project into a curved sequence, E_1 to E_5, on the surface of the Earth, within the time sequence, t_1 to t_5, during which a plate on the surface of the Earth undergoes a rotation R about the point C. See text for details.

involved in the evolution of the curvature of some islands and of some island chains. Stone and Cameron (1970) have discussed the evolution of the curvature of the Aleutian chain. Kawai et al. (1969) have discussed the evolution of the curvature of the island of Honshu.

It may be seen from Fig. 17-10-2-3 that the results will be the same whether (1) the point S moves from S_1 to S_5 during the sequence t_1 to t_5; or (2) the plate on the surface of the Earth undergoes a translation from E_1 to E, in addition to the rotation R, during the sequence t_1 to t_5, over a single point of activity S_1 on the surface of the tectonosphere.

Summarizing, the geometry and mechanics of the world-rift system, although complex when analyzed in terms of modern concepts of continental drift and plate tectonics, are more straightforward when interpreted in terms of the tectonospheric Earth model described in Chapter 6. This simplification arises from the fact that continental drift and plate tectonics consider only "surficial" plate phenomena to a depth of only about 100 km, whereas the tectonospheric Earth model considers all surface and subsurface phenomena to a depth of 1000 km.

<u>The</u> <u>Driving</u> Mechanism <u>for</u> <u>Continental</u> <u>Drift</u> <u>and</u> <u>Polar</u> <u>Wandering</u>
<u>as</u> <u>Interpreted</u> <u>by</u> <u>the</u> <u>Tectonospheric</u> <u>Earth</u> Model.

Continental drift and polar wandering are motivated, according to the tectonospheric Earth model, by the same basic
driving mechanism that motivates other aspects of the surficial
behavior. This driving mechanism, as described in Chapter 6,
has existed since the Earth began 4.6 b. y. ago. The model
postulates, further, that continental masses have existed for
at least 3.6 b. y. (Chapter 7) and that continental drift has
been possible during this period of time.

It is unlikely (and unnecessary), according to the model,
that <u>all</u> continents have ever been joined simultaneously into
a single supercontinental configuration. It is possible, according to the model, that each present contiental mass might
have been attached to one or more other present continental
masses at some time during the past 3.6 b. y. But it is not
likely (nor necessary), under this concept, that all were ever
joined <u>simultaneously</u> into a single supercontinent.

Thus, one speaks, in this concept, not of "<u>a</u> predrift configuration", nor of "<u>the</u> pre-Mesozoic configuration". One speaks,
rather of "the most likely Upper Jurassic configuration", or
of "the most likely Lower Paleozoic configuration", or of any
one of the many possible Precambrian configurations.

Because all continents have been more-or-less continually

789

in motion at all times during the past 3.6 b. y., a myriad of pre-Mesozoic configurations is possible, each differing from each previous and subsequent configuration by at least a few centimeters. At a rate of a few cm/yr, continents may drift hundreds of km per million years, or roughly once around the Earth every few billion years.

Here it is necessary to recall the constraints placed upon continental drift and upon supercontinental configurations by the basic orogenic—cratonic evolution of the continental nuclei, as described in Chapter 7.

Openings between continental masses are allowable, according to the model, in "predrift configurations", but overlaps are not allowed. That is, if an overlap exists in a close—fitting reconstruction, the configuration must be opened, in this concept, until the overlap no longer exists. The result then shows, according to the model, a possible "predrift" configuration, i. e., without overlaps for the particular geologic age being studied.

"Fits" at depths other than sea—level are not necessarily better than those at sea—level, because differential uplifts and subsidences of the various continents in the past have severely decreased the feasibility of using any given present depth as a reliable global horizontal datum for paleoanalyses.

Summarizing, continents fall, according to the model, into

three categories of drifting: (1) those continents that are undergoing individual motion, i. e., as single continents such as present-day North America, South America, Australia, and Antarctica; (2) those experiencing group motion, i. e., as a conglomeration of several continents as a unit such as Eurasia; and (3) those in the process of separating, such as Euroafrica, or of coalescing, such as Hinduasia.

In all these cases of continental drift and related phenomena, according to the model, the driving mechanism is the same basic driving mechanism described in Chapter 6. The details of how this mechanism moves plates, continents, and geoblocks is discussed in the next chapter where plate tectonics is discussed.

REFERENCES

Anderle, R. J. and Smith, S. J., 1968. Observation of 27th and 28th order gravity coefficients based on Doppler observations. J. Astronaut. Sci., 15: 1-4.

Anon., 1971a. On to the Caribbean. Geotimes, 16(1): 27-30.

Anon., 1971b. Deep sea drilling project: Leg 14. Geotimes, 16(2): 14-17.

Anon., 1971c. Geodynamics Project: Development of a U. S. program. Trans. Am. Geophys. Union, 52: 396-405.

Anon., 1971d. Deep sea drilling project: Leg 16. Geotimes, 16(6): 12-14.

Artemjev, M. E. and Artyushkov, E. V., 1971. Structure and isostasy of the Baikal rift and the mechanism of rifting. J. Geophys. Res., 76: 1197-1211.

Beck, M. E., 1970. Paleomagnetism of Keweenawan intrusive rocks, Minnesota. J. Geophys. Res., 75: 4985-4996.

Bell, J. S., 1971. Sea-floor spreading through gravity sliding. Geol. Soc. Am. Abs. Programs, 3: 368.

Beloussov, V. V., 1970. Against the hypothesis of ocean floor spreading. Tectonophysics, 9: 489-511.

Blackett, P. M. S., Bullard, E. C., and Runcorn, S. K. (Editors), 1965. A symposium on continental drift. Phil. Trans. Roy. Soc. A, 258.

Bosshard, E. and MacFarlane, D. J., 1970. Crustal structure

of the western Canary Islands from seismic refraction and gravity data. J. Geophys. Res., 75: 4901-4918.

Briden, J. C., 1969. Intercontinental correlations based on paleomagnetic evidence for recurrent continental drift. In: Gondwana Stratigraphy, UNESCO, Paris, pp. 421-439.

Bullard, E. C., 1964. Continental drift. Quart. J. Geol. Soc. Lond., 120: 1-39.

Bullard, E. C., Everett, J. E., and Smith, A. S., 1965. The fit of the continents around the Atlantic. Phil. Trans. Roy. Soc. Lond., A, 258: 41-51.

Carey, W. S., 1958. A tectonic approach to continental drift. In: W. S. Carey (Editor), Continental Drift: A Symposium. Univ. of Tasmania, Hobart, pp. 177-355.

Cook, K. L., 1970. Active rift system in the Basin and Range province. Tectonophysics, 8: 469-511.

Crawford, A. R. and Wilson, J. T., 1969. Continental drift in the Arctic and Indian oceans and before the Mesozoic era. Geol. Soc. Am. Abs. Programs, 7: 41.

Dewey, J. F. and Bird, J. M., 1971. Origin and emplacement of the ophiolite suite: Appalachian ophiolites in Newfoundland. J. Geophys. Res., 76: 3179-3206.

Dietz, R. S. and Holden, J. C., 1970. East Indian basin (Wharton basin) as pre-Mesozoic ocean crust. Geol. Soc. Am. Abs. Programs, 2: 537.

793

Dietz, R. S. and Sproll, W. P., 1970. Fit between Africa and Antarctica: A continental drift reconstruction. Science, 167: 1612-1612.

Dietz, R. S., Holden, J. C., and Sproll, W. P., 1969. Geotectonic evolution and subsidence of Bahama Platform. Geol. Soc. Am. Abs. Programs, 7: 48.

Elders, W. A., Meidav, T., Rex, R. W., and Robinson, P. W., 1970. The Imperial Valley of California: The product of oceanic spreading cneters acting on a continent. Geol. Soc. Am. Abs. Programs, 2: 545.

Erickson, B. H., Naugler, F. P., and Lucas, W. H., 1970. Emperor fracture system zone: A newly discovered feature in the central North Pacific. Nature, 225: 53-54.

Fairbridge, R. W., 1969. Polar migration, sea-floor spreading, atolls, and climate changes. Tectonophysics, 7: 545-546.

Francheteau, J., Harrison, C. G. A., Sclater, J. S., and Richards, M. L., 1969. Magnetization of Pacific seamounts: A preliminary polar curve for the northeastern Pacific. Trans. Am. Geophys. Union, 50: 634.

Goldreich, P. and Toomre, A., 1969. Some remarks on polar wandering. J. Geophys. Res., 74: 2555-2567.

Gough, D. L. and Porath, H., 1970. Long-lived thermal structure under the southern Rocky Mountains. Nature, 226: 837-839.

Griggs, D. T., 1970. The sinking lithosphere and the focal mechanism of deep earthquakes. In: Symposium on the Nature of the Solid Earth, in Honor of Professor F. Birch, McGraw-Hill, New York (in press).

Grim, P. J., 1970. Bathymetric and magnetic anomaly profiles from a survey south of Panama and Costa Rica. ESSA Tech. Memo ERLTM-AOML, 2.

Hanks, T. C. and Whitcomb, J. H., 1971. Comments on paper by J. W. Minear and M. N. Toksoz, "Thermal regime of a downgoing slab and new global tectonics". J. Geophys. Res., 76: 613-616.

Hart, P. J. (Editor), 1969. The Earth's Crust and Upper Mantle. Am. Geophys. Union, Washington, 735 pp.

Hasebe, R., Fujii, N., and Uyeda, S., 1970. Thermal processes under island arcs. Tectonophysics, 10: 335-355.

Hatherton, T., 1969. Similarity of gravity anomaly patterns in asymmetric active regions. Nature, 224: 357.

Heirtzler, J. R. and Vogt, P. R., 1971. Marine magnetic anomalies and their bearing on polar wander and continental drift. Trans. Am. Geophys. Union, 52: I 220 - I 223.

Hurley, P. M., 1969. Some observations on the geological history of Laurasia. Geol. Soc. Am. Abs. Programs, 7: 112.

Ichiye, T., 1971. Continental breakup by nonstationary mantle convection generated with differential heating of the crust. J. Geophys. Res., 76: 1139-1153.

Irving, E., 1966. Paleomagnetism of some carboniferous rocks from New South Wales and its relation to geological events. J. Geophys. Res., 71: 6025–6051.

Irving, E. and Robertson, W. A., 1969. Test for polar wandering and some possible implications. J. Geophys. Res., 74: 1026–1036.

Isacks, B., Oliver, J., and Sykes, L. R., 1968. Seismology and the new global tectonics. J. Geophys. Res., 73: 5855–5899.

Jacob, K. L., 1970. P-residuals and global tectonic structures investigated by three–dimensional seismic ray tracings with emphasis on Longshot data. Trans. Am. Geophys. Union, 51: 359.

Jeffreys, H., 1970. Imperfections of elasticity and continental drift. Nature, 225: 1007–1008.

Julian, B. R., 1970. Regional variations in upper mantle structure in North America. Trans. Am. Geophys. Union, 51: 359.

Kane, M. F., 1971. Differential rotational inertia of continental plates. Trans. Am. Geophys. Union, 52: 355.

Kawai, N., Hirooka, K., and Nakajima, T., 1969. Paleomagnetic and potassium–argon age informations supporting Cretaceous-Tertiary hypothetic bend of the main island Japan. Paleogeog., Paleoclim., Paleoecol., 6: 277–282.

Knopoff, L., 1969. Continental drift and convection. In: P. J. Hart (Editor), The Earth's Crust and Upper Mantle. Am.

Geophys. Union, Washington, pp. 683-689.

Knopoff, L. and Shapiro, J. N., 1969. Comments on the interrelationships between Grüneisen's parameter and shock of isothermal equation of state. J. Geophys. Res., 74: 1439-1450.

Knopoff, L., Heezen, B. C., and MacDonald, G. J. F. (Editors), 1969. The World Rift System: Internat. Upper Mantle Project Sci. Rept. 19. Tectonophysics, 8, 586 pp.

Larson, E. E. and La Fountain, L., 1970. Timing of the breakup of the continents around the Atlantic as determined by paleomagnetism. Earth and Planetary Sci. Letters, 8: 341-351.

Lee, W. H. K. and MacDonald, G. J. F., 1963. The global variation of terrestrial heat flow. J. Geophys. Res., 68: 6481-6492.

LePichon, X., 1968. Sea floor spreading and continental drift. J. Geophys. Res., 73: 3661-3697.

Luyendyk, B. P., 1971. Comments on paper by J. W. Minear and M. N. Toksoz, "Thermal regime of a downgoing slab and new global tectonics". J. Geophys. Res., 76: 605-606.

Lyustikh, E. N., 1969. Problem of convection in the Earth's mantle. In: P. J. Hart (Editor), The Earth's Crust and Upper Mantle. Am. Geophys. Union, Washington, pp. 689-692.

Macdougall, D., 1971. Deep sea drilling: Age and composition of an Atlantic basaltic intrusion. Science, 171: 1244-1245.

Markowitz, W. and Guinot, B. (Editors), 1968. Continental

797

Drift, Secular Motion of the Pole, and Rotation of the Earth. Reidel, Dordrecht, 107 pp.

Markowitz, W., Stoyko, N., and Fedorov, E. P., 1964. Longitude and latitude. In: H. Odishaw (Editor), Research in Geophysics, vol. 2. M. I. T. Press, Cambridge, pp. 149-162.

Martin, R., 1971. Cretaceous-Lower Tertiary rift basin of Baffin Bay: Continental drift without sea-floor spreading. Geol. Soc. Am. Abs. Programs, 3: 393.

Maxwell, A. E., 1969. Recent deep sea drilling results from the South Atlantic. Trans. Am. Geophys. Union, 50: 113.

McBirney, A. R., 1970. Cenozoic igneous events of the Circum-Pacific. Geol. Soc. Am. Abs. Programs, 2: 749-751.

McElhinny, M. W., 1970. Formation of the Indian Ocean. Nature, 228: 977-979.

McElhinny, M. W. and Luck, G. R., 1971. Paleomagnetism and Gondwanaland. Science, 168: 830-832.

McKenzie, D. P., 1971. Comments on paper by J. W. Minear and M. N. Toksoz, "Thermal regime of a downgoing slab and new global tectonics". J. Geophys. Res., 76: 607-609.

McKenzie, D. P. and Sclater, J. G., 1968. Heat flow inside the island arcs of the northwestern Pacific. J. Geophys. Res., 73: 3173-3179.

Meservey, R., 1969. Topological inconsistency of continental drift on the present sized Earth. Science, 166: 609-611.

Minear, J. W. and Toksoz, M. N., 1970a. Thermal regime of a downgoing slab and new global tectonics. J. Geophys. Res., 75: 1397-1419.

Minear, J. W. and Toksoz, M. N., 1970b. Thermal regime of a downgoing slab. Tectonophysics, 10: 367-390.

Mohr, P. A., 1971. Ethiopian rift and plateaus: Some volcanic petrochemical differences. J. Geophys. Res., 76: 1967-1984.

Moores, E. M., 1970. Patterns of continental fragmentation and reassembly: Some implications. Geol. Soc. Am. Abs. Programs, 2: 629.

Murray, C. G., 1970. Magma genesis and heat flow: Differences between mid-ocean ridges and African rift valleys. Earth Planet. Sci. Letters, 9: 34-38.

Noltimier, H. C., 1970. A proposed structural evolution of the Caribbean and Central America since the Jurassic. Trans. Am. Geophys. Union, 51: 825.

Oliver, J., Sykes, L., and Isacks, B., 1969. Seismology and the new global tectonics. Tectonophysics, 7: 527-541.

Pan, C., 1971. The wandering of the pole and the Earth's evolution cycle. Trans. Am. Geophys. Union, 52: 355.

Rodolfo, K. S., 1971. Contrasting geometric adjustment styles of drifting continents and spreading sea floors. J. Geophys. Res., 76: 3272-3281.

Runcorn, S. K. (Editor), 1962. Continental Drift. Academic, New York, 338 pp.

Salop, L. I. and Scheinmann, Y. M., 1969. Tectonic history and structures of platforms and shields. Tectonophysics, 7: 565–597.

Sclater, J. G. and Francheteau, J., 1970. The implications of terrestrial heat flow observations on current tectonic and geochemical models of the crust and upper mantle of the Earth. Geophys. J., 20: 493–509.

Shor, G. G., Kirk, H. K., and Menard, H. W., 1971. Crustal structure of the Melanesian area. J. Geophys. Res., 76: 2562–2586.

Smith, A. G. and Hallam, A., 1970. The fit of the southern continents. Nature, 225: 139–144.

Stacey, F. D., 1969. Physics of the Earth. Wiley, New York, 324 pp.

Stone, D. B. and Cameron, C. P., 1970. Paleomagnetic investigations in Alaska. Alaska Sci. Conf. 20th, 1969, Proc., pp. 311–318.

Sykes, L. R., Kay, R., and Anderson, D. L., 1970. Mechanical properties and processes in the mantle. Trans. Am. Geophys. Union, 51: 874–879.

Tatsch, J. H., 1969. Sea-floor spreading, continental drift, and plate tectonics unified into a single global concept by the application of a dual primeval planet hypothesis to the Earth. _Trans. Am. Geophys. Union_, _50_: 672.

Toksoz, M. N., Minear, J. W., and Julian, B. R., 1971. Temperature field and geophysical effects in a downgoing slab. _J. Geophys. Res._, _76_: 1113-1138.

Utsu, T., 1967. Anomalies in seismic wave velocity and attenuation associated with a deep earthquake zone. _J. Fac. Sci. Hokkaido Univ. Ser._ _7_, _3_: 1-25.

Van Andel, T. H. and Moore, T. C., 1970. Magnetic anomalies ans seafloor spreading rates in the northern South Atlantic. _Nature_, _226_: 328-330.

Vine, F. J., 1969a. Spreading of the ocean floor: New evidence. _Science_, _154_: 1405-1415.

Vine, F. J., 1969b. Sea-floor spreading and continental drift. _Geol. Soc. Am. Abs. Programs_, _7_: 231-232.

Vinogradov, A. P. et al., 1969. The structure of the mid-oceanic rift zone of the Indian Ocean and its place in the world rift system. _Tectonophysics_, _8_: 377-401.

Vogt, P. R., Higgs, R. H., and Johnson, G. L., 1971. Hypotheses on the origin of the Mediterranean basin: Magnetic data. _J. Geophys. Res._, _76_: 3207-3228.

Wegener, A., 1915. _Die Entstehung der Kontinente und Ozeane._ Sammlung Vieweg, Braunschweig, 94 pp.

Yungul, S. H., 1971. Magnetic anomalies and the possibilities of continental drifting in the Gulf of Mexico. _J. Geophys. Res._, 76: 2639-2642.

Zietz, I., 1971. Magnetic anomalies over the continents. _Trans. Am. Geophys. Union_, 52: I 204 - I 209.

Chapter 18

PLATE TECTONICS

The plate-tectonics concept encompasses the structural
phenomena involved in the development and behavior of lithospheric
plates. Most of the phenomena that have contributed to the
evolution of the surficial part of the Earth's tectonosphere
are embodied in the plate-tectonics concept. These phenomena
may be discussed under four basic headings: (1) present geodynamics; (2) ancestral geodynamics; (3) thermomagmatic activity;
and (4) causal mechanisms.

803

Present Geodynamics Associated with Plate Tectonics.

Present geodynamics falls into four categories of phenomena:
(1) the present configuration of lithospheric plates; (2) the
tectonic activity at plate interfaces; (3) the tectonic activity
within plates; and (4) the surficial manifestations of the Earth's
internal behavior.

The Present Configuration of Lithospheric Plates. The
initial concept of plate tectonics divided the Earth's surface
into a relatively few large plates (See, e. g., Morgan, 1968;
Heirtzler et al., 1968). Closer scrutiny revealed that the
actual number of plates has to be quite large in order to account for the observed behavior of the Earth along inter-plate
regions and to explain the fact that some plates are moving
transversally at the interfaces, that some are moving conver-
gently, and that others are moving divergently with respect to
contiguous plates.

Tectonic Activity at Plate Interfaces. The tectonic behavior
of lithospheric plates at their interfaces presents an extremely
complex problem of geometry and mechanics, and the nature of
the unbalanced forces creating these complexities defies simple
definition. At diverging plate interfaces, the complexities
involve problems associated with (1) the emplacement of the

magmatic effusives required to "create" new crustal material;
(2) the geometry of the magnetic basement; (3) the chemistry of
the crustal rocks; (4) the nature of the unbalanced forces causing
uplifts and subsidences at the interfaces; and (5) the mechanics
of producing thermomagmatic activity at the interfaces as well
as at relatively great distances from the interfaces.

The cyclical but aperiodic nature of uplifts and subsidences
at the interfaces poses a particularly acute problem from the
standpoint of causal relationships. Large-scale diapiric intrusions observed at great distances from the interfaces defined
by plate tectonics are not understood. Large uplifts parallel
to some interfaces apparently are not predictable from present
concepts of transform faulting as defined by the plate-tectonics
concept.

Tectonic activity at <u>converging</u> <u>plate</u> <u>interfaces</u> is even
more complex than is that at diverging interfaces. The complexity
of the geology at most converging plate interfaces defies resolution in terms of geometrical, mechanical, thermal, and chemical
phenomena predicted by the plate-tectonics concept. Even greater
complexities are introduced into the problem when present tectonic behavior is considered in conjunction with ancestral tectonic behavior at these interfaces. Because many of these erstwhile interfaces are now uplifted, subsided, or metamorphosed
beyond recognition, their present whereabouts remains unknown.

Most known interfaces now lie within oceanic areas or at the ocean—continent interfaces, but many erstwhile interfaces must also now lie hidden within continental areas, such as the Ural suture in Eurasia and the Transarctic Range in Antarctica.

Tectonic Activity within Plates. Although most intraplate regions are tectonically quiescent, not all plate deformations occur at the interfaces identified by the plate—tectonics concept. Most tectonic activity occurring within oceanic plates apparently is related to associated behavior at diverging plates. Intraplate tectonic activity where such association does not exist includes the activity that created aseismic ridges such as the Carnegie, Cocos, Sierra Leone, Bermuda, Guinea rises. These and similar aseismic ridges apparently resulted from tectonic activity. Many prominent tectonic features, such as these, although far removed from plate interfaces, contain features suggesting an "interface" origin. But the geometry, mechanics, and present location of the required motivating interface are completely obscure.

Intraplate regions of continental plates generally are as quiescent as are the intraplate regions of oceanic plates. But, as in the case of oceanic plates, most continental plates contain intraplate "anomalies" reminiscent of the "fossil" features i oceanic intraplate regions. Examples of these anomalies within

continental intraplate regions include earthquakes, widespread
effusive activity, uplift, subsidence, and rifting.

Surficial Manifestations of the Earth's Internal Behavior.
Summarizing, tectonic activity at plate interfaces as well as
within intraplate regions is a surficial manifestation of the
Earth's internal behavior. Initially, the relationship between
the observed tectonic behavior on the surface of the Earth and
the deep-seated behavior was considered to be more-or-less direct.
But present observational evidence shows (1) that the actual
association between the surficial manifestations and the causal
activity deep within the Earth is extremely complex; (2) that
the basic geometry, mechanics, and chemistry involved in this
association are poorly understood; and (3) that the details
of the causal mechanism linking these phenomena remains to be
defined.

<u>Ancestral Geodynamics Associated with Plate Tectonics.</u>

Ancestral geodynamics may be separated spatially and temporally into 5 categories of phenomena: (1) polar wandering;
(2) Meso-Cenozoic plate configurations and movements; (3) pre-
Mesozoic plate configurations and movements; (4) orogenic belts
and igneous activity; and (5) vertical movements.

<u>Polar Wandering.</u> Polar wandering must be integrated into
a study of ancestral geodynamics because the polar axis of the
Earth has not remained fixed at all times in the past (Chapter
14). Shifts in the direction of the polar axis serve to complicate an analysis of the geometry and mechanics of the ancestral
plate configurations and movements.

<u>Meso-Cenozoic Plate Configurations and Movements.</u> Recent
plate configurations and movements can be analyzed without regard to polar wandering because the polar axis apparently has
remained more-or-less fixed during Meso-Cenozoic times. Most
major present plate interfaces have been determined. The number
and configurations of the minor plates, however, remain undetermined. This indeterminacy is particularly applicable in
suturing activities between Meso-Cenozoic plates and remnants
of older plates as well as in rifting activities. Meso-Cenozoic
rifting has been and still is creating new plate interfaces.

Most early Mesozoic plate interfaces have been obliterated by subsequent orogenic activity, and sutures within continental areas are more difficult to recognize than are those within oceanic areas. Tentative locations of such sutures may be determined through techniques of paleomagnetic correlation. These include polarity-reversal and magnetic-anomaly techniques.

Pre-Mesozoic Plate Configurations and Movements. Ancestral plate configurations and movements pose a far more complex problem than do those of the Meso-Cenozoic. Because the oldest oceans date back only to the Jurassic, i. e., back only through the past 4% of the Earth's life, the techniques of paleomagnetism are of little help in analyzing the pre-Mesozoic plate configurations and movements in oceanic areas. With the oceans covering 70% of the Earth's surface, this poses a serious problem of global analyses in the pre-Mesozoic eras. Explanations for the earlier eras, extending through 96% of the Earth's life, must be sought, therefore, not in the vast oceanic areas but in the relatively small continental areas.

On the assumption that pre-Mesozoic tectonic behavior resembles Meso-Cenozoic behavior, continental areas may be searched for sutures and other indications of pre-Mesozoic interfaces, such as geosynclines and mountain ranges (Chapter 12), mineralization patterns within continents (Chapter 10), and the unique

809

orogenic-cratonic structure of the continents (Chapter 7). In
short, analytical techniques and indirect approaches, used in
conjunction with a working model of the behavior of the Earth's
tectonosphere, may serve to supply the details of the pre-Mesozoic
plate configurations and movements, particularly in those continental areas containing orogenic belts and associated igneous
activity.

Orogenic Belts and Igneous Activity. The relationships
among orogenic belts, metamorphism, igneous activity, and plate
tectonics apparently are fundamental, global, unified, deep-
seated, and long-lived.

One example of the nature of these relationships may be
seen in the behavior within thermomagmatic belts. In both young
and old belts, thermomagmatic activity is closely associated
with seismic activity, as evidenced by the unique spectra of
rock suites in belts of all ages.

Unanswered questions in orogenic belts concern the nature
of the unbalanced forces that have produced cyclical but aperiodic
uplifts and subsidences within the geosynclinal and mountain-
building sequences (Chapter 12). These unbalanced forces generally appear to have operated, within both continental and
oceanic areas, along extended rectilinear belts. The unique
pattern of these belts suggests that these unbalanced forces are

810

global because rectilinear patterns on the surface of a sphere suggest forces conentrated within central planes, i. e., within planes passing through the center of the sphere. This, in turn, puts closely-bounded constraints on the geometrical and mechanical nature of the driving mechanism producing the unbalanced forces that cause the observed uplifts and subsidences within orogenic belts.

A further constraint in relating the various phenomena of ancestral geodynamics is introduced by the observation that many mountain systems comprise parallel chains aggregating a width of 1000 to 2000 km. These large dimensions are difficult to understand in terms of the plate-tectonics concept of "subduction" by lithospheric plates less than 100 km thick, i. e., an order of magnitude smaller than the width of the mountain systems supposedly produced by the subducting lithospheric plates. A further complication arises from the observation that many of these mountain systems develop on bases of continental crust rather than on bases of oceanic crust as expected from the plate-tectonics concept.

Certain mountain systems develop on subsiding continental rocks contrary to the predictions of the basic plate-tectonics concept. In other systems, large ultamafic stocks have intruded the older Paleozoic and Mesozoic basements, producing systems that are the antithesis of the andesitic effusives expected from

811

subducting lithospheric plates.

The problem of understanding the complexities of geosynclinal development and mountain building is compounded by the
observation that orogenic systems of one age closely resemble
those of other ages. The Hercynian system of southern Europe
and northwestern Africa, for example, resembbles the younger
Mediterranean system.

Some igneous activity, in both oceanic and continental
areas, appears completely unrelated to that expected in terms
of the plate-tectonics concept. Oceanic examples of this lack
of association include (1) the Hawaiian, Canary, and Guinea
islands; and (2) widespread igneous activity, spreading, and
crustal formation, far removed from predictable interfaces.
Continental examples of this lack of association between the
predicted and observed phenomena include (1) the plateau basalts
on almost all continents; (2) the alkalic intrusives in other
parts of these continents; (3) the widespread Cretaceous effusives around the Gulf of Mexico; and (4) extensive volcanism,
along the west coast of the Americas and on other continents,
as much as 1500 km inland from the plate interfaces identified
by the plate-tectonics concept.

Vertical Movements. Observational evidence indicates that
vertical movements of plates occur within plates as well as at

plate interfaces. These vertical movements, often accompanied by thermomagmatic activity, occur in both oceanic and continental areas.

Some vertical movements within plates have exceeded several km during the past 4% of the Earth's existence. At plate interfaces, these vertical movements may be as much as an order of magnitude greater, e. g., the subsidences of 30 to 40 km along the Tethys belt of Eurasia.

The subsurface mass transfers associated with vertical movements of these magnitudes obviously involve large volumes of "fluid" Earth material that, in turn, requires large unbalanced forces. These forces are not predictable from the plate-tectonics concept.

Thermomagnatic Activity Associated with Plate Tectonics.

The exact relationship between thermomagnatic activity and crustal-plate behavior can best be understood by examining some of the phenomena associated with the origin and behavior of the lithosphere: its generation, its disappearance, and its transformation from one form to another.

Thermomagnatic Activity Associated with the Generation of Lithospheric Plates. The lithosphere comprises the tectonosphere above the asthenosphere, or low-viscosity zone, that lies at depths varying from zero to approximately 100 km. The physical properties of the lithosphere vary both horizontally and vertically, and these variations appear to be complex functions of the chemical composition of the lithosphere, of its thermal state, and of its stress field.

The lithosphere is distinguishable from the underlying asthenosphere primarily by its state variables, i. e., by its geometrical, mechanical, thermal, and chemical characteristics. These state variables are poorly understood at great depths, at the litho-asthenospheric interface, and in those regions where lithospheric material is being either generated or destroyed.

Discontinuities in the lithosphere occur along mid-ocean ridges and along the world-rift system. The structural features associated with these belts of lithospheric discontinuity are predominantly tensional, and the effusives found there are primarily basic volcanics with some ultrabasics. Major areal variations occur symmetrically with respect to the belts of lithospheric discontinuity. Linear variations occur <u>along</u> these belts.

<u>Thermomagmatic Activity Associated with the Disappearance of Lithospheric Plates</u>. Lithospheric discontinuities that occur along the seismically active belts associated with island arcs and with some continental margins are further associated with the disappearance of lithospheric material. The geometrical, mechanical, thermal, and chemical details of the processes through which the lithosphere disappears are not understood. Particularly undefined are (1) the nature of the litho-asthenospheric interface where the lithospheric plates apparently are disappearing; and (2) the nature of the subasthenospheric portion of the tectonosphere, to depths of at least 700 km, through which the lithospheric material appears to be disappearing and within which earthquakes may occur to these depths.

<u>Thermomagmatic Activity Associated with the Transformation of Lithospheric Material</u>. One of the most unique characteristic of the lithosphere is the broad spectrum of types of crustal material, ranging from purely "oceanic" to purely "continental"

(Chapter 5). Transformations and transitions apparently occur, within this spectrum of types, along both tensional and compressional segments of belts of lithospheric discontinuity. Much of the "oceanic" crust now found along segments of these belts was once in shallower water and was possibly of "continental" composition; some presently continental material apparently once existed as oceanic material.

Many areas of intermediate oceano-continental material are not assoicated with obvious oceano-continental interfaces defined by the plate-tectonics concept, e. g., the crust of island arcs, of many deep-ocean rises, and of many shallow-ocean basins. Almost all areas of oceano-continental type material are associated with segments of the belts of lithospheric discontinuity. This association is discussed in the last section of this chapter.

Causal Mechanisms for the Geodynamic and Thermomagmatic Behavior
Associated with Plate Tectonics.

The details of the unbalanced forces that contribute to
the geodynamic and thermomagmatic activity associated with plate
tectonics are not understood. An approach to such an understanding
is possible by closely scrutinizing the observational evidence
from the standpoint of identifying a single, global, deep-seated,
long-lived mechanism that has the geometrical, mechanical, thermal,
and chemical attributes required to produce the phenomena that
comprise plate tectonics. These phenomena may be analyzed from
the standpoint of (1) the nature of the unbalanced forces; (2)
the rheid framework within which these forces operate; (3) the
contibutions from the thermochemical behavior and the spatio-
temporal variations within Earth materials.

a. The Nature of the Unbalanced Forces Involved in Geodynamic and Thermomagmatic Phenomena. The nature of the unbalanced horizontal and vertical forces that have been moving parts
of at least the upper 30% of the Earth during at least the past
200 million years are not understood. These unbalanced forces
need not be large, because an unbalanced force, no matter how
small, can do work such as that required to move parts of the
tectonosphere to depths of at least 700 and possibly 1000 km.
But the existence of these unbalanced forces at these great

depths must be explained. Such an explanation must embody planetary geometry and mechanics extending to depths of at least 700 km and back in time at least 200 m. y. It must also relate these geometrical and mechanical attributes of the long-lived deep-seated unbalanced forces to the thermal and chemical environments in which they have been effective at these great depths during these extended periods of time.

Present plate-tectonics concepts do not provide the required geometrical, mechanical, thermal, and chemical details associated with the unbalanced forces involved. Some theoretical analyses involving "mathematical singularities" associated with these unbalanced forces may provide partial answers. Similarly, the determination of generalized "density profiles" of the Earth may, in some cases, provide some insight into an averaged "gravity profile" to drive matter toward the Earth's surface. Other help may be provided from studies embodying minimum shear-velocity boundary zones of partial melting, phase-transformation-induced density discontinuities, vanishing-shear-mode instabilities, density inversions and fusions, and other such theoretical analyses.

All these and other analyses and studies may contribute toward the determination of a successful theory of geodynamics expaining the origin of the unbalanced forces that (1) move large plates, (2) allow lateral inhomogeneities, and (3) sustain mountains over long periods of time. But such analyses, unless embodied into a single geometrical, mechanical, thermal, and chemical system, will probably fall short of providing a complete answer for the geodynamic behavior of the Earth's tectonosphere during at least the past 200 m. y.

b. The Rheid Framework within Which the Unbalanced Forces of the Tectonosphere Operate. The rheid framework within which these unbalanced forces have been able to operate effectively for at least the past 200 m. y. must also be considered in any analysis of plate tectonics. An understanding of this framework requires knowledge of the coefficients of viscosity of the thermal expansivity, of the shear elastic constants, and of the thermal diffusivity at the pressures and temperatures to a depth of at least 700 km over a long period of time. The acquisition of this knowledge will require techniques not now in use.

c. Contributions from the Thermochemical Behavior and the Spatio-Temporal Variations within Earth Materials. Analyses regarding the thermochemical behavior of the Earth's interior will permit a better understanding of energy transfer and thermal structure associated therewith. This, in turn, will permit a determination of the ultimate origin of the unbalanced forces within the tectonosphere, as well as how these forces are influenced

819

by the spatio-temporal variations within the material comprising the various levels of the tectonosphere.

Summarizing, present evidence and techniques do not permit the identification of the causal mechanism that motivates plate and block movements and configurations within the lithospheric, asthenospheric, and subasthenospheric levels of the Earth's tectonosphere.

<u>Geophenomena</u> That <u>Are</u> <u>Not</u> <u>Explainable</u> <u>within</u> <u>the</u> <u>Concept</u> <u>of</u>
<u>Plate</u> <u>Tectonics</u>.

When the predictions of the concept of plate tectonics are
compared with observational evidence, several critical questions
remain unanswered. These may be discussed under the headings
of geodynamic and thermomagmatic phenomena.

A. <u>Salient</u> <u>Geodynamic</u> <u>Phenomena</u> <u>Unpredictable</u> <u>from</u> <u>the</u>
<u>Concept</u> <u>of</u> <u>Plate</u> <u>Tectonics</u>. Primary questions of geodynamics
left unanswered by the concept of plate tectonics fall into
5 categories: (1) vertical motions; (2) seismicity; (3) unbalanced forces; (4) small continental masses; and (5) anomalous
horizontal movements. Following are specific problem areas
within these categories:

    1. <u>Geodynamic</u> <u>Problems</u> <u>of</u> <u>Vertical</u> <u>Motions</u>.

        a. Unpredictable vertical movements within frac-
ture zones.

        b. Major uplifts and subsidences <u>within</u> certain
continental plates.

        c. Unexpected large vertical movements within
small ocean basins such as the Mediterranean, Black, and Caspian
seas.

    2. <u>Geodynamic</u> <u>Problems</u> <u>of</u> <u>Seismicity</u>.

        a. The position of aseismic ridges such as the

Walvis, Carnegie, Cocos, Tehuantepec, and Nasca.

   b. The local concentrations of earthquakes <u>within</u> certain regions of continental plates.

   3. <u>Geodynamic Problems of Unbalanced Forces</u>.

   a. Differences in tectonic style along rift zones.

   b. Tectonic events within the oceanic positions of plates apparently unrelated to the plate stress patterns prevailing at the time of their origin. These include: (1) the Bermuda, Sierra Leone, and Guinea rises in the Atlantic; (2) the volcanic islands of the Bight of Guinea that continue into the Cameroon volcanic belt on the continent; (3) the Naturaliste plateau and a group of large shoal blocks west of Australia; and (4) the Ninety-East Ridge of the Indian Ocean.

   4. <u>Geodynamic Problems of Small Continental Masses</u>.

   a. The role of small continental masses, such as the Seychelles, which generally have been regarded as "trails" behind drifting continents but whose positions within young oceanic crust makes this unlikely.

   b. The origin of down-dropped continental masses in some small ocean basins such as the Caribbean, Mediterranean, and other segments of the extended Tethys.

   5. <u>Geodynamic Problems of Anomalous Horizontal Movemen</u>

   a. The puzzle in the general region of the Philippine Sea, where numerous small troughs, ridges, and basins form a complex pattern that is wholly different from that of normal oceanic plates and their margins.

       b. Similar problems on the continents, such as the location of a plate boundary somewhere in Alaska.

       c. The relationship of pre-Mesozoic orogenic events to the concept of plate tectonics.

B. <u>Salient Thermomagmatic Phenomena Unpredictable from the Concept of Plate Tectonics</u>. Primary questions of thermomagmatics left unanswered by the concept of plate tectonics fall into 3 categories: (1) heat-flow anomalies; (2) intrusive and extrusive activity; and (3) distinctions between oceanic and continental behavior. Following are specific problem areas within these categories:

       1. <u>Thermomagmatic Problems of Heat-Flow Anomalies</u>.

       a. Major variations in heat flow along rift zones.

       b. The causal nature of events such as the pan-African thermal event of about 500 m. y. ago.

       2. <u>Thermomagmatic Problems of Intrusive and Extrusive Activity</u>.

       a. The position of seamount swarms and areas of major volcanism such as Hawaii, the Canaries, the Cape Verdes, and the Kelvin - New England seamount group.

       b. The appearance and disappearance of volcanic

823

activity, in space and time, in view of the fact that the geographic distributions of volcano swarms do not appear to bear any relationship to the overall geometry of the plates.

b. The absence of evidence for the existence of a Mesozoic Darwin Rise.

d. The lack of apparent relationship between plate movements and the guyots and volcanic plateaus of an extensive region in the southwest Pacific.

e. The variable nature of the Hercynian orogeny.

f. The position of the plateau basalts that cover very large continental areas without any apparent connection with plate tectonics.

3. Thermomagmatic Problems of Oceanic versus Continental Behavior.

a. The fact that the plate-tectonics concept has a reasonable record for predicting events or phenomena in oceanic areas but an unimpressive record for predicting events or phenomena within continental areas.

The Geodynamic and Thermomagmatic Behavior of the Earth as Interpreted by the Tectonospheric Earth Model.

An examination of the concept of plate tectonics shows that certain phenomena of the Earth's geodynamic and thermomagmatic behavior are left unanswered by that concept. These unexplained phenomena may be analyzed in two salient areas of non-concordance between the predictions of plate tectonics and observational evidence: (1) geodynamic phenomena; and (2) thermomagmatic phenomena.

Specific questions in certain problem areas of these two categories, as summarized in the previous section, are interpreted in this section in terms of the tectonospheric Earth model. The same numbering format that was used in the previous section, for listing the questions left unanswered by the concept of plate tectonics, will be used in this section, for summarizing the answers to those questions as interpreted by the concept of the tectonospheric Earth model.

A. Geodynamic Phenomena.

1. Vertical Motions.

a. Vertical movements within the fracture zones are a predicted consequence of the tectonospheric Earth model's basic driving mechanism which, in simplest terms, consists essentially of the resultant of two factors: (1) the potential

825

energy inherent in the geogenetically disequilibrated shape
of the primordial Earth and (2) the selectively channeled energy
from the preferential flow of heat, and possibly of volatiles,
outward from the subtectonospheric fracture system described in
Chapter 6.

b. Events of some magnitude that apparently
are completely unrelated to plate tectonics, such as major uplifts and subsidences within certain continental plates, are
associated with the predicted behavior of the 8 basic geoblocks
of the tectonospheric Earth model (Chapter 6).

c. Large vertical movements in areas such as the
Mediterranean, Black, and Caspian seas are associated with the
predicted behavior of the three mutually orthogonal wedge-belts
of activity of the tectonospheric Earth model, in this case
with the Gibraltar-Bengal segment of the tectonospheric equator
of the model.

2. Seismicity.

a. Ridges, such as the Walvis, Carnegie, Cocos,
Tehuantepec, and Nasca, are aseismic because they are "fossil"
with respect to the current wedge-belts of activity of the model
defined by the basic subtectonospheric points, Galapagos, Gibraltar
Bengal, Kermadecs, Aleutians, and Bouvet, lying roughly 1000 km
beneath the Earth's surface.

b. The local concentration of earthquakes within

certain regions of continental plates are associated with segments of those mutually orthogonal wedge-belts of activity that
happen to lie beneath continental areas. Most such earthquakes
are associated with current wedge-belts of activity; others
are associated with fossil wedge-belts as discussed in Chapter 8.

3. Unbalanced Forces.

a. Differences in tectonic style along rift
zones are a predictable consequence of the selectively channeled
energy from the preferential flow of heat, and possibly of volatiles, outward from the subtectonospheric fracture system described in Chapter 6.

b. Tectonic events within the oceanic portions
of surficial plates are not related to the stress patterns prevailing at the time of their origin but are surficial manifestations of the behavior of the deeper-lying geoblocks and their
associated wedge-belts of activity.

4. Small Continental Masses.

a. Small contiental masses, such as the Seychelles,
are not "trails" behind drifting continents but are surficial
manifestations of the wedge-belts of activity, the Seychelles
being associated with the Bouvet-Bengal arc of the ortho-meridional wedge-belt of activity.

b. The down-dropped continental masses in some
small ocean basins are the remnants of down-dropped wedge-belts

827

of activity.

     5. <u>Anomalous</u> <u>Horizontal</u> <u>Movements</u>.

        a. The Philippine Sea puzzle of numerous small troughs, ridges, and basins forming a complex patterns wholly different from that of normal oceanic plates and their margins is identifiable with the surficial manifestations of the intersection of two wedge-belts of activity (Chapter 17).

        b. The suspected location of a plate boundary somewhere in Alaska is identifiable with the Aleutians point of the subtectonospheric fracture system, where four geoblocks of the model meet to create unusual surficial manifestations, as they do also at the five other basic subtectonospheric points of the model: Galapagos, Gibraltar, Bengal, Kermadecs, and Bouvet.

        c. Pre-Mesozoic orogenic events are not relatable directly to the post-Mesozoic surficial plate behavior of the concept of plate tectonics, but these orogenic events are relatable to the subsurficial plate and block behavior of the tectonospheric Earth model. The tectonospheric Earth model concept, unlike the plate-tectonics concept, makes allowances for differential motions between the surficial plates and the subsurficial plate-block complex.

B. <u>Thermomagmatic</u> <u>Phenomena</u>.

1. <u>Heat</u> <u>Flow</u> <u>Anomalies</u>.

a. Major variations in heat flow along rift zones are a predictable consequence of a combination of: (1) the basic plate-block structure of the tectonosphere as described in Chapter 6; and (2) the two factors of the driving mechanism of the model: (a) the potential energy inherent in the geogenetically disequilibrated shape of the primordial Earth and (b) the selectively channeled energy from the preferential flow of heat, and possibly of volatiles, outward from the subtectonospheric fracture system.

b. Events such as the pan-African thermal event of about 500 m. y. ago may be related to the plate-tectonics concept, but their cause is more easily relatable to the 8 basic tectonospheric geoblocks and the 3 mutually orthogonal wedge-belts of activity of the tectonospheric Earth model.

2. <u>Intrusive</u> and <u>Extrusive</u> <u>Activity</u>.

a. The positions of seamount swarms and of areas of major volcanism such as Hawaii, the Canaries, the Cape Verdes, and the Kelvin – New England seamount group are identifiable with the wedge-belts of activity of the model. Hawaii is identifiable with the Kermadecs-Aleutians segment of the current prime-meridional wedge-belt. The Canaries are identifiable with the Galapagos point of the subtectonosphere, i. e., the intersection of the tectonospheric equator and the prime-meridian of the model. The Cape Verdes are identifiable with the Gibraltar-Bouvet segment of the current prime-meridional wedge-belt. The Kelvin – New England seamount group is identfiable with the Mesozoic position of the Gibraltar point of the subtectonosphere.

      b. The cyclical but aperiodic nature of volcanic activity is not related to the geometry of the surficial plates of the plate-tectonics concept but it is related to the 8 deeper-lying tectonospheric geoblocks and to the 3 mutually orthogonal wedge-belts of activity along the interfaces of the geoblocks to depths of 1000 km.

      c. The Mesozoic Darwin Rise, if it existed, is identifiable with the surficial manifestations of the Mesozoic position of the Kermadecs-Aleutians wedge-belt of activity, the associated surficial plates of the Pacific having rotated 78° counterclockwise since the Mesozoic (Chapter 15).

      d. The guyots and volcanic plateaus of an extensive region in the southwest Pacific and other areas of the Earth are not related to surficial plate movements. Rather, they are surficial manifestations of the 8 deeper-lying tectonospheric geoblocks and the 3 mutually orthogonal wedge-belts along their interfaces.

      e. The Hercynian orogeny and its variable character are not related to the surficial plates of the plate-tectonics conecpt. Rather, they are related to the Mesozoic position of the 3 mutually orthogonal wedge-belts of activity. The world-wide Hercynian orogeny occurred, according to the tectonospheric Earth model, along the active segments of the Early Mesozoic position of these wedge-belts.

   f. Plateau basalts covering very large areas of the continents are not identifiable with the interfaces of the _surficial_ plates considered in the plate-tectonics concept. Rather, these extensive basaltic flows are associated with the deeper-seated behavior of the 8 geoblocks and the 3 mutually orthogonal wedge-belts of the tectonospheric Earth model. Basaltic effusives flow, according to the model, from any wedge-belt that is in a "tensional" state (Chapter 10). When such a tensional segment of a wedge-belt lies beneath an oceanic lithospheric plate, the basaltic effusives are "oceanic basalts"; when beneath a continental lithospheric plate, they are like the Deccan, Antrim, Laki, Columbia, Keweenawan, Stromberg, New Jersey, Parana, Western Australia, Antarctic, and other continental basaltic flows, floods, and traps. The thicknesses of some of these flows (5 km) and their areas (as much as a million km^2 on most continents) poses problems for the concept of plate tectonics. These problems are not encountered within the implications of effusive activity predicted by the tectonospheric Earth model and described in Chapter 10.

   3. _Oceanic_ versus _Continental_ Behavior.

a. The plate-tectonics concept has a reasonable record for predicting events or phenomena in oceanic areas but an unimpressive one for doing so in continental areas, because the plate-tectonics concept consider only the relatively recent behavior of only the _surficial_ plates. The tectonospheric Earth model concept, on the other hand, considers the behavior of the surficial plates as well as of the subsurficial plates and blocks, to a depth of 1000 km, over the entire 4.6 b. y. that the Earth has existed.

REFERENCES

Anon., 1971a. On to the Caribbean. Geotimes, 16(1): 27-30.

Anon., 1971b. Deep sea drilling project: Leg 14. Geotimes, 16(2): 14-17.

Anon., 1971c. Geodynamics Project: Development of a U. S. program. Trans. Am. Geophys. Union, 52: 396-405.

Anon., 1971d. Deep sea drilling project: Leg 16. Geotimes, 16(6): 12-14.

Artemjev, M. E. and Artyushkov, E. V., 1971. Structure and isostasy of the Baikal rift and the mechanism of rifting. J. Geophys. Res., 76: 1197-1211.

Beck, M. E., 1970. Paleomagnetism of Keweenawan intrusive rocks, Minnesota. J. Geophys. Res., 75: 4985-4996.

Bell, J. S., 1971. Sea-floor spreading through gravity sliding. Geol. Soc. Am. Abs. Programs, 3: 368.

Beloussov, V. V., 1970. Against the hypothesis of ocean floor spreading. Tectonophysics, 9: 489-511.

Bird, J. M. and Dewey, J. F., 1970. Lithospheric plate: Continental margin tectonics and the evolution of the Appalachian orogen. Bull. Geol. Soc. Am., 81: 1031-1060.

Blackett, P. H. S., Bullard, E. C., and Runcorn, S. K. (Editors), 1965. A symposium on continental drift. Phil. Trans. Roy. Soc. A, 258: 1-40.

Blackwell, D. D. and Roy, R. F., 1971. Geotectonics and

Cenozoic history of the western United States. <u>Geol</u>. <u>Soc</u>. <u>Am</u>.
<u>Abs</u>. <u>Programs</u>, <u>3</u>: 84-85.

Bosshard, E. and MacFarlane, D. J., 1970. Crustal structure of the western Canary Islands from seismic refraction and
gravity data. <u>J</u>. <u>Geophys</u>. <u>Res</u>., <u>75</u>: 4901-4918.

Briden, J. C., 1969. Intercontinental correlations based
on paleomagnetic evidence for recurrent continental drift. In:
<u>Gondwana</u> <u>Stratigraphy</u>, UNESCO, Paris, pp. 421-439.

Bullard, E. C., 1964. Continental drift. <u>Quart</u>. <u>J</u>. <u>Geol</u>.
<u>Soc</u>. <u>Lond</u>., <u>120</u>.

Bullard, E. C., Everett, J. E., and Smith, A. S., 1965.
The fit of the continents around the Atlantic. <u>Phil</u>. <u>Trans</u>.
<u>Roy</u>. <u>Soc</u>. <u>Lond</u>. <u>A</u>, <u>258</u>: 41-51.

Carey, W. S., 1958. A tectonic approach to continental
drift. In: W. S. Carey (Editor), <u>Continental</u> <u>Drift</u>: <u>A</u> <u>Sympo-
sium</u>. Univ. of Tasmania, Hobart, pp. 177-355.

Carr, M. J., Stoiber, R. E., and Drake, C. L., 1971. A
model for the upper mantle below Japan. <u>Trans</u>. <u>Am</u>. <u>Geophys</u>.
<u>Union</u>, <u>52</u>: 279.

Conolly, J. R., 1969. Western Tasman Sea floor. <u>New</u> <u>Zea-
land</u> <u>J</u>. <u>Geol</u>. <u>Geophys</u>., <u>12</u>: 310-343.

Cook, K. L., 1970. Active rift system in the Basin and
Range province. <u>Tectonophysics</u>, <u>8</u>: 469-511.

Dewey, J. F. and Bird, J. M., 1970a. Mountain belts and

the new global tectonics. J. Geophys. Res., 75: 2625-2647.

Dewey, J. F. and Bird, J. M., 1970b. Plate tectonics and geosynclines. Tectonophysics, 10: 625-638.

Dewey, J. F. and Bird, J. M., 1971. Origin and emplacement of the ophiolite suite: Appalachian ophiolites in Newfoundland. J. Geophys. Res., 76: 3179-3206.

Dietz, R. S. and Holden, J. C., 1970. East Indian basin (Wharton Basin) as pre-Mesozoic ocean crust. Geol. Soc. Am. Abs. Programs, 2: 537.

Dietz, R. S. and Sproll, W. P., 1970. Fit between Africa and Antarctica: A continental drift reconstruction. Science, 167: 1612-1614.

Dietz, R. S., Holden, J. C., and Sproll, W. P., 1969. Geotectonic evolution and subsidence of Bahama Platform. Geol. Soc. Am. Abs. Programs, 7: 48.

Elders, W. A., Rex, R. W., Meidav, T., Robinson, P. T., and Biehler, S., 1971. A plate tectonic model for the Salton Trough. Geol. Soc. Am. Abs. Programs, 3: 116-117.

Elsasser, W. M., 1971. Sea-floor spreading as thermal convection. J. Geophys. Res., 76: 1101-1112.

Emery, K. O., Uchipi, E., Phillips, J. D., Bowin, C. O., and Bunce, E. T., 1970. Continental rise off eastern North America. Am. Assoc. Geol. Bull., 54: 44-108.

Erickson, B. H., Naugler, F. P., and Lucas, W. H., 1970.

Emperor fracture system zone: A newly discovered feature in the central North Pacific. Nature, 225: 53-54.

Fink, K. L., 1971. Evidence in the eastern Caribbean for Mid-Cenozoic cessation of sea-floor spreading. Trans. Am. Geophys Union, 52: 251.

Francheteau, J., Harrison, C. G. A., Sclater, J. S., and Richards, M. L., 1969. Magnetization of Pacific seamounts: A preliminary polar curve for the northeastern Pacific. Trans. Am. Geophys. Union, 50: 634.

Gough, D. L. and Porath, H., 1970. Long-lived thermal structure under the southern Rocky Mountains. Nature, 226: 837-839.

Griggs, D. T., 1970. The sinking lithosphere and the focal mechanism of deep earthquakes. In: Symposium on the Nature of the Solid Earth, in honor of Professor F. Birch. McGraw-Hill, New York, in press.

Hall, J. K., 1970. Arctic Ocean geophysical studies: The Alpha cordillera and Mendeleyev ridge. Columbia Univ. CU-2-70.

Hamblin, W. P. and Petersen, M. S., 1971. A possible sub-duction zone in western U. S. Geol. Soc. Am. Abs. Programs, 3: 385.

Hamilton, W., 1970. The Uralides and the motion of the Russian and Siberian platforms. Bull. Geol. Soc. Am., 81: 2553-2576.

Hanks, T. C. and Whitcomb, J. H., 1971. Comments on paper by J. W. Minear and M. N. Toksoz, "Thermal regime of a downgoing slab and new global tectonics". J. Geophys. Res., 76: 613–616.

Hart, P. J. (Editor), 1969. The Earth's Crust and Upper Mantle. Am. Geophys. Union, Washington, 735 pp.

Hasebe, R., Fujii, N., and Uyeda, S., 1970. Thermal processes under island arcs. Tectonophysics, 10: 335–355.

Hatherton, T., 1969. Similarity of gravity anomaly patterns in asymmetric active regions. Nature, 224: 357.

Hayes, D. E. and Pitman, W. C., 1970. Magnetic lineations in the North Pacific. Geology of the North Pacific Basin. Geol. Soc. Am. Mem. 126.

Heirtzler, J. R. and Vogt, P. R., 1971. Marine magnetic anomalies and their bearing on polar wander and continental drift. Trans. Am. Geophys. Union, 52: I 220 – I 223.

Heirtzler, J. R., Dickson, G. O., Herron, E. M., Pitman, W. C., and LePichon, X., 1968. Marine magnetic anomalies, geomagnetic field reversals, and motions of the ocean floor and continents. J. Geophys. Res., 73: 2119–2136.

Hey, R. N. and Morgan, W. J., 1971. Parallel seamounts in the northeast Pacific. Trans. Am. Geophys. Union, 52: 236.

Holcombe, T. L., 1971. Evolution of the sea-floor relief of the Burma Rise and adjoining provinces. Trans. Am. Geophys. Union, 52: 250.

Ichiye, T., 1971. Continental breakup by nonstationary mantle convection generated with differential heating of the crust. J. Geophys. Res., 76: 1139-1153.

Isacks, B. and Molnar, P., 1969. Mantle earthquake mechanisms and the sinking of the lithosphere. Nature, 223: 1121-1124.

Isacks, B., Oliver, J., and Sykes, L. R., 1968. Seismology and the new global tectonics. J. Geophys. Res., 73: 5855-5899.

Jeffreys, H., 1970. Imperfections of elasticity and continental drift. Nature, 225: 1007-1008.

Julian, B. R., 1970. Regional variations in upper mantle structure in North America. Trans. Am. Geophys. Union, 51: 359.

Kane, M. F., 1971. Differential rotational inertia of continental plates. Trans. Am. Geophys. Union, 52: 355.

Karig, D. E., 1970. Ridges and basins of the Tonga-Kermadec island arc systems. J. Geophys. Res., 75: 239-254.

Karig, D. E., 1971. Origin and development of marginal basins in the western Pacific. J. Geophys. Res., 76: 2542-2561.

Kawai, N., Hirooka, K., and Nakajima, T., 1969. Paleomagnetic and potassium-argon age informations supporting Cretaceous-Tertiary hypothetic bend of the main island Japan. Paleogeog. Paleoclim. Paleoecol., 6: 277-282.

Knopoff, L., 1969. Continental drift and convection. In: P. J. Hart (Editor), The Earth's Crust and Upper Mantle. Am. Geophys. Union, Washington, pp. 683-689.

Knopoff, L., Drake, C. L., and Hart, P. J. (Editors), 1968. The Crust and Upper Mantle of the Pacific Area. Am. Geophys. Union, Washington, 522 pp.

Knopoff, L., Heezen, B. C., and MacDonald, G. J. F. (Editors), 1969. The World Rift System. Internat. Upper Mantle Project Sci. Rept. 19. Tectonophysics, 8, 586 pp.

Larson, R. L. and Chase, C. G., 1970. Relative velocities of the Pacific, N. America, and Cocos plates in the Middle America region. Earth Planet. Sci. Letters, 7: 425-428.

LePichon, X., 1968. Sea floor spreading and continental drift. J. Geophys. Res., 73: 3661-3697.

LePichon, X. and Heirtzler, J. R., 1968. Magnetic anomalies in the Indian Ocean and sea-floor spreading. J. Geophys. Res., 73: 2101-2117.

Luyendyk, B. P., 1971. Comments on paper by J. W. Minear and M. N. Toksoz. "Thermal regime of a downgoing slab and new global tectonics". J. Geophys. Res., 76: 605-606.

Lyustikh, E. N., 1969. Problem of convection in the Earth's mantle. In: P. J. Hart (Editor), The Earth's Crust and Upper Mantle. Am. Geophys. Union, Washington, pp. 689-692.

Macdougall, D., 1971. Deep sea drilling: Age and composition of an Atlantic basaltic intrusion. Science, 171: 1244-1245.

Martin, R., 1971. Cretaceous – Lower Tertiary rift basin

of Baffin Bay: Continental drift without sea-floor spreading. Geol. Soc. Am. Abs. Programs, 3: 393.

Maxwell, A. E., Von Herzen, R. P., Hsu, K. J., Andrews, J. E., Saito, T., Percival, S. F., Milow, E. D., and Boyce, R. E., 1970. Deep sea drilling in the South Atlantic. Science, 168: 1047-1059.

McBirney, A. R., 1970. Cenozoic igneous events of the Circum-Pacific. Geol. Soc. Am. Abs. Programs, 2: 749-751.

McElhinny, M. W., 1970. Formation of the Indian Ocean. Nature, 228: 977-979.

McElhinny, M. W. and Luck, G. R., 1971. Paleomagnetism and Gondwanaland. Science, 168: 830-832.

McKenzie, D. P., 1971. Comments on paper by J. W. Minear and M. N. Toksoz, "Thermal regime of a downgoing slab and new global tectonics". J. Geophys. Res., 76: 607-609.

Menard, H. W., 1967. Transitional types of crust under small ocean basins. J. Geophys. Res., 72: 3061-3073.

Minear, J. W. and Toksoz, M. N., 1970a. Thermal regime of a downgoing slab and new global tectonics. J. Geophys. Res., 75: 1397-1419.

Minear, J. W. and Toksoz, M. N., 1970b. Thermal regime of a downgoing slab. Tectonophysics, 10: 367-390.

Mitchell, A. H. and Reading, H. G., 1969. Continental margins, geosynclines, and ocean floor spreading. J. Geol., 77: 629-646.

840

Mohr, P. A., 1971. Ethiopian rift and plateaus: Some volcanic petrochemical differences. J. Geophys. Res., 76: 1967–1984.

Moores, E. M., 1970. Patterns of continental fragmentation and reassembly: Some implications. Geol. Soc. Am. Abs. Programs, 2: 629.

Morgan, W. J., 1968. Rises, trenches, great faults, and crustal blocks. J. Geophys. Res., 73: 1959–1982.

Morgan, W. J., 1971. Island chains, aseismic ridges, and hot spots. Trans. Am. Geophys. Union, 52: 371.

Noltimier, H. C., 1970. A proposed structural evolution of the Caribbean and Central America since the Jurassic. Trans. Am. Geophys. Union, 51: 825.

Oliver, J., Sykes, L., and Isacks, B., 1969. Seismology and the new global tectonics. Tectonophysics, 7: 527–541.

Oliver, J. and Isacks, B., 1967. Deep earthquake zones, anomalous structures in the upper mantle, and the lithosphere. J. Geophys. Res., 72: 4259–4275.

Oriel, S. S., 1971. Southern Idaho thrust belt. Geol. Soc. Am. Abs. Programs, 3: 400–401.

Phillips, J. D. and Luyendyk, B. P., 1970. Central North Atlantic plate motions over the last 40 million years. Science, 170: 727–729.

Pitman, W. C., 1971. Sea-floor spreading and plate tectonics.

Trans. Am. Geophys. Union, 52: I 130 – I 135.

Price, R. A. and Mountjoy, E. W., 1970. Geologic structure of the Canadian Rocky Mountains between Bow and Athabasca
Rivers: A progress report. Geol. Assoc. Canada Spl. Paper 6:
7-25.

Price, R. A. and Mountjoy, E. W., 1971. The Cordilleran
foreland thrust and fold belt in the southern Canadian Rockies.
Geol. Soc. Am. Abs. Programs, 3: 404-405.

Rayleigh, Lord, 1916. On convection currents in a horizontal layer of fluid when the higher temperature is on the
under side. Phil. Mag., 32: 529-546.

Robinson, G. D., 1971. Montana disturbed belt segment of
Cordilleran orogenic belt. Geol. Soc. Am. Abs. Programs, 3:
408-409.

Rona, P. A., Brakl, J., and Heirtzler, J. R., 1970. Magnetic anomalies in the northeast Atlantic between the Canary
and Cape Verde islands. J. Geophys. Res., 75: 7421-7425.

Runcorn, S. K. (Editor), 1962. Continental Drift. Academic, New York, 338 pp.

Salop, L. I. and Scheinmann, Y. M., 1969. Tectonic history and structures of the platforms and shields. Tectonophysics,
7: 565-597.

Scholz, C. H., Barazangi, M., and Sbar, M., 1971. Cenozoic evolution of the Basin and Range province. Trans. Am.

Geophys. Unn., 52: 350.

Sclater, J. G. and Francheteau, J., 1970. The implications of terrestrial heat flow observations on current tectonic and geochemical models of the crust and upper mantle of the Earth. Geophys. J., 20: 493-509.

Sclater, J. G., Hawkins, J. W., and Chase, C. G., 1971. Crustal extension between the Tonga and Lau ridges: Petrologic and geophysical evidence. Trans. Am. Geophys. Union, 52: 194.

Shor, G. G., Kirk, H. K., and Menard, H. W., 1971. Crustal structure of the Melanesian area. J. Geophys. Res., 76: 2562-2586.

Smith, A. G. and Hallam, A., 1970. The fit of the southern continents. Nature, 225: 139-144.

Stacey, F. D., 1969. Physics of the Earth. Wiley, New York, 324 pp.

Stone, D. B. and Cameron, C. P., 1970. Paleomagnetic investigations in Alaska. Alaska Sci. Conf. 20th., 1969, Proc.: 311-318.

Sykes, L. R., Kay, R., and Anderson, D. L., 1970. Mechanical properties and processes in the mantle. Trans. Am. Geophys. Union, 51: 874-879.

Takeuchi, H. and Sakata, S., 1970. Convection in a mantle of variable viscosity. J. Geophys. Res., 75: 921-927.

Talwani, M., Windisch, C. C., and Langseth, M. G., 1971.

Reykjanes ridge crest: A detailed geophysical study. *J. Geophys. Res.*, *76*: 473-517.

Tatsch, J. H., 1963. Certain volcanological implications of applying a dual primeval planet model to the Earth. *Trans. Am. Geophys. Union*, *44*: 892.

Tatsch, J. H., 1964. Distribution of active volcanoes: Summary of preliminary results of three-dimensional least-squares analysis. *Geol. Soc. Am. Bull.*, *75*: 751-752.

Tatsch, J. H., 1966. Certain correlations between seismological observations in the African continent and deductions made from applying a dual primeval planet model to that region of the Earth. *Trans. Am. Geophys. Union*, *47*: 490.

Tatsch, J. H., 1967. Global geomagnetic evidence supporting the existence of a geophysical equator predicted as a consequence of applying a dual primeval planet model to the Earth. *Trans. Am. Geophys. Union*, *48*: 59.

Tatsch, J. H., 1969. Sea-floor spreading, continental drift, and plate tectonics unified into a single global concept by the application of a dual primeval planet hypothesis to the Earth. *Trans. Am. Geophys. Union*, *50*: 672.

Tatsch, J. H., 1970. Global seismicity patterns as interpreted in accordance with a dual primeval planet hypothesis. *Geol. Soc. Am. Abs. Programs*, *2*: 153.

Toksoz, M. N., Minear, J. W., and Julian, B. R., 1971.

Temperature field and geophysical effects in a downgoing slab. *J*. *Geophys*. *Res*., 76: 1113-1138.

Torrance, K. E. and Turcotte, D. L., 1971. Structure of convection cells in the mantle. *J*. *Geophys*. *Res*., 76: 1154-1161.

Utsu, T., 1967. Anomalies in seismic wave velocity and attenuation associated with a deep earthquake zone. *J*. *Fac*. *Sci*. *Hokkaido Univ*. *Ser*. 7, 3: 1-25.

Utsu, T. and Okada, H., 1968. Anomalies in seismic wave velocity and attenuation associated with a deep earthquake zone, 2. *J*. *Fac*. *Sci*. *Hokkaido Univ*. *Ser*. 7, 3: 65-84.

Vine, F. J., 1969a. Spreading of the ocean floor: New evidence. *Science*, 154: 1405-1415.

Vine, F. J., 1969b. Sea-floor spreading and continental drift. *Geol*. *Soc*. *Am*. *Abs*. *Programs*, 7: 231-232.

Vogt, P. R. and Ostenso, N. A., 1970. Magnetic and gravity profiles across the Alpha cordillera and their relation to Arctic sea-floor spreading. *J*. *Geophys*. *Res*., 75: 4925-4937.

Vogt, P. R., Schneider, E. D., and Johnson, G. L., 1969. The crust and upper mantle beneath the sea. In: P. J. Hart (Editor), *The Earth's Crust and Upper Mantle*. Am. Geophys. Union, Washington, pp. 556-617.

Vogt, P. R., Anderson, C. N., Bracey, D. R., and Schneider, E. D., 1970. North Atlantic magnetic smooth zone. *J*. *Geophys*.

Res., 75: 3955-3966.

Watanabe, T., Epp, D., Uyeda, S., Langseth, M., and Yasui, M., 1970. Heat flow in the Philippine Sea. Tectonophysics, 10: 205-224.

Wegener, A., 1915. Die Entstehung der Kontinente und Ozeane. Sammlung Vieweg, Braunschweig, 94 pp.

Young, J. C., 1970. Caribbean Island arc. Geotimes, 15(9): 17-19.

Yungul, S. H., 1971. Magnetic anomalies and the possibilities of continental drifting in the Gulf of Mexico. J. Geophys. Res., 76: 2639-2642.

Zietz, I., 1971. Magnetic anomalies over the contients. Trans. Am. Geophys. Union, 52: I 204 - I 209.

Chapter 19

ASTEROIDS, METEORITES, AND TEKTITES

An analysis of the evolutionary sequences through which the
asteroids, meteorites, and tektites originated serves to explain
why certain similarities and differences developed among these
solar-system "rocks". An analysis of these similarities and
differences, in turn, permits a better understanding of the
evolutionary sequences through which the Earth's tectonosphere
developed from the primordial solar system.

This chapter reconstructs the behavior of the asteroids,
the meteorites, and the tektites within the 4.6 b. y. span during
which the Earth's tectonosphere has been developing. From this,
it relates the salient geometrical, mechanical, thermal, and
chemical aspects of the evolutionary behavior of the asteroids,
meteorites, and tektites to the present state variables and
behavioral patterns of the Earth's tectonosphere.

Asteroids.

The asteroids form a belt of planetary fragments lying between the orbits of Mars and Jupiter. Most evolutionary hypotheses for the asteroids use the Titius-Bode Law (See, e. g., Roy, 1967) to postulate that, between the oribts of Mars and Jupiter, where the asteroids are found, there can be placed one or, possibly, two planetary orbits (See, e. g., Tatsch, 1962; Stacey, 1969). This suggests a catastrophic origin for the asteroids from one or two planetary systems.

The exact number of asteroids is unknown because only those larger than 1 km in diameter are detectable. Some analyses show that the "population" of the asteroids increases strongly with decreasing size (See, e. g., Blanco and McKuskey, 1961). This suggests that the total mass of asteroidal material is indeterminate.

Analyses used to estimate the ages of meteorites show that the hypothesized meteoritic "parents" were differentiated from a common source. In some hypotheses, this common source was asteroidal and had a history of chemical differentiation prior to the asteroidal fragmentations that produced the meteorites. Analyses of the meteorites show that this asteroidal differentiation occurred roughly 4.5 b. y. ago, or very nearly at the beginning of the solar system.

The paucity of small meteorites suggests that meteorites

848

are the products of relatively recent asteroidal collisions, because older small meteorites would have been destroyed by the mechanics of the Poynting-Robertson effect (Chapter 1). If a primary asteroidal fragmentation had occurred very early in the history of the solar system, say 3.6 to 4.6 b. y. ago, then all primary products of that fragmentation smaller than 50 cm would have fallen, long since, into the Sun from the Poynting-Robertson effect.

Analyses, using the Poynting-Robertson effect combined with cosmic-ray-exposure data, suggest that, within the delicate balance of the solar-system mechanics and the laws of probability, all the meteorites could have evolved: (1) from one or two fairly large, planetary-sized bodies; or (2) from many small asteroidal-sized bodies; or (3) from a sequential combination of (1) and (2) (See, e. g., Tatsch, 1962; Stacey, 1969).

Meteorites.

Meteorites are extra-terrestrial rock-like objects that can be divided into three categories: siderites, siderolites, and aerolites. The siderites, or iron meteorites, consist almost entirely of iron alloyed with nickel. The siderolites are mixtures of nickel-iron and heavy basic silicates such as olivine and pyroxene. The aerolites, or stony meteorites, consist almost entirely of heavy basic silicates.

The gradations of meteorites within and between these three categories is such that they form a complete spectrum of compositions that fits fairly well onto the "basic" end of the spectrum of terrestrial rocks.

The close association of the terrestrial and meteoritic compositional spectrums is shown in Fig. 19-2-1. The interspectrum gradations are such that some terrestrial dunites resemble some meteoritic aerolites, and some meteoritic siderites resemble nickeliferous core-type materials.

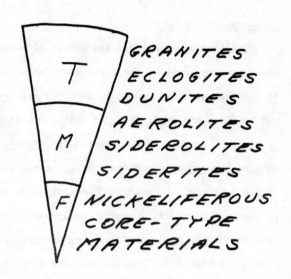

GRANITES
ECLOGITES
DUNITES
AEROLITES
SIDEROLITES
SIDERITES
NICKELIFEROUS
CORE-TYPE
MATERIALS

Figure 19-2-1. The close association of the spectrums of
terrestrial rocks, T, of meteroties, M, and of nickeliferous
core-type materials, F.

851

<u>Tektites.</u>

Tektites are small rounded pieces of silica-rich glass
found in many areas of the Earth's surface. These bottle-green
to blackish vitreous bodies include the <u>moldavites</u> of Bohemia
and Moravia; the <u>australites</u> of southern Australia; the <u>queens-
townites</u> of northeastern Tasmania; the <u>billitonites</u> of the East
Indies; the <u>rizalites</u> of the Philippines; the <u>indochinites</u> of
Cambodia, Annam, and Siam; and possibly some of the <u>glassites</u>
of the Libyan desert. Geographically most tektites appear to
be concentrated approximately along the great-circular extended
Tethys (Chapter 15). Smaller concentrations have been found
in most parts of the world.

The largest known tektite weighs only a few kilograms;
most are much smaller. The most common tektite forms are rough
spheroids, ovoids, pearshapes, buttons, lenses, dumb-bells,
spindles, and some irregular shapes. The surfaces of most tektites are pitted, grooved, and molded in a manner suggesting
flight through the atmosphere, but none has ever been observed
to fall from the sky.

For each geographical region of the larger groups (Europe,
Australia, Tasmania, East Indies, Philippines, Indo-China, and
Libya), the ages of the tektites correlate with the ages of the
geological formations in which found. But the compositions of
tektites are unrelated to those geological formations. The

852

ages (since last molten) range from 0.3 to 35 million years and fall into 4 or 5 distinct groups, suggesting that the known tektites were produced in 4 or 5 events during the past 35 m. y. Older tektites, now geologically lost, might have been produced in earlier "events" spanning the remaining 92% of the life of the Earth.

All tektites resemble all other tektites in that they have a "melted" appearance. Tektites, in this respect, resemble certain "glasses" found on the Moon (Chapter 2).

In some evolutionary hypotheses, all tektites are assumed to have originated on the Moon; in others, all are assumed to have originated on the Earth (See, e. g., O'Keefe, 1963; Geiss and Goldberg, 1963).

The Celestial Mechanical Aspects of Asteroidal Evolution as
Interpreted by a Dual Primeval Planet Hypothesis.

The asteroidal system, from a celestial-mechanical standpoint, provides one of the most well-behaved geometrico-mechanical
complexes of the solar system. The salient geometrical, mechanical, thermal, and chemical aspects of this system have been
analyzed (See, e. g., Tatsch, 1960, 1962, 1962-1971).

Only one of these aspects need be considered here: the
asteroids owe their existence to their genetic environment.
That is, the asteroids exist, according to the dual primeval
planet hypothesis, because Aster and Aster Prime (Chapter 1)
were created within the threshold interval between the terrestrial planets (Mercury, Venus, Earth, and Mars) and the Jovian
planets (Jupiter, Saturn, Uranus, and Neptune). The asteroidal
belt exists, in this concept, not as an "accident" but as a
well-behaved and expected consequence of solar-system evolution.

The evolution of the asteroidal belt from the hypothesized
planet Aster, A, and its prime, A', is shown schematically in
Fig. 19-10-1-1. As in the case of other planetary evolutionary sequences (Chapter 1), the center of mass, C, of the asteroidal system continues its orbit about the Sun, shown in the
direction, S, and along the orbital path A'A (Cf., e. g., Chapter 2, Fig. 2-11-3, showing the evolution of the Moon and the
tectonosphere from 6 of the octants of Earth Prime).

854

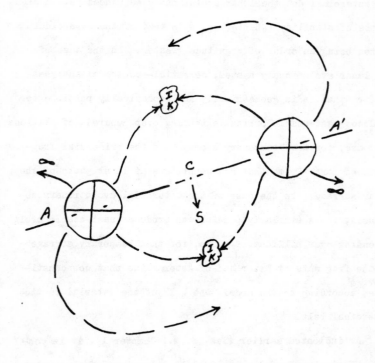

Figure 19-10-1-1. Schematic sketch showing the evolution of

the asteroidal belt according to the dual primeval planet hypo-

thesis. A = Aster. A' = Aster Prime. C = center of mass. S =

direction of Sun. IK = regions of initial collisions of the

octants of Aster and Aster Prime.

855

Within the dual primeval planet hypothesis concept, both the asteroidal and the lunar evolutionary sequences have a high degree of stability (Chapter 2), with that of the asteroidal system being an order of magnitude greater. In the case of the lunar evolutionary system, celestial-mechanical analyses confirm that it is geometrically and mechanically possible to produce quasi-stable orbits, extending over hundreds of millions of years, for the "temporary storage" of the primordial fragments of Earth Prime that now constitute at least parts of the lunar surface. In the case of the asteroidal evolutionary sequences, it is similarly possible to produce quasi-stable orbits, extending over billions of years, for the "temporary storage" of the fragments of Aster and of Aster Prime that now constitute, according to the model, the bulk of the material in the asteroidal belt.

As indicated earlier (See, e. g., Chapter 1), it is possible that the present asteroidal belt developed from two complete Aster systems in "parallel" orbits between Mars and Jupiter; i. e., Aster I and Aster II, each with its own "prime". This possibility does not change the results of any analysis, such as this, which is done with a first-order dual primeval planet hypothesis.

The Relationship of Asteroids to Meteorites and to Other Rocks
of the Solar System as Interpreted by the Dual Primeval Planet
Hypothesis.

Asteroids, according to the dual primeval planet hypothesis,
bear a primo-primal ("first cousin") relationship to the terrestrial, lunar, and meteoritic rocks. The asteroidal belt, according to this concept, constitutes what remains of hypothetical Aster and Aster Prime (or of two such "asteroidal" planets,
Aster I and Aster II) that formed between the Martian and Jovian
orbits in the primordial solar-system evolutionary sequences
(Chapter 1).

The hypothetical planet Aster and its prime both separated
into their octantal parts, as described in the previous section.
These octantal parts, in turn, formed the asteroids through
further fragmentations as they continued to orbit the Sun, in
an extension of the initial collisions shown in Fig. 19-10-1-1
of the previous section.

Fig. 19-10-5-1 shows schematically how one octant of Aster
might have looked just after the initial separation, i. e.,
just prior to the first of the many collisions that have been
fragmentizing these octants into the asteroids during the past
3.6 to 4.6 b. y.

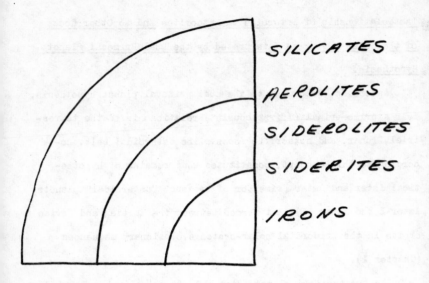

Figure 19-10-5-1. Schematic sketch showing how an octant of
the hypothetical planet Aster might have looked, compositionally, 3.6 to 4.6 b. y. ago, i. e., just prior to the subsequent
collisions that fragmentized these octants into the asteroids.

REFERENCES

Alfven, H. and Arrhenius, G., 1969. Two alternatives for the history of the Moon. Science, 165: 11-17.

Alfven, H. and Arrhenius, G., 1970. Mission to an asteroid. Science, 167: 139.

Anders, E., 1963. Meteorites and the early history of the solar system. In: R. Jastrow and A. G. W. Cameron (Editors), Origin of the Solar System. Academic, New York and London, pp. 95-142.

Anders, E., 1971. How well do we know "cosmic" abundances? Geochim. Cosmochim. Acta, 35: 516-522.

Blanco, V. M. and McCuskey, S. W., 1961. Basic Physics of the Solar System. Addison-Wesley, Reading, pp. 179 et seq.

Bogard, D. D., 1971. Noble gases in meteorites. Trans. Am. Geophys. Union, 52: I 429 - I 435.

Burnett, D. S., 1971. Formation times of meteorites and lunar samples. Trans. Am. Geophys. Union, 52: I 435 - I 440.

Dodd, R. T., 1971. Chondrites. Trans. Am. Geophys. Union, 52: I 447 - I 453.

Dohnanyi, J. S., 1971. Micrometeoroids. Trans. Am. Geophys. Union, 52: I 459 - I 464.

Garlick, G. D., Naeser, C. W., and O'Neil, J. R., 1971. A Cuban tektite. Geochim. Cosmochim. Acta, 35: 731-734.

Gehrels, T., 1971. Asteroids and comets. Trans. Am. Geophys. Union, 52: I 453 - I 459.

Geiss, J. and Goldberg, E. D. (Editors), 1963. Earth Science and Meteoritics. North-Holland, Amsterdam, 312 pp.

Gentner, W., Glass, B. P., Storzer, D., and Wagner, G. A., 1970. Fission track ages and ages of deposition of deep-sea microtektites. Science, 168: 359-361.

Guth, J. H., 1970. Asteroid landing. Science, 170: 1431-1432.

Heide, F., 1964. Meteorites. Univ. of Chi., Chicago and London, 144 pp.

Herzog, G. F. and Anders, E., 1971. Absolute scale for radiation ages of stony meteorites. Geochim. Cosmochim. Acta, 35: 605-611.

Keays, R. R., Ganapathy, R., Laul, J. C., Anders, E., Herzog, G. F., and Jeffreys, P. M., 1970. Trace elements and radioactivity in lunar rocks: Implications for meteorite infall, solar-wind flux, and formation of the Moon. Science, 167: 490-493.

Keays, R. R., Ganapathy, R., and Anders, E., 1971. Chemical fractionations in meteorites - IV. Abundances of fourteen trace elements in L-chondrites; implications for cosmothermometry. Geochim. Cosmochim. Acta, 35: 337-363.

Mason, B., 1962. Meteorites. Wiley, New York and London, 274 pp.

Moore, C. B. (Editor), 1962. Researches in Meteorites. Wiley, New York and London, 227 pp.

Morgan, J. W., 1969. Uranium and thorium in tektites. Earth Planet. Sci. Letters, 7: 53-63.

Nairn, F., 1966. Spatial distribution and motion of the known asteroids. Am. Inst. Aero. Astro. Paper No. 66-149.

O'Keefe, J. A. (Editor), 1963. Tektites. Univ. of Chi., Chicago and London, 228 pp.

O'Keefe, J. A., 1970. Apollo 11: Implications for the early history of the solar system. Trans. Am. Geophys. Union, 51: 633-636.

O'Keefe, J. A., 1970. Tektite glass in Apollo 12 sample. Science, 168: 1209-1210.

O'Keefe, J. A., 1971. Tektites from the Earth. Science, 171: 313-314.

Ronca, L. B., 1965. A geological model of Mare Homorum. Icarus, 4: 390-395.

Roy, A. E., 1967. Bode's law. In: S. K. Runcorn (Editor), International Dictionary of Geophysics, 2 vols. Pergamon, Oxford, p. 146.

Stacey, F. D., 1969. Physics of the Earth. Wiley, New York, 324 pp.

Stanyukovich, K. P. and Bronshten, V. A., 1962. The role of external cosmic factors in the evolution of the Moon: Collisions of meteorites and asteroids. In: A. V. Markov (Editor), The Moon. Univ. of Chi., Chicago, pp. 304-337.

Tatsch, J. H., 1960. The evolution of the solar system in accordance with a dual primeval planet model. U. S. Embassy, Caracas, Venezuela, private publication, given limited distribution to approximately 200 selected astronomers and geophysicists.

Tatsch, J. H., 1962. The evolution of the retrograde motion of certain planetary satellites in accordance with a dual primeval planet model, outline of a doctoral dissertation, submitted to graduate advisor February 4, 1962, about 75 pages.

Tatsch, J. H., 1962-1971. The celestial-mechanical aspects of the solar-system evolutionary sequences as interpreted by a dual primeval planet hypothesis. Unpublished analyses.

Turekian, K. K. and Clark, S. P., 1969. Inhomogeneous accumulations of the Earth from the primitive solar nebula. Earth Planet. Sci. Letters, 6: 346.

Urey, H. C., 1951. The origin and development of the Earth and other terrestrial planets. Geochim. Cosmochim. Acta, 1: 207.

Urey, H. C., 1952. The Planets, Their Origin and Development. Yale, New Haven, 178 pp.

Urey, H. C., 1956. Diamonds, meteorites, and the origin of the solar system. Astrophys. J., 124: 625.

Urey, H. C., 1959. Primary and secondary objects. J. Geophys. Res., 64: 1721-1727.

Urey, H. C., 1960. Criticism of the melted Moon theory.

J. _Geophys_. _Res_., 65: 358-359.

Urey, H. C., 1968. Mascons and the history of the Moon.
Science, 162: 1408-1410.

Urey, H. C., 1971. Tektites from the Earth. _Science_,
171: 312-313.

Wasson, J., 1971. Differentiated meteorites. _Trans_. _Am_.
Geophys. _Union_, 52: I 441 - I 447.

Wood, J. A., Dickey, J. S., Marvin, U. B., and Powell,
B. N., 1970. Lunar anorthosites. _Science_, 167: 602-604.

Wood, J. A., Marvin, U. B., Powell, B. N., and Dickey,
J. S., 1970. Mineralogy and petrology of the Apollo 11 lunar
sample. _Smithsonian Astrophys_. _Obs_. _Special Report_ 307, 99 pp.

Chapter 20

THE INTEGRATED EARTH AND ITS FUTURE

In an analysis of the evolution of the Earth's tectonosphere,
a discussion of the integrated Earth can be written from any
one of a myriad of viewpoints, e. g., (1) from a philosophical
standpoint emphasizing the role of the integrated Earth within
an integrated universe; or (2) from a celestial-mechanical standpoint emphasizing how the geometrico-mechanical behavior of the
integrated Earth will continue to be as well-behaved in the
future as it has been during the past 4.6 b. y.; or (3) from an
ecological standpoint emphasizing how the tectonospheric Earth
model can be used to help mankind.

This chapter uses the last of these viewpoints as the basis
for summarizing several approaches whereby the model can help
man enjoy the Earth's benefits, avoid its disasters, and pre-
serve its landscape in a planned manner.

864

The Ecological Problems of the Earth and Its People.

In the search for solutions to the ecological problems of the Earth and its people, we must consider alternatives that will permit man to enjoy the benefits of industrialization without creating irreparable damage to his environment. It is clear, for example, that the Earth cannot sustain the present exponential growth rate in the use of its limited material and energy resources.

The destruction and abuse of the environment from the exploration, drilling, and digging for fossil and non-fossil substances does not now involve extensive areas. But larger and larger areas will be required as the needs for fuels and metals continue to increase. Any techniques or approaches that can reduce these areas and the damage to the environment within these areas will be a boon to mankind and its posterity.

Long before atomic fuels can satisfy our energy requirements, depletions in present oil and gas reserves will require the burning of increasing amounts of coal. Coal mining, particularly strip mining, leads to extensive land pollution. This condition is aggravated in those areas where the coal reserves lie in steep topography.

Oil shales provide an alternate energy source, but the thermo-chemical behavior of shales presents problems for the environment. Salient among these problems are: (1) each ton

865

of shale produces only about a barrel of oil; (2) large pollution-belching processing plants are required to remove the oil from the shale; and (3) the post-extractive shale residue is 10% greater than that of the unprocessed shale.

Summarizing, the energy and metals industries pose severe ecological problems for the future of mankind. These problems can best be solved through a constructive approach that recognizes and considers both aspects of this problem: (1) man must have energy and other mineral resources to sustain his wellbeing; and (2) he must seek ways to get these resources with minimum damage to the environment.

In this connection, the extreme conservationist will have to realize that the cessation of the exploration for and the processing of natural resources may conserve the environment but it tends to destroy mankind. Furthermore, the extreme explorationist will have to realize that the unlimited exploitation of the Earth for natural resources destroys the environment in which man hopes to enjoy these natural resources.

This will require thoughtful planning, constructive effort, and timely action. The tectonospheric Earth model may be used as an "ecological tool" to supplement this planning, effort, and action.

The Tectonospheric Earth Model as an Ecological Tool.

If man had detailed knowledge of the Earth's geometrical, mechanical, thermal, and chemical behavior, he could take action necessary to conserve his environment while fulfilling his needs for natural resources. Man could utilize, for example, a detailed knowledge of the central causal mechanism to analyze the origin, evolution, migration, and present accumulation of energy and metal deposits. Also, he could utilize a detailed knowledge of the origin, evolution, and destructive effects of the Earth's stress pattern to analyze the causes of earthquakes or, perhaps, their prediction and control. Likewise, he could utilize a detailed knowledge of the origin, evolution, and variation in the Earth's thermal pattern to analyze the cause of volcanoes, their prediction, and control.

Therefore, with a detailed understanding of the tectonosphere and its behavior to a depth of 1000 km during the past 4.6 b. y., man can evaluate the Earth's present behavior in terms of natural resources and environmental catastrophes.

A basic ecological tool that can help this evaluation is provided by the tectonospheric Earth model.

Earthquake Prediction and Control Procedures Derived from the
Tectonospheric Earth Model.

Earthquakes, according to the tectonospheric Earth model,
are motivated by a single, global, deep-seated, and long-lived
mechanism. An understanding of the geometrical, mechanical,
thermal, and chemical aspects of this mechanism provides an
approach to earthquake prediction and control.

The details of this approach were described in Chapter 8
and need not be considered here except to repeat the conclusion
of that chapter: although the exact time and location of earthquakes are difficult to predict, the model provides a basis
for establishing closely-constrained temporal and spatial bounds
within which future earthquakes are apt to occur.

Fig. 20-3-1 is a modified Venn diagram (See, e. g., Venn,
1888; Kendall, 1947; Miller and Kahn, 1962) showing regions
of high probability of seismic activity predicted by the tectonospheric Earth model. Continental profiles are shown in a
Mollweide projection, and representative regions of severe earthquakes are shown: California (CA), Chile (CH), Fiji Islands
(FI), Japan (JP), Java Sea (JS), Kuriles (KU), New Hebrides
(NH), Philippines (PL), Peru (PE), and Yugoslavia (YU). The
radial surficial projections of the 6 basic points of the subtectonospheric fracture system also are indicated: Aleutians
(AL), Bengal (BN), Bouvet (BV), Kermadecs (KR), Galapagos (GA),

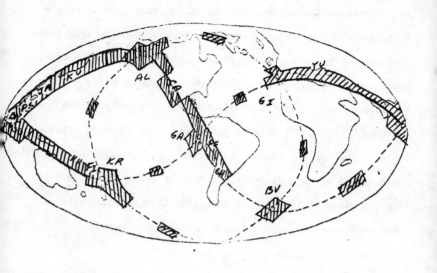

Figure 20-3-1. First-order unadjusted Venn diagram showing
regions of high probability of seismic activity predicted by
the tectonospheric Earth model. See text for details.

and Gibraltar (GI).

Adjustments are not included in Fig. 20-3-1 for errors in projections through the 1000 km of heterogeneous materials that separate the 6 basic points of the subtectonospheric fracture system from the surface of the Earth shown on the sketch. These and other adjustments are always made when using the tectonospheric Earth model for actual earthquake analyses and predictions. The adjustments in these cases are based on predictable variations within the geometrical, mechanical, thermal, and chemical aspects of tectonospheric behavior that affect the time and place of earthquakes according to the tectonospheric Earth model.

<u>Selective</u> <u>Exploration</u>, <u>Drilling</u>, <u>and</u> <u>Mining</u> <u>Procedures</u> <u>Derived</u>
<u>from</u> <u>the</u> <u>Tectonospheric</u> <u>Earth</u> <u>Model</u>.

Ambiguous evidence makes it difficult to define the source
of minerals (See, e. g., Goldschmidt, 1954; Levorsen, 1967;
Krauskopf, 1971).

The processes involved in the origin, evolution, migration,
enrichment, and emplacement of mineral deposits appear to constitute a multistage sequence encompassing thermomagmatic and
geodynamic activity, including: (1) subsurficial magmatic intrusion, cooling, and differentiation; (2) surficial volcanism
and other extrusive activity; (3) a complete spectrum of metamorphic activity; and (4) erosion, weathering, and leaching
by surficial and subsurficial hydraulic activity.

These processes, under the tectonospheric-Earth-model concept, may be expressed as aA + bB, where A represents thermomagmatic activity, B represents geodynamic activity, and $\underline{a}$ and
$\underline{b}$ are constants such that $a + b = 1$. In processes (1) and (2)
above, the value of $\underline{b}$ is roughly zero; in processes (3) and
(4), the value of $\underline{a}$ is roughly zero.

The tectonospheric Earth model provides an approach to
"selective" exploration, drilling, and mining procedures. This
approach, in turn, provides benefits of two types: (1) a more-
economical means for finding fossil and non-fossil minerals;
and (2) a means for saving the environment by eliminating needless

871

digging and drilling in mineralogically uneconomical areas.

The tectonospheric-Earth-model approach to mineral exploration is based on a straightforward premise: if one knew today where all the Earth's minerals were 4.6 b. y. ago, and if he could define the geometrical, mechanical, thermal, and chemical behavior of the Earth during the past 4.6 b. y., then he should be able to predict the present locations of the mineral deposits.

Because most fossil substances are an order of magnitude "younger" than are non-fossil substances, the problems of predicting the present locations of fossil substances are simpler than are those of predicting the present locations of non-fossil substances. Also, most petroleums are found in rocks that fall into a few closely-bounded temporal sequences, and almost all coal is found in rocks that fall into a single temporal sequence, the Carboniferous.

Most petroleums are found in rocks that are considerably younger than are the rocks that contain the coals; that is, 58% of the petroleums are found in Tertiary rocks, 18% in Cretaceous rocks. The remaining petroleums are found deposited in rocks of various ages, including the Precambrian, but excluding the Pleistocene.

The tectonospheric Earth model, by utilizing these and other distribution anomalies, provides predictions that serve

872

as valuable supplements to present exploration techniques: (1) a 3-dimensional solution for the present locations of Tertiary and Cretaceous petroleum deposits; (2) spatio-temporally adjusted 4-dimensional solutions for the present locations of the Permian, Carboniferous, Devonian, Silurian, Ordovician, and Cambrian petroleum deposits; and (3) a similar 4-dimensional solution for the present location of the coal measures (Carboniferous).

Fig. 20-4-1 shows a first-order unadjusted Venn diagram of regions for which the tectonospheric Earth model predicts Tertiary petroleum deposits. The indicated regions, superimposed upon a Mollweide projection of the Earth, show the unadjusted first-order concomitant residues from the simultaneous application of three predictions of the model: (1) pre-Tertiary proto-petroleum source materials; (2) Tertiary thermo-mechanical energy sources for converting proto-petroleum source materials to petroleum; and (3) Tertiary sediments suitable for the origin, evolution, and migration of petroleum.

Fig. 20-4-1 does not embody any structural, stratigraphic, or composite entrapment criteria. These entrapment criteria are used in actual analyses and predictions for petroleum. The results show: (1) the locations of most presently known or suspected Tertiary petroleum deposits; and (2) the locations of some presently unknown and unsuspected Tertiary petroleum deposits.

873

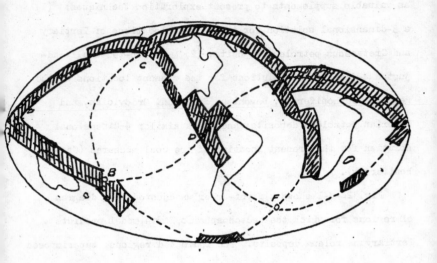

Figure 20-4-1. Venn diagram superimposed upon Mollweide Earth
to show the unadjusted first-order concomitant residues from
the simultaneous application of three predictions of the tectonospheric Earth model: (1) pre-Tertiary proto-petroleum source
materials; (2) Tertiary thermo-mechanical energy sources for
converting proto-petroleum source materials to petroleum; and
(3) Tertiary sediments suitable for the origin, evolution, and
migration of petroleum. See text for details.

It is not necessary for the purpose of this book to adjust Fig. 20-4-1 for the entrapment criteria. Nor is it necessary in this "qualitative" analysis to adjust the predictions for the differential horizontal movements that have occurred between the Earth's surficial features and the subtectonospheric fracture system (Chapter 6) since the petroleum deposits were entrapped.

In higher-order analyses and predictions for petroleum, these adjustments, amounting to as much as 140 km for some early Tertiary deposits, are made for the regions being analyzed. More-complex adjustments must be made in analyses involving "older" petroleum deposits and in those involving metallic deposits. These adjustments may amount to 300 km for Cretaceous petroleum deposits; 500 km for Permian deposits.

These adjustments, because of the "limit-stop" aspect of the orogenic-cratonic evolution of the continents (Chapter 7), rarely exceed 3500 km, except that the adjustments for some of the oldest Archean metal deposits may amount to 4900 km under the tectonospheric-Earth-model concept (Chapter 6). No single adjustment under this concept ever amounts to more than a few km, because individual adjustments are made incrementally in a spatio-temporal frame extending over 4.6 b. y. to a depth of 1000 km. The details of the mechanical, geometrical, thermal, and chemical aspects of these incremental adjustments are beyond the scope of this book, but these are included in a forthcoming book specifically directed toward the solution of problems of mineral exploration by using the tectonospheric Earth model.

Summarizing, the tectonospheric Earth model provides an approach to "selective" exploration, drilling, and mining procedures by suggesting (1) a more-economical means for finding fossil and non-fossil minerals; and (2) a means for saving the environment by eliminating needless digging and drilling in mineralogically uneconomical areas.

REFERENCES

Goldschmidt, V. M., 1954. Geochemistry. Oxford, Clarendon, 730 pp.

Kendall, M. G., 1947. The Advanced Theory of Statistics, vol. 1. Griffin, London, 457 pp.

Krauskopf, K. B., 1971. The source of ore metals. Geochim. Cosmochim. Acta, 35: 643–659.

Levorsen, A. I., 1967. Geology of Petroleum, 2d. ed. Freeman, San Francisco, 724 pp.

Miller, R. L. and Kahn, J. S., 1962. Statistical Analysis in the Geological Sciences. Wiley, New York and London, 483 pp.

Venn, J. A., 1888. The Logic of Chance. Macmillan, London, 188 pp.